Greek Historiography

Greek Historiography

EDITED BY
Simon Hornblower

CLARENDON PRESS · OXFORD
1994

Oxford University Press, Walton Street, Oxford OX2 6DP

Oxford New York Toronto
Delhi Bombay Calcutta Madras Karachi
Kuala Lumpur Singapore Hong Kong Tokyo
Nairobi Dar es Salaam Cape Town
Melbourne Auckland Madrid

and associated companies in
Berlin Ibadan

Oxford is a trade mark of Oxford University Press

Published in the United States
by Oxford University Press Inc., New York

British Library Cataloguing in Publication Data
Data available

Library of Congress Cataloging in Publication Data
Greek historiography / edited by Simon Hornblower.
Includes bibliographical references.
1. Historiography—Greece. 2. Greece—History—To 146 B.C.—
Historiography. I. Hornblower, Simon.
DF211.G75 1994   938'.0072—dc20   93-45956
ISBN 0-19-814931-X

1 3 5 7 9 10 8 6 4 2

Typeset by Joshua Associates Ltd., Oxford
Printed in Great Britain on acid-free paper by
Bookcraft (Bath) Ltd., Midsomer Norton

Preface

The papers in this book were, with one exception, given at the Oxford ancient history seminar on 'Greek Historiography' held on the Tuesdays of Michaelmas Term 1991 at Oriel College. The exception is my own paper on 'Narratology and Narrative Techniques in Thucydides', which was given at the Oxford Philological Society in February 1992. Professor Badian's paper at the original seminar was called 'Herodotus and Thucydides on Macedon' and was in two parts. The Herodotean part appears in the present volume; the Thucydidean part is included, as 'Thucydides and the *Arche* of Philip', in Badian's new collection of essays on fifth-century Greek history, *From Plataea to Potidaea* (Baltimore, 1993).

The title of the book, *Greek Historiography*, is the title of the seminar series. But the comments of the book's anonymous referees helped to convince me that, for a book, that title would be misleadingly wide without some attempt by me as editor to put into context the ancient writers discussed, to fill in gaps in the coverage of the volume, to pull the threads together, and to identify some common problems addressed by the various papers. Hence the long Introduction. Without it, a more accurate title for the book would have been 'Aspects of' or 'Papers on' Greek Historiography. (For the shape of the Introduction see further p. 1.)

I am grateful to the contributors for participating in the original seminar, and making their papers available for publication in the present book. I much regret that Tony Andrewes, who had originally agreed to speak at the seminar, did not live to do so.

I must also thank my college, Oriel, for the hospitality which it extended to my seminar and to its speakers; and my Oriel pupil Paola De Carolis for her help with the practical and social sides. Finally, thanks to Richard Rutherford and Peter O'Neill for help with reading the proofs.

SIMON HORNBLOWER

Oxford
February 1993

Contents

Notes on the Contributors

ERNST BADIAN is John Moors Cabot Professor of History at Harvard University and an Honorary Fellow of University College, Oxford. His books include *Studies in Greek and Roman History* (Oxford, 1964) and he wrote the chapter on Alexander the Great in the *Cambridge History of Iran*, ii (1985). His most recent book is *From Plataea to Potidaea: Studies in the History and Historiography of the Pentecontaetia* (Baltimore, 1993).

JOHN DAVIES is Rathbone Professor of Ancient History in the University of Liverpool. His books include *Athenian Propertied Families* (Oxford, 1971) and *Democracy and Classical Greece* (revised edn., London, 1993); and he has edited (with L. Foxhall) *The Trojan War: Its Historicity and Context* (Bristol, 1984). He has written on all periods of Greek history in the *Cambridge Ancient History*, of which he is also an editor.

PETER DEROW is Fellow and Tutor in Ancient History of Wadham College, Oxford. He has written on Polybius and Roman Republican history in journals and in the *Cambridge Ancient History* and is currently at work on a book about Rome and the Greeks.

P. M. FRASER was until recently Reader in Hellenistic History at Oxford University and a Fellow of All Souls College. He was Director of the British School at Athens from 1968 to 1971. His books include *Ptolemaic Alexandria* (Oxford, 1972). He is co-editor of the *Lexicon of Greek Personal Names* (Oxford), of which volume 1 appeared in 1987.

JOHN GOULD was until recently H. O. Wills Professor of Greek at the University of Bristol. His books include *The Development of Plato's Ethics* (London, 1955), *Herodotus* (London, 1989), and (with D. M. Lewis) a revision of A. W. Pickard-Cambridge's *Dramatic Festivals of Athens* (Oxford, 1988). His book on Greek Tragedy will appear soon.

SIMON HORNBLOWER is Fellow and Tutor in Ancient History of Oriel College, Oxford. His books include *Thucydides* (London, 1987) and *A Commentary on Thucydides*, i: *Books I–III* (Oxford, 1991). He has written on the literary sources for the fourth century BC in the *Cambridge*

Ancient History and is now working on volume 2 of his three-volume Thucydides commentary.

KENNETH S. SACKS is Professor of Classics and Chairman of the Department of History at the University of Wisconsin-Madison. His books include *Polybius and the Writing of History* (Berkeley, 1981) and *Diodorus and the First Century BC* (Princeton, NJ, 1990).

ANTONY SPAWFORTH was Assistant Director of the British School at Athens from 1978 to 1981 and is now a Senior Lecturer in the Classics Department at the University of Newcastle upon Tyne. He is the author with P. A. Cartledge of *Hellenistic and Roman Sparta: A Tale of Two Cities* (London, 1989) and co-editor with S. Hornblower of the third edition of the *Oxford Classical Dictionary* (in preparation).

List of Abbreviations

ANRW	*Aufstieg und Niedergang der römischen Welt* (*Festschrift J. Vogt*), ed. H. Temporini and W. Haase (Berlin and New York, 1972–)
Ath. Pol.	*Athenaion Politeia* (*Athenian Constitution*) attributed to Aristotle
Beloch, *GG*	K. J. Beloch, *Griechische Geschichte*², 4 vols. in 8 (Strasburg, Leipzig, and Berlin, 1912–27)
BICS	*Bulletin of the Institute of Classical Studies*
CAH	*Cambridge Ancient History*
CID	*Corpus des inscriptions delphiques*, ed. G. Rougemont *et al.* (1977–)
D–K	H. Diels and W. Kranz, *Die Fragmente der Vorsokratiker*⁶, 3 vols. (Berlin, 1952)
E/K	L. Edelstein and I. Kidd, *Posidonius*; see Bibliography under Kidd
FdeD	*Fouilles de Delphes*
FGrHist	F. Jacoby, *Die Fragmente der griechischen Historiker*, 15 vols. (Berlin, 1923–30; Leipzig, 1940–58)
FHG	C. and T. Müller, *Fragmenta Historicorum Graecorum* (Paris, 1841–72)
Fornara	C. W. Fornara, *Archaic Times to the end of the Peloponnesian War* (Cambridge, 1983)
GGM	C. Müller (ed.), *Geographi Graeci Minores*, 3 vols. (Paris, 1855–61)
Harding	P. Harding, *From the End of the Peloponnesian War to the Battle of Ipsus* (Cambridge, 1985)
Hercher, *Epist. Gr.*	R. Hercher (ed.), *Epistolographi Graeci* (Paris, 1873; repr. Amsterdam, 1965)
Hp.	Hippocrates, ed. E. Littré (Paris, 1839–61)
HCP	F. W. Walbank, *Historical Commentary on Polybius*, 3 vols. (Oxford, 1957–79)
HCT	A. W. Gomme, A. Andrewes, and K. J. Dover, *Historical Commentary on Thucydides* (Oxford, 1945–81)
HM	N. G. L. Hammond, *A History of Macedonia* (Oxford:

	i, 1972; ii (with G. T. Griffith), 1979; iii (with F. W. Walbank), 1988)
IDélos	*Inscriptions de Délos*, ed. F. Durrbach (Paris, 1926–37)
IG	*Inscriptiones Graecae*
ML	R. Meiggs and D. Lewis, *A Selection of Greek Historical Inscriptions to the End of the Fifth Century BC*, rev. edn. (Oxford, 1988)
NH	Pliny the Elder, *Natural History*
OCD	N. G. L. Hammond and H. H. Scullard (eds.), *The Oxford Classical Dictionary*² (Oxford, 1970; a 3rd edn. is in preparation)
OGIS	W. Dittenberger, *Orientis Graecae Inscriptiones Selectae*, 2 vols. (Leipzig, 1903–5)
PEleph.	*Elephantine Papyri*, ed. O. Rubensohn (Berlin, 1907)
POxy.	*The Oxyrhynchus Papyri*, ed. B. P. Grenfell, A. S. Hunt, *et al.*; in progress (London, 1898–)
RE	A. Pauly and G. Wissowa, *Real-Encyclopädie der classischen Altertumswissenschaft*, 83 vols. (Stuttgart, 1894–1980)
Ptol. Alex.	P. M. Fraser, *Ptolemaic Alexandria*, 3 vols. (Oxford, 1972)
SCI	*Scripta Classica Israelica*
SEG	*Supplementum Epigraphicum Graecum*
*Syll*³	W. Dittenberger, *Sylloge Inscriptionum Graecarum*³, 4 vols. (Leipzig, 1915–24)
TLS	*Times Literary Supplement*
Tod	M. N. Tod, *Greek Historical Inscriptions*, 2 vols. (Oxford, 1946², 1948)
UCPCPh	*University of California Publications in Classical Philology*
VS	Philostratus, *Lives of the Sophists* (*Vitae sophistarum*)
Wehrli	F. Wehrli, *Die Schule des Aristoteles: Texte und Kommentar*, 10 vols. (Basle and Stuttgart, 1944–69)

I

Introduction*

SIMON HORNBLOWER

I. GENERAL INTRODUCTION

A word is necessary about the shape of this Introduction. After Section I, the present opening section, I summarize the papers in this book (Section II), and thereafter I often refer forward to them. Then the chronologically organized Section III sketches the general story of Greek historiography, thus I hope providing the larger frame into which the individual papers fit, and introducing the reader to some names taken for granted in the individual chapters. Section IV is thematic, and addresses itself to such problems as 'intertextuality'—the awareness shown by one writer of another's work, and the difficulties of assuming direct influences—and the legitimate use of arguments from silence. What can we infer from a historian's non-mention of a particular item? These issues are (see further below, pp. 69ff.) raised in one form or another by all the chapters in the book.

II. SUMMARY OF THE PAPERS

Peter Derow's paper on historical explanation discusses the attitudes of four Greek historians—Hecataeus of Miletus, Herodotus, Thucydides, and Polybius—to questions of causation. At one end is the naïve rationalizing of Hecataeus, who nevertheless deserves credit for beginning the process of historical explanation; at the other is the professionalism and methodological sophistication of Polybius. In between are Herodotus and Thucydides. Herodotus often stressed psychological motives, above all revenge and greed; but he did sometimes recognize that people do things as a response to circumstances, such as the growth of someone

* This Introduction has been much improved by the thorough and invaluable comments of Richard Rutherford and Rosalind Thomas, to both of whom I offer warm thanks. The usual exemption clause applies.

In the notes, Hdt. = Herodotus and Th. = Thucydides.

else's power, thus at 1.46 Croesus is explicitly said to react to the increase of Persian power. Thucydides' agents also react, more noticeably and more often, to circumstances of this sort. Thus in his famous chapter 1.23 the Spartans are said to fear the growth of Athenian power, hence the two Peloponnesian wars. But unlike Herodotus, Thucydides offers full, perceptive, and neutral analysis of those circumstances, what Derow calls the 'who, what, where and when'. It was Polybius, in the second century BC, who added the 'how and why': in fact, he 'defined the historian's task as explanation'. It is an important part of Derow's thesis (see most explicitly p. 84) that Polybius, at least, was engaged in conscious dialogue with the predecessors whom he mentions in what Derow aptly calls Polybius' 'bibliography': this raises questions of intertextuality to which we shall return in Section IV below (p. 54).

Derow's Herodotus leaves a place for predetermination or 'fate' as facts of life; but fate is not accorded explanatory power in Herodotus' scheme: on the whole, human affairs call for human explanations. John Gould's paper, on Herodotus and religion, begins with an insistence, very reminiscent of Derow, that Herodotus is 'cautious' in admitting non-human causation. But Derow is talking about fate, Gould is concerned with the gods; and Gould believes that Herodotus assumed (see e.g. Hdt. 9. 100) a massive intervention by the gods in human affairs, while at the same time treating those gods in a conspicuously cautious, even inhibited, way. Gould's examination of the literary dynamics of that caution has implications for Herodotus' method generally, as well as in the particular area of religion. One manifestation of caution is for Herodotus to demand additional weight of evidence; another is to decline to identify the particular god concerned; a third is to use *oratio obliqua* or indirect speech as a distancing device; finally there is Herodotus' technique of offering alternative possibilities. These are all features of presentation which, as narratologists would say (see below, Ch. 5), establish a particular relationship or mode of interaction with the reader or hearer. Gould shows that Herodotus' nuanced language is designed to flag real uncertainty about the gods, rather than a conviction that they were irrelevant. The second half of Gould's paper looks at Herodotus' treatment of alien religions, and shows that Herodotus perceived them in largely *ritual* terms. In this he was a true Greek, but his neglect of the iconography and ideology of religion was a weakness (or rather seems a weakness, from a modern perspective) in an interpreter of foreign religions.

Ernst Badian's chapter, on Herodotus and Alexander I of Macedon, resembles the first part of John Gould's in that it makes its points by drawing attention to some purposeful stylistic devices ('subtle art') used by Herodotus. Badian's concern is however politics not religion. He shows how Herodotus, by the way he presented and arranged his material, exposed the past misbehaviour of, but avoided giving offence to, a powerful dynasty, namely the family of Alexander I of Macedon. (That is politics; but the reader of this book will be reminded of Gould's 'religion' chapter at the point where Badian, in passing, compares Herodotus' political tact to his care not to give offence to the gods, 'Egyptian no less than Greek': p. 120). The underlying truth for which Badian argues is that Alexander medized, i.e. favoured Persia, and that Herodotus wished us to know this shameful fact but did not wish to assert it bluntly. (We may, reading Badian, be tempted to contrast Herodotus' round condemnation at 8. 73 of the neutrality-amounting-to-medism of a less powerful group in Herodotus' day, Alexander's ancestors the Argives.) Badian's story begins with the marriage of Alexander's sister to an ascertainable Achaemenid, and his father Amyntas' offer of allegiance to Persia. It continues with four episodes portraying Alexander as ostensibly loyal, but actually treacherous, to the Greek cause. By a brilliant narrative device Herodotus chooses to give the full genealogical proof of Alexander's Greek descent in precisely the context (at the end of book 8) of what Badian calls Alexander's 'most conspicuous act of medism'. (This is in effect a piece of 'narrative delay' of a kind further discussed in Ch. 5 below: Herodotus has already, at 5. 22, *introduced* us to the theme of Alexander as Greek.) Finally, Badian shows that Herodotus was concerned to save not just Macedonian but Athenian faces, because the Athenian grant of *proxeny* (diplomatic honours) to Alexander was a reward for mediating Athens' own, subsequently embarrassing, submission to Persia in the final years of the sixth century.

My own chapter explores, explicitly and in detail, more narrative devices of the kind Gould and Badian argue for in Herodotus; but my concern is Thucydides. (The summary which follows takes for granted some technical and semi-technical words and phrases defined fully in the paper.) Essentially I attempt to apply the insights of the new art or science of narratology, developed for Proust by Genette and for Homer and Euripides by de Jong, to the supposedly austere and objective historian Thucydides. And I end the paper by asking how that or any other historian's use of such devices differs from a poet's or a novelist's. In the main and central part of the paper I discuss, with reference to

Thucydides, problems of focalization and inferred motivation; techniques of opening and closure; narrative displacement (anticipation and delay, and the effects thereby produced; cp. what was said above about Badian); use of iterative verbs and expressions; distancing devices, i.e. ways of expressing reserve or disbelief (cp. above on Gould) and their converse, devices for reinforcing or compelling belief; and other techniques by which the narrator 'buttonholes' or interacts with the narratee, such as: presentation through negation; 'if . . . not' and other counterfactual locutions; and the use of evaluative and affective words and of attributive discourse. All these devices, I suggest, help us to understand the art by which Thucydides' 'objective' manner is achieved. Finally I discuss the issue of narrative voice, as it bears on historical discourse.

We move on to the fourth and third centuries with P. M. Fraser's subject, who is Theophrastus (*c.*370–288 BC). He is best known as the author of the *Characters*, but later in this Introduction I shall try to uphold his general claim to the title of historian (p. 33). The particular focus of Fraser's paper is the 'geographical aspect of botany', and this represents a move away from the more familiar type of topic handled by modern ancient historians. But it is a miserable and pre-Braudelian definition of history which would exclude the material which Fraser adduces and discusses in the botanical connection, not merely because of its intrinsic interest, but because Fraser's enquiry raises fundamental questions about the extent, and date of acquisition, of Greek knowledge of (i) the territories conquered by a later King Alexander of Macedon than Badian's, in fact Alexander III, the Great; and (ii) the 'Greek West'; i.e. Italy and Sicily. Despite the exotic subject-matter of Fraser's paper, there are immediate echoes of the earlier chapters of this book. For instance Theophrastus, like Herodotus, has (p. 171) his areas of reticence or 'taciturnity' on matters we have to suppose he knew about very well (a topic to which I return below, p. 71); and Theophrastus, like Herodotus before him and Arrian after him, uses distancing devices like 'they say' and *oratio obliqua* to signal items which he declines personally to authenticate. (See pp. 180, and 181 for Egypt and Cyrene in particular.) The years around 300 BC were like the age of the opening up of the Americas, a period when reports of both real and imaginary marvels flooded in from the east (see for example p. 178 for Theophrastus' knowledge of the banana-tree). In such a period, a cautious historian like Theophrastus must have needed all his scholarly training in the testing of informants (bematists, explorers, and the like), and all the stylistic repertoire available for the expressing of reservations. As for the west, here too we find such distancing devices as 'they say' (see pp. 184f. for Adriatic plane-trees, and the fir trees of Corsica). Paradoxically—given that Greeks had known Italy for centuries—it turns out that Theophrastus' information on the west was actually sketchier than about the east.

Theophrastus' botanical works are perhaps an unexpected text to act as a reminder that the historians of antiquity, including Thucydides, had to face the problem of how to evaluate oral tradition. John Davies' chapter attacks the 'oral tradition' problem by examining not an author but an event (or pseudo-event) of archaic Greek history. The event is the First Sacred War, a major episode if it happened, but which (it has recently been argued) may not have happened at all: *not* a combination of features with which historians of any period are accustomed to deal. (A further disconcerting feature is that the war, unlike, say, the Trojan War from which Herodotus distances himself in his first few chapters, was not identified as problematic in antiquity. Herodotus and therefore Thucydides ignored it, certain other sources imply or accept it.) The war was, the sceptics say, an invention of the fourth century. That century was admittedly the period from which we derive our fullest information about the war; hence the positioning of Davies's chapter in this roughly chronologically arranged book. Since 1978, when the sceptical case was first made, the choice between belief and rejection has been ostensibly stark, but Davies exploits two sophisticated modern methodologies to show a way between the horns of the dilemma, so that by the end of his paper the historicity of the war has again become something we can warily admit ('a plausible hypothesis, but no more'). The first methodological advance is that which has imposed a less gullible scholarly attitude to fabricated documents and decrees (a species of what has been called 'invented tradition'; see below, p. 39). The second is the increase in our understanding of Greek oral tradition generally, or rather 'semioral' tradition: the hyphenation is necessary because one important new insight is that cultures may retain markedly oral features even after the advent of widespread literacy. This coexistence of features helps to explain why traditions may continue to be unstable even after they have begun to be written down: certainly, Greek ideas about the past were reworked in response to a 'continually shifting present'. The reconstruction of an event is different from the reconstruction of the consciousness of an event. Against this theoretical background Davies elaborates a kind of stratigraphy, or scientific separation of the layers, of the tradition about the war. He shows that although the fourth century was a particularly busy phase in the manufacture of that tradition, there is an irreducible residue dating from the archaic period itself and ultimately deriving from local legends. By his treatment of this test case, Davies deconstructs Greek history; the challenge will be to reconstruct it with the tools he has provided.

Unlike Davies, for whom oral tradition is crucial, Kenneth Sacks is concerned with a line of written tradition, the sources used by the first-century BC historian Diodorus Siculus. But both chapters show that traditions were slippery and that the present (i.e. *their* present, in the sense of the ancient moment from which the remoter past was viewed) could not help contaminating the past. This is not an invitation to post-modernist despair at the impossibility of any reconstruction at all; rather, it means that we must reckon with the kind of interference which scientists call 'noise'. Thus fourth-century accounts of the sixth century are not just information about the sixth century, but (see Davies, above) tell us about the fourth and fifth as well as the sixth. Equally, the Ephorus we find in Diodorus is not a clear, interference-free, Ephoran signal, scrupulously (or a less charitable view, mindlessly) transmitted by Diodorus, but is simultaneously informative about the transmitting author, his attitudes, and his period, in fact about 'Diodorus and the First Century BC'. (The quotation marks enclose the title of Sacks's 1990 book, the starting-point for his chapter in the present volume.) Sacks argues that Diodorus was more independent and creative than has sometimes been allowed, and that ethical, philosophical and political judgements are often intrusions by Diodorus himself. In particular, Diodorus' narrative incorporates his distinctive views on both Athenian and Roman imperialism. Rome, it appears, elicited no enthusiasm from Diodorus; on the contrary he was reserved and even critical. So much for Diodorus' attitudes; Sacks also puts his accuracy under the lens and argues that on one key issue of early second-century history Diodorus is right and Livy wrong, or rather Diodorus more accurately reflects the original source, Polybius. Sacks neatly reconciles his 'independence' thesis about Diodorus with his 'accuracy' thesis by concluding that Diodorus had his own views but stopped short of altering the narrative to fit them.

Finally, Antony Spawforth discusses the Persian-Wars tradition in the Roman empire. This picks up one main thread of Davies's paper ('invented tradition' about a war, or rather elaborated or encrusted tradition, because there is no doubt about the reality of the early fifth-century Persian Wars, unlike the First Sacred War). At the same time Spawforth follows into the imperial period a path of inquiry taken for a slightly earlier period by Sacks, in that both are interested in educated Greek reactions to Rome at the high noon of its power. But Spawforth's Greeks, unlike Sacks's Diodorus, use the past not as a mechanism for the articulation of discontent, but as a way of expressing, through various types of civic discourse, attitudes perfectly compatible with loyalty to Rome. To the categories of evidence treated in earlier chapters of this book, mainly literary texts and oral tradition, Spawforth adds full-length treatments of inscriptions and monuments. He begins by showing that the inscription in honour of Nero, impudently erected on the architrave of the Parthenon, attests not only an obvious and traditional Greek pride in the Persian Wars but a Greek response to a theme positively favoured by Roman imperial ideology. He goes on to explain this paradox (*Roman* adoption of a *Greek* achievement) by demonstrating in detail the equation between Persia the ancient, and Parthia the modern, enemy. Rome is now, what Greece itself no longer has the muscle to be, the 'champion of Hellas against the Orient'. Spawforth even shows (p. 244) that Greek medism, a major theme of Badian's paper, still had vitality in the first century AD: he describes how some Argives, in a Roman context, feel embarrassed by their medism of more than half a millennium earlier. Greeks were flattered by Roman adoption of their own most splendid historical achievement, which thus served as a symbol of unity. The absorption of Herodotus' great subject into Roman imperial ideology is a suitable closure for a book on Greek historiography.

III. THE STORY OF GREEK HISTORIOGRAPHY

In the first and best work of European literature, we already find a pre-occupation with the past, and with the urge to transmit it to future generations. In book 9 of Homer's *Iliad*, Achilles accompanies himself on his lyre and sings of the 'glorious deeds of men' (lines 189, 194). In book 22, Hector, who is near death, prays that he may not die ingloriously but after doing some great deed for posterity to hear of (lines 304–5). Similarly Helen in book 6 (357–8) observes to Hector that Zeus has ordained an evil fortune for herself and Paris, 'so that in times to come we may be sung about by people yet unborn'. In an earlier book (3. 125–8) we learn that the first historian of the Trojan Wars is a woman, namely Helen herself: the goddess Iris finds her in her room working at a great web, 'weaving into it many contests between the horse-taming Trojans and the bronze-clad Achaeans'. That is, Helen is busy producing an illustrated record of the calamities she has herself helped to bring

about. Homeric scholars have recently been much interested in the kind
of poetic self-reference here implied:[1] the poet is in fact making a definite
claim on posterity, and 'Hector's prayer', it has been neatly said,[2] 'is fulfilled every time the poem is recited or read'.

The poet naturally puts a high value on fellow exponents of verbal
skills. At the vengeful end of the *Odyssey*, the collaborating bard
Phemius is more honourably treated than other suspect members of
Odysseus' household (22. 330–80[3]). And in the list of craftsmen at
Odyssey 17. 384–5, the bard has a whole line to himself, whereas the
prophet, the doctor, and the builder are all crammed into the preceding
line.[4] Phoenix describes a great leader as a 'speaker of words and a doer
of deeds', in that order (*Iliad* 9. 443). With this compare Thucydides'
description (1. 139. 4) of Pericles as 'a man of great ability in words and
action'. The echo of the epic poet of the seventh century BC by the fifth-
century prose reporter of the Peloponnesian War is clear; indeed the need
for posterity to know, and the compulsion to preserve glorious deeds, are
themes which echo throughout the archaic period, down to Herodotus
and Thucydides. But much was to happen on the way, and we cannot
jump from Homer to Thucydides as rapidly as that: 'whatever else it may
have been, the epic was *not history*', wrote Moses Finley in a celebrated
article on 'Myth, Memory and History', first published in 1965. Finley's
concern was a slightly different one from ours, namely the problem of
reconstructing the Dark Ages, rather than the problem whether the epics
carry the germ of a historical attitude. But Finley's paper has a bearing on
that second problem too. He did not deny that Homer might contain
some kernel of historical fact, but was concerned to insist that epic was
timeless, lacking 'dates and a coherent dating scheme'.[5] Put in so stark
and positivist a way, Finley's point is undeniable. But it surely remains of
great interest and importance, for future developments, that the poet is
aware both that there will be a posterity (see above), and that he himself

---

[1] O. Taplin, *Homeric Soundings: The Shape of the Iliad* (Oxford, 1992), 243, cp. 97f.,
284. Note also Hecuba in Euripides' *Trojan Women* 1244–5 with S. Barlow's commentary
(Warminster, 1986). Cassandra at line 365 is also relevant.

[2] R. B. Rutherford, 'What's New in Homeric Studies?', *JACT Review* 11 (1992), 15–17
at 17; Taplin (above, n. 1), 243. See also C. Segal, 'Bard and Audience in Homer', in
R. Lamberton and J. J. Keaney (eds.), *Homer's Ancient Readers: The Hermeneutics of
Greek Epic's Earliest Exegetes* (Princeton, NJ, 1992), 3–29.

[3] See Heubeck's note on that section in J. Russo, M. Fernandez-Galiano, and A. Heubeck, *A Commentary on Homer's Odyssey*, iii: *Books XVII–XXIV* (Oxford, 1992), 278.

[4] See Russo's note in Russo *et al.* (above, n. 3), 38.

[5] For the 7th-cent. dating see Taplin (above, n. 1), 33–5. Finley's essay is republished in
his *Use and Abuse of History*[2] (London, 1975), 11–33, see esp. 14f.

belongs to a much later generation than that of the heroes he describes. (Thus he speaks of 'mortals of the kind who exist now', *Iliad* 5. 304 and frequently.) It is worth our staying with Homer a little longer, because of his use of and attitude to the past.

Homeric interest in the human past is all-pervasive. Two major concerns of the *Iliad* are warfare and genealogy. Typically, a minor hero is killed in battle. We are told who his parents and sometimes who his remoter ancestors were, and now we are told how he died and at whose hands. (Simple examples are Agamemnon's killing of Deikoon son of Pergasos, and Aeneas' retaliatory killing of Krethon and Orsilochos, sons of Diokles, grandsons of Ortilochos, and great-grandsons of the river-god Alpheios: *Iliad* 5. 533–60. Aeneas in book 20. 215–40 and Glaukos in 6. 153–206 have even longer genealogies.) As with Thucydides' echo of the Homeric Phoenix, we find that tastes were not so different in the classical period. There is not much explicit genealogizing in Thucydides, but consider 7. 69, Nikias' 'old-fashioned' appeal to his troops by their fathers' names, πατρόθεν, and by the fame of their ancestors. Given in full, this speech might have looked distinctly Homeric: compare *Iliad* 10. 68 (from the *Doloneia*, a 'Homeric' book if not actually by Homer) where Agamemnon tells Menelaus to exhort every man calling him by his own and his father's name, πατρόθεν ἐκ γενεῆς ὀνομάζων ἄνδρα ἕκαστον. More explicit is a passage of Plato, who tells us through the speaker who gave his name to his dialogue *Hippias* that in the fifth century the Spartans (the greatest fighters of Greece) enjoyed hearing tales about the genealogies of heroes *and men*—and about the foundations of cities: *Hippias Major* 285d. (See below for Hippias of Elis and his chronological interests.)

The second part of that sentence of Plato, about the 'foundations of cities', takes us to the subject-matter of the *Odyssey*: the return of a hero after his withdrawal and resulting disasters, followed by the (re-)establishment of a community. The idea of the walled settlement with temples, the 'sacred city', is important not just in the *Iliad* (where 'sacred Troy' is basic to the poem) but also to the *Odyssey* in a general way.[6] And the particular community of the Phaiakians, with its king Alkinoos and its princess Nausikaa, was identified in antiquity with the historical island of Corcyra (Corfu): even the realist Thucydides casually mentions a sacred precinct of Alkinoos on Corcyra (3. 70. 4) alongside one to Zeus.[7] Returning heroes other than Odysseus had, like him, their *nostoi* or

---

[6] S. Scully, *Homer and the Sacred City* (Ithaca, NY, 1990) with Rutherford (above, n. 2), 16.   [7] See further below, p. 65.

'epics of return'; P. M. Fraser (below, p. 183) discusses the traces left by one such example, the tradition about the great Achaean warrior Diomedes, who is so prominent in *Iliad* book 5, where he even wounds Aphrodite. Diomedes settled in the west, as we are told by (among other sources) a scholiast or ancient commentator on Thucydides 1. 12—but not by Thucydides himself. Or rather, not by Thucydides in so many words, though he did in that chapter record in unspecific terms the colonizing of the west by Peloponnesians and the Dark Age colonizing of the east (Ionia) by the Athenians. Thucydides was reticent; but such precise foundation legends, which certainly reflect a historical early Greek presence in Italy and Sicily, were taken seriously by other historians: Thucydides' contemporary Hellanicus is the first Greek writer explicitly to give the story that the Trojan Aeneas founded Rome (*FGrHist* 4 F 84; cp. Th. 6. 2. 3 and 4. 120. 1: Trojan and Achaean legends). We are now in an area where myth and history, 'true' and 'false' history,[8] overlap, but at the very least the Aeneas legend says something important about the Roman self-image: it has been wittily said that the point of the Aeneas story is to demonstrate to the world that the Romans were neither quite Greeks nor quite Etruscans.[9]

The Iliadic world of genealogy and the post-Iliadic or Odyssean world of foundation legends intersected. There is a remarkable genealogical inscription surviving on stone, giving the pedigree of one Heropythos of Chios. It has been conjectured that the pedigree begins with the Athenian foundation (cp. above) of Ionian Chios.[10] Certainly a genealogy of the famous Athenian aristocratic family of Miltiades, preserved by yet another fifth-century writer, Pherekydes, includes the name Oulios at an early point (*FGrHist* 3 F 2): this unusual name was connected with the Athenian colonization of or migration to Ionia.[11] And it may be that some local generation count lies behind the claim of Thucydides' Melians that their community was founded (by Sparta) 700 years before 416 BC, the date of the Melian Dialogue: 5. 112. 1.[12]

---

[8] E. Gabba, 'True History and False History in Classical Antiquity', *JRS* 71 (1981), 50–62.

[9] A. Momigliano, *CAH* vii²(2) (1989), 56, cp. 111.

[10] H. T. Wade-Gery, *The Poet of the Iliad* (Cambridge, 1952), 8; R. Thomas, *Oral Tradition and Written Record in Classical Athens* (Cambridge, 1989), 156, 159, 190. Wade-Gery fig. 1 gives a photograph of the Heropythos pedigree, with a text; and R. Thomas uses the inscription on the front cover of the paperback version of her book.

[11] Thomas (above, n. 10), 164–5. For the interesting name Oulios see O. Masson, 'Le Culte ionien d'Apollon Oulios, d'après des données onomastiques nouvelles', *Journal des savants* (1988), 173–83 = *SEG* XXXVIII 1996 *bis*.

[12] Andrewes, *HCT* iv (1970), 180–1.

In view of all this, Oswyn Murray is perhaps too sweeping (think of Plato's Spartans, above) when he writes that 'comparative absence of genealogies is one of the characteristics of Greek tradition',[13] though even he makes an exception for royal genealogies like the Spartan king lists in Herodotus (7. 204; 8. 131, cp. 7. 11 for Persia. But the last example—Xerxes' citation of his own ancestry—is a reminder that a historian, like a poet, may insert lists to add atmosphere and solemnity, like the epic lists in Milton, studied by Barbara Everett[14]). Instances like Oulios, Heropythos, and even the Spartan kings—who are in an obvious sense coterminous with the Spartan state, founded as it was by Dorian newcomers—encourage us to take it as a working hypothesis that genealogies often took their start from the moment of civic or colonial foundation. John Gould's general approach seems preferable to Murray's: in his recent book on Herodotus, he speaks (below, n. 35, at his p. 40) of Herodotus' 'ability to think readily and fluently in terms of kinship and generations, and his use of these structures to map the past' (see further below). By introducing Herodotus we have anticipated illegitimately; but Gould's stress on genealogy is right, as is his implication on the same page that facility with genealogy (a special case of list-making) is characteristic not just of Herodotus, who names nearly a thousand individuals, but of the oral tradition he inherited. However, as we shall see (below, n. 22), genealogical traditions were unstable and gappy until people like Hecataeus and Pherekydes began to systematize them. For the moment I am merely concerned to make the point that, though we can hardly expect support amounting to proof, Greek colonization may have been a kind of starting pistol.

Genealogies were surely not the only kind of record-keeping encouraged by the colonizing act. The impulse to mark—perhaps by some ritual and solemnly recorded innovation—the founding of a new city, far from the motherland or *metropolis*, must have been strong; though even successful colonies (and there were many failures) may have been less good at keeping the proud record up to date. Herodotus on Cyrene (4. 145–205) is, however, an obvious example of a historian

[13] O. Murray, 'Herodotus and Oral History' in H. Sancisi-Weerdenburg and A. Kuhrt (eds.), *Achaemenid History*, ii: *The Greek Sources* (Leiden, 1987), 93–115 at 98. For genealogies in Herodotus see E. Vandiver, *Herodotus and Heroes: The Interaction of Myth and History* (Frankfurt am Main, 1991), 29–60.

[14] B. Everett, *Poets in their Time: Essays on English Poetry from Donne to Larkin* (Oxford, 1991), ch. 3, 'The End of the Big Names: Milton's Epic Catalogues'. With Xerxes' catalogue compare Sophocles, *OT* 266–8, where Oedipus is unconsciously giving his own ancestry.

exploiting such a line of tradition which goes well beyond the initial 'big bang' of seventh-century colonization and takes in Cyrene's sixth-century time of troubles. Or take the material about early Sicily, which introduces Thucydides' book 6 and the story of the great Sicilian Disaster of 415–413. This may indeed go back, as K. J. Dover acutely demonstrated forty years ago, to the local Syracusan historian Antiochus of Syracuse (*FGrHist* 555).[15] But equally, some of Thucydides' Sicilian foundation-dates in that same section may derive from the archives, if that is not too formal a concept, of the temple of Apollo Archegetes ('Apollo the Founder') at Sicilian Naxos.[16] That was the temple from which Thucydides says (6. 3. 1) *theoroi* or sacred ambassadors sacrificed before setting out (to Old Greece, presumably). To say so much is not to return to the old view, discredited by the great modern historiographer F. Jacoby, that Greek historiography had its origins in local historiography.[17]

It is then an important truth that two great concerns of the Homeric epic, the past of individual men and the past of the cities of men, are still reflected in later Greek historiography. It is, in fact, not just the manner (see my chapter below, p. 148) but the matter of Herodotus and Thucydides that looks back to Homer. (See further below for *war* as another epic concern bequeathed to classical historiography.)

If the Homeric poems took their shape in the seventh century, they precede by only about a century and a half the first prose historian with any claim to that title, Hecataeus of Miletus. It may seem bold to claim this for Hecataeus, but as Peter Derow puts it below (p. 73), he was the first to 'identify the past as a field of critical study'. This is a good place to say that I am concerned in this Introduction with *Greek* historiography, not with Jewish or other claims to primacy over Greek; and the same goes for the other contributions to this volume. That is not to deny the

[15] K. J. Dover, 'La colonizzazione della Sicilia in Tucidide', *Maia* 6 (1953), 1–20, Ger. tr. 'Über die Kolonisierung Siziliens bei Thukydides' in H. Herter (ed.), *Thukydides* (Darmstadt, 1968), 344–68. For the kind of thing (lists of victors, office-holders, and so forth) recorded in early epigraphically preserved lists, see L. H. Jeffery, rev. A. W. Johnston, *The Local Scripts of Archaic Greece* (Oxford, 1990), 59–61.

[16] O. Murray (above, n. 13), 98 and n. 8, referring to a forthcoming article on 'Thucydides and Local History'. But innocent belief in the efficiency and extensiveness of ancient Greek archives will not survive a reading of R. Thomas, *Literacy and Orality in Ancient Greece* (Cambridge, 1992), ch. 7.

[17] F. Jacoby, *Atthis: The Local Chronicles of Ancient Athens* (Oxford, 1949), 184. Jacoby was also concerned, see esp. *Atthis*, 1–8, to combat the related view of Wilamowitz (*Aristoteles und Athen*, i (Berlin, 1893), 260–90, esp. 280; cp. *Glaube der Hellenen*, i. 40 n. 1) that local Athenian history or Atthidography, for which see pp. 23 and 34 below, had its origins in chronicles kept by the *exegetai* or official interpreters of religious matters.

chronological priority over Greece of the source (tenth century BC?) of the sixth-century 'Deuteronomist' author of some of the early narrative parts of the Old Testament, in particular the remarkably lively 'Succession narrative' in parts of 2 Samuel, chapters 9 and following, and the beginning of 1 Kings. (It includes the story of David's shameful acquisition of Bathsheba, the wife of Uriah the Hittite.) Nor is it to deny the importance of this early but anonymous Jewish achievement, which has something approximating to a secular principle of causation, and comes close to being, in a modern phrase, 'contingency-oriented'; that is, the author of the 'Succession narrative' sees the historical process as 'contingent' rather than as basically independent of human action. But although Jews and Greeks reacted in similar ways to the Persian presence which overshadowed them both, it is not plausible to postulate direct influence, in either direction, before the fifth century BC at earliest. And in the last analysis it is not until Herodotus that we find a whole work (as opposed to particular and fairly short biblical episodes) governed by the contingency-oriented approach.[18]

Hecataeus wrote a work of genealogy or mythography, and a geographical description or *periegesis* of his world (*FGrHist* 1). Both works survive only in 'fragments' (quotations by later writers), and we have many more fragments of the *periegesis* than of the other work. So Hecataeus' interest in genealogy may have been greater than his surviving output would suggest. Famously, his younger rival Herodotus scoffed at Hecataeus' own genealogical pretensions (2. 143): there was a god in my family sixteen generations back, Hecataeus had claimed. This may remind us of the Homeric Krethon and Orsilochos and their descent from the river-god Alpheios (above). It also looks forward to Alexander the Great's divine descent from Thetis via Achilles, through Achilles' son Neoptolemus, the ancestor of Olympias, Alexander's mother. Alexander took this seriously, and we shall see that ties of shared descent or kinship (συγγένεια) mattered more and more as the classical world melted into the Hellenistic.[19] Ancient antipathies persisted, as well as ancient

---

[18] R. Lane Fox, *The Unauthorized Version: Truth and Fiction in the Bible* (London, 1991), 188–90; R. N. Whybray, *The Succession Narrative* (London, 1968). On 'contingency-oriented' history see C. Meier, 'Historical Answers to Historical Questions: The Origins of History in Ancient Greece', in D. Boedeker (ed.), *Arethusa* 20 (1987), an interesting special number on *Herodotus and the Invention of History*, 41–57 at 44. Shared Greek and Jewish reaction to Persia: this was a favourite theme of the late A. Momigliano, see for instance his posthumous *The Classical Foundations of Modern Historiography* (Berkeley and Los Angeles, 1990), 17. I am grateful to Dr E. W. Nicholson for discussing these Old Testament topics with me.

[19] S. West doubts whether we can believe Herodotus' information about Hecataeus,

kinship: it is possible that the duel between Tlepolemos and Sarpedon in *Iliad* 5 reflects and symbolizes that eternal rivalry between Rhodes and Lycia which came most conspicuously to the surface after the Peace of Apamea in 189 BC (see Polybius 22. 5 etc.). But the nature of our sources (inscriptions tend to record friendly rather than hostile relations) means that more emphasis is placed on shared descent than on inherited enmities: the guest-friendship of Diomedes and Glaukos in *Iliad* 6 seems to indicate a tie between Argos and Lycia.[20] For the formulation of such ties, genealogical interest and skill was necessary: after all, as Jan Vansina puts it, 'genealogies are among the most complex sources in existence'.[21] This will come as no surprise to the student of the victory odes of Pindar, with their mythical pedigrees and their stress on inherited athletic ability. Pindar's very first poem, the Tenth Pythian, asserts Heraclid links between Sparta and Thessaly, while the Sixth Olympian, which deals with the family of the Iamidai, goes geographically very far afield to make its points.

But the transition from Homeric-style oral genealogies to the written genealogies of the fifth and later centuries was not smooth. It has recently been demonstrated that city and family traditions tend at the oral stage to concentrate on the remoter and even legendary past, and were naturally accurate on the 'narrow band' constituted by the *very* recent past; but the middle distance, so to speak, was neglected. It took the efforts of people like Hecataeus and Pherekydes to produce order from the chaos of family memory, and above all it took Herodotus (below) to force attention on the *recent* past.[22] Nevertheless, even after allowing for gaps in the traditions, and 'telescoping' of genealogies before Pherekydes and others got busy, one germ of Greek historiography was surely contained

either in book 2 or elsewhere: 'Herodotus' Portrait of Hecataeus', *JHS* 111 (1991), 144–60. For Hellenistic stress on kinship see esp. *SEG* xxxviii. 1476 and L. Robert, *Documents d'Asie Mineure* (Paris, 1987), 173–86; E. J. Bickerman, 'Origines gentium', *CP* 47 (1952), 65–81. The idea features in Thucydides more than is often recognized, see my forthcoming article 'Thucydides and συγγένεια'; J. Alty, 'Dorians and Ionians', *JHS* 102 (1982), 1–14 at 3. For kinship ties as the subject-matter of 'false' history in the Hellenistic period see Asklepios of Myrleia, cited by Walbank at *HCP* on 9. 1. 4.

[20] Rhodes–Lycia enmity: G. S. Kirk, *The Iliad: A Commentary*, ii: *Books 5–8* (Cambridge, 1990), 122. Argos–Lycia relationship: ibid. 180.

[21] J. Vansina, *Oral Tradition as History* (London, 1985), 182.

[22] For the Iamidai see A. Andrewes, 'Sparta and Arcadia in the Early Fifth Century', *Phoenix* 6 (1952), 1–5 at 1. For oral neglect of what one might call the middle past, as opposed to very recent events on the one hand and the legendary past on the other, see R. Thomas, *Oral Tradition*, ch. 3, and now her *Literacy and Orality*, 108–13; note also S. West, *JHS* 111 (1991), 151 n. 40.

in the ability and motivation to learn long, stirring, and complex pedigrees of Homeric and other pre-literate types.

Returning to Hecataeus, his divine pretensions may have struck Herodotus as preposterous, but Hecataeus was not the last scholar to have grand ideas about his own social pedigree: the eminent late sixteenth-century humanist Joseph Justus Scaliger and his father Julius Caesar Scaliger were convinced (wrongly, it seems) that they descended from the Della Scala rulers of Verona. 'Whatever the truth of that,' commented Wilamowitz of the younger Scaliger, 'he would still be a prince, every inch a prince.'[23] Herodotus was less charitable than this.

Hecataeus' genealogical book is a serious loss to us, but his geography or *periegesis* was in any case his more tangibly influential work. Herodotus was certainly familiar with it, as were later geographers and lexicographers (Strabo; Stephanus of Byzantium); and it was arguably a source for Thucydides in for instance his treatment of Epirus and North Greece, in sections where that austere writer becomes most noticeably ethnographic. I postpone this question until later (see below, p. 58).

Hecataeus' home town of Miletus was a Greek *polis* with a local Carian element (for which see Hdt. 1. 146). The same was true of Herodotus' home town of Halicarnassus, a little to the south. Both men thus grew up in a Greek-speaking, but not straightforwardly Greek, enclave of the Persian empire. This is surely relevant to the width of their intellectual horizons and their interest in ethnography; but there the similarities stop.

What was the relation between Herodotus and Hecataeus, and why did Herodotus write what seems (even allowing for the pathetic state of the surviving fragments of Hecataeus) to have been a different and richer sort of work? A problem of intertextuality greets us at the outset of Greek historiography, because the extent of Herodotus' debt to Hecataeus is unclear. That he regarded Hecataeus as his predecessor in a general way is not disputed, and is important; but the degree of acquaintance and consultation is not obvious or agreed. For Jacoby, who perhaps exaggerated the division between ethnography and history, Herodotus started as a Hecataean-style ethnographer and would have stayed one had it not been for the subsequent and direction-changing impact of Athens. In particular, Herodotus came under the wand of Pericles. An alternative view is that the irrigation which resulted in the blossoming of historiography, above all but not only in the achievement of Herodotus, was

[23] U. von Wilamowitz-Moellendorff, *History of Classical Scholarship* (London, 1982), 49.

provided by that greatest of 'Great Events', the Persian War. This view, with its correct insistence that Herodotus' attention to the recent past was a breakthrough, has powerful attractions—provided we do not over-look the Athenian and Hecataean dimensions (not just Hecataeus the ethnographer but Hecataeus the genealogist); and provided we do not force the chronological evidence for some shadowy figures, so as to produce prose 'Herodotuses before Herodotus'. (Simonides had already narrated Persian War events in *verse*, a generation before Herodotus: *POxy.* 3965.)[24]

Herodotus survives gloriously complete. This, for us, is one of his most important differences from Hecataeus. There are in our texts nine 'books' or subdivisions of Herodotus' work, but the warrant for this is no better than for the twenty-four books of Homer's *Iliad*:[25] that is, the nine Muses are not more organizationally respectable than the letters of the Greek alphabet.

Book-divisions will recur in this Introduction, and deserve a word to themselves. They are a fourth-century phenomenon, though an argu-ment has been made for an unfulfilled 'pentadic' plan (two matching five-book slices, cp. the 'decades' or ten-book slices of Livy) underlying the eight books of Thucydides' unfinished work of the fifth century.[26] But there are, for Thucydides, none of the compositional certainties which entitle us to say that the thirty books of Tacitus' *Annals* and *Histories* were conceived as a unity of 18 + 12 books, and represent five 'hexads' or six-book building blocks.[27] The decisive Tacitean consideration is an argument from 'closure' (for this term see below, p. 138): Tacitus closes each of the first three books, that is the first half-hexad, of the *Annals*

[24] General Hecataean influence on Herodotus is conceded even by S. West, *JHS*, 1991, 144–60 at 159, though she warns against seeing Hecataean reminiscences or anti-Hecataean polemic in Herodotus at every turn. The debate is an old one; for a detailed and attractive statement of the other view which upholds Herodotus' detailed dependence on Hecataeus see H. Diels, 'Herodot und Hekataios', *Hermes* 22 (1887), 411–44: still well worth reading. Jacoby's classic study of Herodotus is reprinted as *Griechische Historiker*, 7–164, see esp. p. 23 for Athens and p. 76 for Hecataeus. For the 'Great Event' view see R. Drews, *Greek Accounts of Eastern History* (Washington, DC, 1973), with which R. Thomas, *Literacy and Orality*, 108–13, sympathizes. Against 'Herodotuses before Herodotus' see my *Thucydides* (London, 1987), 19 n. 14. For Simonides see below, n. 163.

[25] On Homer see Taplin (above, n. 1), 285–93; N. J. Richardson, *The Iliad: A Com-mentary*, vi: *Books 21–24* (Cambridge, 1993), 20f. The number of books of Xenophon's *Hellenika* was variously given in antiquity as seven or nine: C. J. Tuplin, *The Failings of Empire: A Reading of Xenophon Hellenica 2.3.11–7.5.27* (*Historia*, Einzelschr. 76; Stutt-gart, 1993), 20.

[26] H. Rawlings, *The Structure of Thucydides' History* (Princeton, NJ, 1981).

[27] R. Syme, *Tacitus* (Oxford, 1958), 253 and app. 35 (disposing of other proposed schemes).

with a resounding allusion to *libertas*.[28] It is not conceivable that this is due to chance. Polybius' 40-book work is more complex, but here too it is possible to conjecture a basically hexadic structure, this time with a book of substantial digression closing each hexad and offering a pause. Book 6 deals with the Roman constitution and army; book 12 deals with historical method (see Derow, p. 85 below); book 18 functions less obviously as a digression, it must be conceded, but it does have some discursive discussion of military matters; and note the explicit statement at 18. 28 that Polybius now resumes the military discussion of book 6. Book 34 (but perhaps it was originally book 24 before Polybius decided to extend the chronological scope of his work by ten books[29]) is again thematically isolated from its context, and concerns geography; while book 40 (originally designed to be 30?) was an index or table of contents. But (to repeat) no such hexadic, pentadic, or other principles of internal division can be detected in Herodotus. They are a feature of a more consciously literate and book-minded culture than was Herodotus' world—a feature, that is, of the fourth and later centuries, not the fifth. (On this view the still partly oral Thucydides, who perhaps did no more than plan a work in two halves, is an in-between figure, a role he plays elsewhere. Thus his speeches occupy a middle ground between the looseness and untechnicality of the fifth century and the systematization and precision of the fourth.[30])

   Herodotus wrote about the Ionian revolt against Persia and the war or

[28] Tac. *Ann*. 1. 81 'libertatis imagine ... servitium'; 2. 88 'liberator haud dubie Germaniae' (cp. 'libertatem popularium' just above); at 3. 76 the artfully achieved reference to the liberators or tyrannicides Brutus and Cassius is another clear allusion, though the word *libertas* is absent. For closure see p. 138 below.

[29] Pol. 3. 4–5 with F. W. Walbank, *Polybius* (Berkeley–Los Angeles–London, 1972), 16f., 23. In my view, geography always introduced the final hexad.

[30] R. Thomas, *Literacy and Orality* (above, n. 16), 103–4, is surely right to allow the possibility that there was an oral aspect even to Thucydides (conversely, Thomas 102 is right to object that it is too simple to see Thucydides' predecessor Herodotus as representative of an 'oral style'). Rarely (5. 1, cp. 1. 8 and 3. 104; 6. 94. 1, cp. 6. 4. 2) Th. explicitly cross-refers ('as I have shown earlier') to his own narrative, surely an originally oral feature, cp. e.g. Hdt. 6. 43 with 3. 80. Another oral feature found in Th. is what Homeric scholars call 'clustering': e.g. ἐκκρούειν, 'to repulse', which is clustered in Th. bk. 4, esp. the last forty chs. For the concept see B. Hainsworth, *The Iliad: A Commentary*, iii: *Books 9–12* (Cambridge, 1993), 27f., 98; Richardson (above, n. 25), 89, 198. The famous claim at Th. 1. 22. 4 is perfectly compatible with recitation of (?) purple passages, see my *Commentary on Thucydides*, i: *Books I–III* (Oxford, 1992), ad loc. and *Thucydides* (above, n. 24), 29 and n. 65. G. Ryle, *Plato's Progress* (Cambridge, 1966), ch. 2, is excellent on the recitation of dialogues (like the Melian Dialogue at Th. 5. 85 ff.?). For oral tradition in Thucydides see E. Badian, *From Plataea to Potidaea* (Baltimore, 1993), 78. For the speeches point see my *Thucydides*, 96–109.

wars between Persia and Greece, a sequence which lasted from 500 to 479 BC. Like Homer before him and Thucydides after him, his main subject was thus a great war. But that programme, Homerically announced in the preface, was systematically attacked only in, roughly, the last five of his nine 'books': in the first four, Herodotus dealt more discursively with what has been called 'the other', that is the non-Greek areas which he had visited as ethnographer or rather tourist.[31] This is a long preamble, and such a distribution of attention would alone make it unlikely (see above) that the impact and trauma of the Persian Wars can suffice by themselves to explain the invention of history by Herodotus: there is some force in Jacoby's view that the ethnographic (and, surely, genealogical) background was still discernible in Herodotus. But the early books are structured around the growth of the Persian empire and are therefore not wholly discrete. Nor is their content such that they should be seen as a kind of historiographic fossil attached to the living organism of the later and more political or 'Thucydidean' books. On the contrary, the work of Momigliano and Gould has taught us to prize, for its own sake, Herodotus' exploitation of oral tradition and 'social memory' for the understanding of other cultures—even though his handling of religion (below, Ch. 3) has at first sight baffling features.

But did Herodotus really visit the non-Greek areas he claims to have visited? And did he really talk to the people he claims to have talked to? Detlev Fehling and O. Kimball Armayor have doubts on both counts.[32] Fehling's argument, an analysis of Herodotus' *Quellenangaben* or source-attributions, is worked out in thorough, sophisticated, and ingenious detail; in some quarters his book has become the—not always relevant or appropriate—standard by which all other, newer, books on Herodotus must be judged.[33] But though there are fairy-tale elements in Herodotus, particularly the early stretches, it is not likely, nor has it been demonstrated, that Herodotus is merely an ingenious teller of lies and fairy-tales—a view which surely makes too little allowance for the difficulty of using oral tradition. There is another objection. We sell the pass

---

[31] F. Hartog, *The Mirror of Herodotus: The Representation of the Other in the Writing of History*, tr. J. Lloyd (Berkeley–Los Angeles–London, 1988); J. Redfield, 'Herodotus the Tourist', *CP* 80 (1985), 97–118: tourists remain conscious foreigners, ethnographers align themselves with the people they are studying.

[32] D. Fehling, *Herodotus and his Sources: Citation, Invention and Narrative Art*, tr. J. Howie (Liverpool, 1989); O. K. Armayor, 'Did Herodotus Ever Go to the Black Sea?', *HSCP* 82 (1978), 45–62; id. *Herodotus' Autopsy of the Fayoum: Lake Moeris and the Labyrinth of Egypt* (Amsterdam, 1985); for other works by Armayor see Fehling, 261.

[33] See for instance S. West, *CR* 41 (1991), 23–5, review of a good book by D. Lateiner, *The Historical Method of Herodotus* (Toronto–Buffalo–London, 1989).

too cheaply if we allow that literary or rather rhetorical stylization of presentation are somehow incompatible with truthful reporting. Thucydides offers an obvious comparison. I have tried to show in Chapter 5 below that Thucydides uses the whole tool-box of rhetoric without, in my view, forfeiting his claim to be believed when he tells us (for instance) that there was a plague at Athens in 429 BC or—to take a still less controversial item, since even Thucydides' plague has actually been doubted in modern times—that the Peace of Nikias guaranteed free access to the common sanctuaries of Greece: 2. 47–54; 5. 18. 2.[34] Why should the same not be true of Herodotus? (In any case, the recent study of oral tradition, and of the development of what has been called social memory, has swung Herodoteans away from the blind alley down which they seemed at one time to be hurtling.[35] Herodotus' awareness of the double tradition about the foundation of Cyrene, pp. 11 f. above, is specially difficult to explain away in Fehling's sceptical mode.) There is an analogy with Homeric studies, though the oral approach has played a different sort of role. At one time (the 1960s) it was hard to find a good up-to-date book in English which treated Homer as poetry. But now the insights of the oralists have been digested, and the epics are treated once again as poetry, though of a special sort.[36]

It is, in fact, scarcely conceivable that any single book will say the last word about an author of the richness of Herodotus. The most satisfying recent approaches have been those which try to explain that richness and complexity (remember those thousand individuals). John Gould, in his

[34] The plague: A. J. Woodman, *Rhetoric in Classical Historiography* (Beckenham, 1988), 39; but see S. Hornblower, *Commentary on Thucydides*, i. 318.

[35] See above all J. Gould, *Herodotus* (London, 1989), esp. 136 n. 16 on Fehling, and ch. 2 generally (note the interesting p. 41 on how unthinkable it would have been for Thucydides to tell a story to make a political point, in the manner of Hdt. 5. 92, the Cypselid chapter); also D. Lateiner (above, n. 33); O. Murray (above, n. 13) and 'The Ionian Revolt', in *CAH* iv² (1988), ch. 8, an important study of Herodotus and the way oral traditions develop; H. Flower, 'Herodotus and Delphic Traditions about Croesus', in M. Flower and M. Toher (eds.), *Georgica: Greek Studies in Honour of George Cawkwell* (London, 1991), 57–77; H. Erbse, *Studien zum Verständnis Herodots* (Berlin, 1992), 153–6, arguing against Fehling on the Arion story at Hdt. 1. 23–4; P. J. Rhodes, *Défense et illustration des historiens grecs* (pamphlets from conference held at Liège, 1992); S. Hornblower, *Thucydides*, ch. 1. On oral tradition see R. Thomas (above, n. 10), and *Literacy and Orality* (above, n. 16).

[36] See above all J. Griffin, *Homer on Life and Death* (Oxford, 1980); see also C. W. Macleod, *Iliad XXIV: Commentary* (Cambridge, 1982), and more recently such sensitive works as Taplin (above, n. 1) and R. B. Rutherford, *Odyssey XIX–XX: Commentary* (Cambridge, 1992). For an outstanding application to the *Iliad* of techniques of modern literary theory, see I. de Jong, *Narrators and Focalizers: The Presentation of the Story in the Iliad* (Amsterdam, 1987), with Ch. 5 below.

brilliant book *Herodotus*,[37] sees reciprocity—requital for both benefits and injuries—as Herodotus' governing principle. This requital operates across geographical space and generational time, and is his way of 'structuring the past' (cp. above on genealogy and kinship). The world of Herodotus is a kind of vast 'Favour Bank', to borrow a concept from Tom Wolfe's 1989 novel *The Bonfire of the Vanities*,[38] where it is wittily used about the Irish community in late 1980s Manhattan. You make your deposits against a rainy day, and eventually you or your descendants (here we recall yet again the importance of lineage and genealogy in Homer and Herodotus) can expect to cash in. The network of favours and benefits is essential to Herodotus' causal scheme: compare Peter Derow (below, p. 76) on the revenge motive which led Darius to invade Scythia: 4. 1. 1. It is also essential to Herodotus' cultural geography: Egypt balances Scythia as a result of an urge for requital or symmetry (this is an approach taken further by François Hartog in his *Mirror of Herodotus* (see n. 31). Herodotus' search for requital, structure, and symmetry had its minus side. John Gould's paper on Herodotean religion (below, Ch. 3) shows that Herodotus, while noting ritual inversions (symmetries, we may call them), such as the non-performance by Persians of this or that Greek act, totally failed to grasp that Persians or other 'others' perceived religion differently.

Much of what has been said so far concerns the Herodotus of the first four books, and parts of books 5 and 6: in those two books only the Ionian revolt (part of book 5) and the battle of Marathon (part of book 6) are strictly part of the Persian Wars. In the final three books (7–9), covering the main Persian War of 480–479, Herodotus is more at the mercy of the kind of (perhaps unconscious) political pressure, leading to bias, which is a feature of Thucydides and Xenophon (below). Thus we can see from Badian's Chapter 4 below, that contemporary political sensitivities meant that Herodotus had to be careful where he trod, in his account of Athenian dealings with King Alexander I of Macedon; and that he played down, without quite covering up, both Macedonian and Athenian guilty secrets. Most of Badian's evidence is from books 7–9.

This raises the general question of Herodotus' politics, and on the whole it has to be said that his attitude to the major Greek states is remarkably fair-minded and positive: attempts to detect a vein of crude anti-Athenianism in Herodotus have not been made out. On the contrary, he was surely an admirer of Athens and its democracy, at a time

---

[37] For Gould see n. 35 above.
[38] *TLS* for 10–16 November 1989, 1237 (review of Gould).

when admiration for Athens was not fashionable (Thuc. 2. 8. 5): there is no getting round the implication of Hdt. 5. 78. 1 which praises Athenian democracy ($\dot{\iota}\sigma\eta\gamma o\rho\acute{\iota}\eta$) explicitly. As for Sparta, it has been well said that a theme of Herodotus' History is that the Persians gradually come to discover what their principal opponents the Spartans are like, and to recognize their good qualities, including courage and love of freedom. An even-handed attitude to Athens and Sparta is implied by passages like 6. 98, which deplores the sufferings brought on Greece as the chief Greek states fought for hegemony—a clear forward reference to the great Peloponnesian War, and perhaps an agonized but coded message to two states which he admired equally but in different ways.[39]

What impact did Herodotus' long work have? This raises general questions to which I return later in this Introduction. But we should not assume that all his contemporaries and successors knew his work equally well, or reacted to it in the same way if they *did* know it. (That is why I have called this section the 'story' rather than e.g. the 'development' of Greek historiography, because that more usual word has too smooth and inevitable a sound.) One probable imitator of his style and Ionic diction has already been mentioned, Antiochus of Syracuse, and I return to his general relation to Herodotus shortly. Theopompus in the fourth century did some kind of an epitome of Herodotus (*FGrHist* T 1; F 1–4), and other fourth-century writers like Nearchos (*FGrHist* 133) were certainly familiar with Herodotus.[40] But (to return to the fifth century) Thucydides' relationship to Herodotus is more problematic. I have argued elsewhere that Thucydides did know Herodotus' work (even this has been doubted), and that he followed Herodotus' factual content closely and pretty well exclusively *in speeches*, though not so slavishly in narrative,

---

[39] F. D. Harvey, 'The Political Sympathies of Herodotus', *Historia* 15 (1966), 254–5 for a succinct disproof of Strasburger's view (*Studien zur alten Geschichte*, ed. W. Schmitt-henner and R. Zoepffel (Hildesheim and New York, 1982), 592–626, originally in *Historia* 4 (1955), 1–25) that Hdt. was at least some of the time anti-Athenian. See also Jacoby cited above (n. 17) for the impact on Hdt. of his experience of Athens. Sparta: see D. M. Lewis, *Sparta and Persia* (Leiden, 1977), 148f.; and for the idea that Hdt. had a contemporary message, and was issuing tragic warnings to Athens and Sparta while admiring both, see C. W. Fornara, *Herodotus: An Interpretative Essay* (Oxford, 1971), esp. 65–91, 'The Archidamian War'. On Hdt.'s composition date see J. Cobet, in I. S. Moxon, J. D. Smart, and A. J. Woodman (eds.), *Past Perspectives: Studies in Greek and Roman Historical Writing* (Cambridge, 1986), 17. On Hdt.'s politics see also Badian, below, p. 121 n. 15.

[40] Antiochus: *FGrHist* 555 F 2 with Dover, *HCT* iv. 20, and in Herter (ed.), *Thukydides* (above, n. 15), 351. On Theopompus' 'epitome' of Herodotus see M. R. Christ, 'Theopompus and Herodotus: A Reassessment', *CQ* 43 (1993), 47–52, conjecturing that it was really just part of the *Philippika*. For Nearchos see below, p. 42.

where he felt freer to draw on alternative traditions.[41] But there was certainly some factual dependence; and we shall see later that Thucydides may have drawn on other written works, not just Hecataeus but Antiochus of Syracuse. It is possible for one historian to take his facts from another but to write a new sort of history with those facts: a well-known early modern example is Gibbon's use of Le Nain de Tillemont. Thus Rudolf Pfeiffer wrote that Tillemont 'provided Gibbon with the essential material for his new concept of Roman history'.[42] Does that dependence lessen Gibbon's stature? Mommsen and Wilamowitz were probably in a minority in thinking so.[43]

Thucydides, then, did something new with material which was partly second-hand. But it was in any case a small part: Thucydides was writing about contemporary events, most of which Herodotus never touched on at all (and the writings of Hecataeus and Antiochus helped him only over certain areas like Epirus and Sicily). Nevertheless Thucydides and Herodotus share with each other and with Homer their overarching thematic conception, a great war.

This was not true in the same way of the fifth-century *local* historians in the period after Herodotus, though they too were perhaps stimulated into activity by the Great Event—unless what stimulated them, in their different ways, was Herodotus' own example. Thus on the one hand Antiochus of Syracuse (*FGrHist* 555) perhaps wrote his important work to supplement Herodotus' patchy account of the west. On the other hand there were local historians who went over Herodotus' Persian War ground for their own purposes. For instance Lysanias of Mallos (*FGrHist* 426) seems, though not an Eretrian, to have combined local

---

[41] 'Thucydides' Use of Herodotus', in J. Sanders (ed.), *ΦΙΛΟΛΑΚΩΝ* (1992), 141–54. If I was right, and Th. followed different principles in speeches and narrative, he was following Homeric precedent, see J. Griffin, 'Words and Speakers in Homer', *JHS* 106 (1986), 36–57, with I. de Jong, 'Homeric Words and Speakers: An Addendum', *JHS* 108, 188–9. On one crucial item, the handling by the two writers of the Theban attack on Plataea, C. R. Rubincam, *LCM* 6 (1981), 47–9, argues that though Th. was certainly correcting Hdt., their disagreement on the numerical facts was not great. Some 'intertextual' awareness of Hdt. by Th. may just show that both were aware of topical issues raised in the years around 430, e.g. the Corinthian loan of ships at Th. 1. 41. 2, cp. Hdt. 6. 89. At Th. 1. 18, cp. Hdt. 5. 92 (Spartan hostility to tyranny), Th. may be in a way prior to Hdt.: see my commentary. And there is a sense in which Th. 3. 68 is actually prior to Hdt. 6. 108: Hdt.'s interest in Athenian–Plataian relations was surely prompted by the Peloponnesian War events described by Th.

[42] R. Pfeiffer, *A History of Classical Scholarship from 1300 to 1850* (Oxford, 1976), 133 f.

[43] Mommsen: G. W. Bowersock, 'Gibbon's Historical Imagination', *The American Scholar* (Winter 1988), 33–47 at 46 f. Wilamowitz (above, n. 23), 83.

history of Eretria with anti-Herodotean coverage of the Ionian revolt. (That is on the evidence of our one and only fragment, F 1!) The Ionian revolt also featured in the writings of the 'Horographer' (local annalist) Charon of Lampsacus (*FGrHist* 262 F 10), but his relation to Herodotus is not clear. He wrote about his native Lampsacus—but also (a separate work) about Spartan political institutions (T 1), perhaps as late as 403 when Sparta had won the Peloponnesian War and the eyes of Greece were turned on Sparta. But by then both Herodotus and Thucydides had both had plenty to say about Sparta, so perhaps Charon was reacting intertextually to them.

There is comparable ambiguity about another figure we have already mentioned, Hellanicus of Lesbos, the first Atthidographer or local historian of Athens. On the one hand he may like Antiochus have aimed in his *Argive Priestesses* (see below) to fill a Herodotean gap. But this gap was a gap in time (the Trojan War to contemporary Greece) rather than a neglected geographical area like Sicily. On the other hand Hellanicus lived in exciting times and we do not need intertextual arguments to explain why he was drawn to write the local history of Athens. Though a citizen of an island subject of the 'tyrant city' Athens, Hellanicus wrote about Athens at a time when (as Jacoby observed) other East Greeks were 'turning to the new sun' represented by Lysander's Sparta. Hellanicus is a very substantial figure indeed, not just a historian of Athens but a mythographer, a genealogist, and a pioneer of chronological research who catalogued the priestesses of Argos.[44]

Perhaps to call Hellanicus a 'pioneer' is unjust to Hippias of Elis, whose document-based list of Olympic victors was a first attempt to lay down a kind of international grid for chronology back to 776: all states of any consequence knew what went on at Olympia and Delphi, and many no doubt had a better idea about that than about the history of the *polis* next door. Hippias' researches transcended particularism. But Hippias, as we saw above when discussing Plato's *Hippias*, used oral tradition as well as records.

Chronology, of a more precise kind than genealogies could come up with, was important, and the Athenians of Hippias' generation seem to

---

[44] L. Pearson, *Early Ionian Historians* (Oxford, 1939), chs. 4 (Xanthus) and 5 (Hellanicus); for Hellanicus see also Pearson, *Local Historians of Attica* (Philadelphia, 1942), ch. 1. The essential studies of Hellanicus are by Jacoby in his *Atthis* (above, n. 17) and in the introd. to *FGrHist* 323a (the point about Hellanicus' time of writing, Lysander, etc. is at p. 20 of the introd. to 323a). For Charon see, in addition to Jacoby's commentary on no. 262, his remarks in *Atthis*, 68 and n. 113. On Antiochus see *Atthis*, 118 as well as the commentary on no. 555.

have acknowledged as much by erecting an inscribed list of their own archaic archons (ML 6=Fornara 23). They did this in the 420s, when Hippias was active. A connection between this list and Hippias is not provable; but it is plausible.[45]

There is one, and only one, Athenian archon date in Herodotus: he tells us (8. 5 1. 1) that the battle of Salamis happened in the archonship of Kalliades. Local Samian or other dates are to be found in his work—thus 3. 59. 4 dates something by a Samian king called Amphikrates. Thucydides (5. 20. 2) harshly considered that such dates were useless because parochial; and some of the literary activities of Charon and Hellanicus for instance may have been undertaken with a view to improving on Herodotus in the chronological department (not that Thucydides thought much of Hellanicus' efforts either). But we must allow for the difficulties Herodotus faced, when weaving together the oral traditions of so many distinct communities and nevertheless imposing chronological coherence on the whole. Thucydides, to whom we must now turn, was not the man to give him credit for this—or for anything else.

Modern historiographers, above all Momigliano (see n. 18 at his pp. 52f.), have reasserted emphatically and rightly the cardinal importance and value of the 'extrapolitical' history of Herodotus. Nevertheless, Thucydides, for any student of historiography, is The Master. No prose writer of the ancient world matched his gift for building an overarching narrative structure, his attention to significant detail, and his emotional sweep. His effects are achieved by the employment of a number of different registers, from the plain and prosaic to the epic and tragic. To say that is to say nothing new. Where modern research has advanced the study of Thucydides is by attending to the rhetorical effects of the plain and prosaic sections as well as to those more obvious set pieces whose emotional power so impressed ancient critics like Dionysius of Halicarnassus (see below, p. 136). Thus we have got used to the idea of a 'rhetoric of numbers' in Thucydides. There has also been study of Thucydides' devices for making us believe what he so authoritatively says: on all this see generally Chapter 5 below.

Here I propose to take up some other topics which shed light on Thucydides' rhetorical presentation of his historical research, while they cast doubt on some of the more ambitious claims made for him as the master of exhaustively accurate and objective history.

---

[45] R. Stroud in *Athens Comes of Age: From Solon to Salamis* (Princeton, NJ, 1978), 33, is sceptical.

Thucydides, as we have said, wanted to write up a great war, the Peloponnesian War of 431 to 404 BC (he got no further than 411 by the end of his work). But he was determined to improve on Herodotus in two ways. He wanted to show that his war was bigger; and he wanted to record it with un-Herodotean precision. Exactness over dates was one part of the new programme. But despite the explicit polemic of Thucydides 5. 20. 2, and his implied polemic at the beginning of book 2 (a whole series of different dating formulae for the beginning of the war), the odd thing is how *loose* Thucydides remains. This is true of his units not only of time but of space.

First, time. The Peloponnesian War began, says Thucydides at the beginning of book 2, in the forty-eighth year of the priesthood of the priestess Chrysis at Argos, when Pythodorus the archon at Athens had two months of his year to run, and so on.[46] It therefore comes as a surprise to notice that Thucydides very rarely dates events *inside* a year, so that scholars interested in the fine tuning of Thucydidean chronology are unsure whether or not he was, like Polybius, operating with a fixed date for the beginning of spring.[47] (It is not certain whether Thucydides reckoned the beginning of spring to be the time of the evening rising of the star Arcturus, but the better view is that Thucydides' spring simply had no fixed beginning at all.) It has even been said, perhaps a little harshly, that 'Thucydides' indicators of time are as infrequent and inexact as those of Herodotus'.[48] So what was the point of that elaborate information about for instance the priestess of Argos? There are I suggest two reasons for it. One is solemnity: this is a kind of drum-roll, a momentous way of signalling the *arche* or beginning of a momentous sequence of events. The other is polemical. Thucydides was impatient of other people's mistakes (see 1. 20) and of their imprecision; he himself was trying to offer an alternative chronological model. That model was a layout organized by campaigning seasons: see 2. 1 'my history has been

[46] For this question see J. D. Smart, 'Thucydides and Hellanicus', in I. S. Moxon *et al.* (eds.), *Past Perspectives*, 19–35. On 6. 2–5 (Sicily dates) see my forthcoming comm.

[47] A. Andrewes, *HCT* v (1981), 148f., cp. iv (1970), 20.

[48] O. Wenskus, *Gnomon* 37 (1990), 577. Major themes of Badian's important collection *Plataea to Potidaea* are that Thucydides' narrative contains much general 'selective omission and disinformation', and that over chronological issues in particular his 'interest in precision appears to be limited'; the quotations are from pp. 184 (from the essay on the *arche* of Philip, which itself argues for some chronological distortion, see 183) and 75 (the introduction to the essay on the chronology of the Pentecontaetia). Perhaps the most important single area of imprecision in all Thucydides (because the Persian factor was ultimately decisive) is the date at which Athens began to support Pissouthnes and Amorges. See Andrewes' classic exposition of the problem in *Historia* 10 (1961), 1–18.

set out by summers and winters', just four words of Greek. In the *next* chapter, the Theban invasion is signposted by traditional dating methods; it is cumbersome by comparison, a point Thucydides makes only indirectly in book 2, but which he makes quite explicit in book 5 (ch. 20). So part of the point of Chrysis and so forth is mocking: this is the way you have to do it if you are going to follow the old relative way by dating of magistrates.

The impression we are left with is of a Thucydides who is neither consistently accurate about chronology nor consistently sloppy, but capricious in his own handling, while bad-tempered about other people's efforts. Capriciousness is as we shall see an impression borne out in other areas where we would expect accuracy.

That brings us to space: Thucydides' units of distance. The stade is the standard ancient Greek unit of measurement. At 4. 3. 2 Thucydides says that Pylos is about 400 stades from Sparta. At first sight an impressively hard figure. The Greek word which I have translated stade is *stadion*, which is the same as our stadium. It just means a running track, and the connection between the two senses comes about because, allegedly, the most famous running track, that at Olympia in western Greece, was exactly a stade long. (Actually it is 192.3 metres.) But Olympia does not provide the necessary fixity because the Olympic stadium was said by Pythagoras to be unusually long.[49] A stade was supposed to be 600 *feet* but this does not solve the problem either, because human feet as we all know come in different sizes, although there is some evidence for attempts at standardization of the unit of measurement. An example of this is the sculpture called the metrological relief in the Ashmolean Museum in Oxford. It is just possible to make Thucydides' 400 stades for Sparta to Pylos work if one takes Thucydides to be working with a stade of 175 metres; the distance will then come to 70 kilometres. This means that the Sparta-to-Pylos distance was reckoned across the Langhada pass, the modern route to Kalamata. But perhaps Thucydides' figure is just wrong, because the easiest as opposed to the quickest route is to take a big loop to the north through southern Arcadia.[50] This would produce a distance of 600 not 400 stades.

At this point we may wish to ask whether Thucydides elsewhere operates with a precise stade, and the surprising answer is that his stade

---

[49] See the still valuable F. Hultsch, *Griechische und römische Metrologie* (Berlin, 1882), 32–3, 48–9; for Pythagoras, Aulus Gellius 1. 1 with L. Holford-Strevens, *Aulus Gellius* (London, 1988), 94.

[50] P. Cartledge, *Sparta and Lakonia* (London, 1979), 240f.

seems to fluctuate disconcertingly. R. Bauslaugh proved this in a short but painstaking and important article in 1979[51] by taking Thucydides' measurements for the distances between ancient places whose exact positions are still known and measurable, and seeing what unit of stade Thucydides' particular statements of stade distances imply he was using. For instance at the beginning of book 6 his distance across the straits of Messina implies a stade of 140 metres (we know it is roughly 2.8 kilometres; Thucydides says 20 stades). At 8. 67 (Kolonos to Athens) he works with 170 metres, and at 4. 66 (the Long Walls from Megara to Nisaia) the stade must be about 225 metres. There are many other examples and it is hard to find two exactly the same. This must make fundamentalist believers in the scientific accuracy of Thucydides queasy. We can evade the implications by saying that it was not Thucydides' own stade that varied, it was the stade of his informants; but if so, what becomes of Thucydides' reputation for accurate checking and critical intelligence? The only general conclusion can be that Thucydides was less than precise about figures for distance, just as he is less chronologically accurate than he seems at first sight to be. But his confident, detailed figures for stades, such as the statement (4. 45. 1) that Crommyon to Corinth is 120 stades, make us *feel* he is doing something reassuringly accurate.

I have looked briefly at chronology and at distance. Much the same can be said about other sorts of figures. One example is money. Even when he gives us a definite monetary figure, it is often useless to us because we do not know all we need to about his monetary assumptions.[52] Thus Thucydides on various occasions talks about Aiginetan obols and Persian Darics, i.e. coins,[53] but he never bothers to translate them into Athenian equivalents. All this is maddening because sometimes Thucydides shows that he had a hard-headed grasp of financial reality, indeed 'money matters very much' is one message of the *Archaeology*, i.e. the first twenty chapters of book 1; and at 7. 28 he says that the Athenians in 413 abolished the imperial tribute and substituted a tax on shipping of five per cent, thinking that would bring in more revenue. This sounds very modern, and is impressive when we remember

[51] R. Bauslaugh, 'The Text of Thucydides iv.8.6 and the South Channel at Pylos', *JHS* 99 (1979), 1–6, esp. the data set out at 5 f. This article underlies my text.

[52] For instance at 3. 70 a fine for cutting down some sacred poles is estimated at one stater per pole, where a stater is a unit of monetary exchange. This can be made sense of in the context only if we assume a stater of *gold*, as editors tend to do. That is all very well but Thucydides does not actually *say* gold. See my *Commentary* on 3. 70. 4.

[53] 5. 47. 3; 7. 28. 4.

that the ancient Greeks had nothing like double-entry bookkeeping. But what evidence did he really have for their reasoning? This is really guess-work or, as it is euphemistically called, inferred motivation, for which see below, p. 137. Why, we can reasonably ask in exasperation, did Thucydides not give us more tribute figures more often? Stylistic dislike of raw financial data might seem an attractive explanation; but look at 4. 57 where Kythera (a special case, agreed) is assessed at four talents tribute, or the statement at 7. 28. 4, just mentioned. So stylistic laws cannot be invoked here. The conclusion, as before, has to be that Thucy-didean accuracy is capricious and unreliable, and that there is much arithmetical vagueness and even 'innumeracy', or rather, perhaps, that there was a Thucydidean rhetoric of numbers.[54]

I am suggesting that Thucydidean 'precision' had its limits. It does not follow that Thucydides was an irresponsibly inaccurate historian: we can expect too much of him if we forget that things were made extremely difficult even for a would-be accurate historian by the extreme relativism and particularism of Greek city-state culture. What Thucydides *did* do was implant in the minds of posterity the idea and ideal of a universally based accurate narrative, even if he himself, with his sliding stades and so on, sometimes leaves us baffled.

Thucydides' other achievement is of course literary, but I have dis-cussed this aspect of Thucydides at some length in Chapter 5 below, to which I hope I may here be allowed to refer. This is, however, a good place to mention the biggest problem about narratological and other such sophisticated literary approaches to Thucydides: they leave aside problems of composition. Homeric scholars like Oliver Taplin (n. 1 above) who detect responsions between speeches, and echoes or other allusions within the narrative, tend (naturally) to be 'unitarians' rather than 'analysts'. Similarly, Thucydidean scholars who think that (for instance) items are deliberately held back from their logical place in one book to another later book, or that the historian makes use of the epic 'technique of increasing precision' (see below, p. 146), are basically

---

[54] Bookkeeping: R. Macve, 'Some Glosses on Ste Croix's "Greek and Roman Account-ing"', in P. Cartledge and D. Harvey (eds.), *CRUX: Essays presented to G. E. M. de Ste Croix on his 75th Birthday* (Exeter, 1985), 233–64. 'Innumeracy': see Lewis cited in my *Thucydides*, 202; and see C. R. Rubincam, 'Qualification of Numerals in Thucydides', *AJAH* 4 (1979), 77–9 at 86, for Thucydidean numbers as a 'rhetorical factor'. There has been some welcome recent work on Thucydides' use of numbers: see Rubincam above and in *LCM*, 1981 (above, n. 41), and her 'Thucydides 1.74.1 and the Use of *ΕΣ* with Numer-als', *CP* 74 (1979), 327–37; S. Berger, 'Seven Cities at Th. 6.20', *Hermes* 115 (1992), 25–9 and cp. below Ch. 5 n. 58 on the Attic slaves who deserted after the Spartan occupa-tion of Decelea.

making unitarian assumptions and treating the work as a literary whole;[55] and yet Thucydides' work is incomplete and contains layers which do seem to have been 'thought' at different times. The problem is perhaps best dealt with in a commentary, on individual passages as they arise, rather than here. But it deserves to be stated squarely. I think and hope that the two positions—admission that Thucydides' text is not a finished whole, and acknowledgment of its complex narrative structure and individual subtleties—are compatible. Even the most uncompromising analyst can hardly deny certain obvious examples of literary patterning, for instance the correspondences between speeches, or the way ἀληθεστάτη πρόφασις ('truest cause') at 1. 23. 6, the beginning of the whole war, is picked up at 6. 6, the beginning of the Sicilian expedition. Other resonantly recurrent expressions are 3. 49=7. 2, 'so close did Mytilene/Syracuse come to destruction', and 1. 110=7. 87, 'few out of many returned'. Or there are explicit back-references like the allusion at 7.71.7 to Pylos in book 4. One can surely be struck by such echoes without, for instance, necessarily subscribing to the 'pentad' theory of the composition of Thucydides (see above, n. 26; on that theory the two occurrences of the phrase 'truest cause' would occur at the beginning of each pentad). Some modern literary critics might put it differently, and say that it does not matter what Thucydides' own authorial intentions were, it is enough that the *text* has these features: we can study Thucydidean patterning without troubling too much whether the author Thucydides elaborated any pentad structure. This kind of view is fashionable but too easy: even allowing for the unfinished character of the work, and thus for the provisional character of Thucydides' decisions about its arrangement, the author Thucydides surely did give some thought to questions of the grand architecture of his work. But even if we lay theoretical considerations aside, the absence of good biographical information means that that 'surely' is not susceptible of proof.

Xenophon's *Hellenika* is often called a continuation of Thucydides, and this is obviously true in a general sort of way, although the fit between Thucydides 8. 109. 2 and Xenophon, *Hellenika* 1. 1. 1 is notoriously not exact. The absence of any formal preface, especially of any methodological announcement at the outset, is a surface disappointment both to professional modern historians and to students of 'openings' (below,

---

[55] See D. M. Lewis, 'Antony Andrewes', *PBA* 80 (1993), 221–31 at 229 on Andrewes' disquiet at the 'new generation of those who insisted on seeing [Thucydides'] work as a completed literary whole to be read from beginning to end'.

p. 138); but Xenophon's silence about what he thought he was doing is in a way an eloquent tribute to Thucydides, as is the adoption of a dating system organized by Thucydidean campaigning seasons. But this system is not consistently carried through, and the absence of much explicit methodology could equally be seen as a distancing device, or a sign that Thucydides' influence was merely superficial. In general, the assessment of Xenophon by Thucydidean canons is an obstacle to understanding.[56]

In the *Hellenika*, Xenophon was primarily interested in Greece and especially the Peloponnese, and his account of 411–362 is angled accordingly. His interest in Sparta has often been remarked, usually with the illicit extension that he was biased in favour of Sparta and Spartans. This is not so; nor is he consistently partial towards Athens, his place of origin. His dislike for Thebes is less controversial.[57] Xenophon's political prejudices do not run on completely straight lines: deviations tend to be explicable by reference to his strong moral and religious prejudices. Thus he withholds approval from the normally pious Spartans when they act in such a way as to incur the anger of the gods. Such attitudes, made explicit on the page, are a return to the god-fearing norms of poetry (Pindar as well as Homer) and of Herodotus, after the godless detour represented by Thucydides' history.[58] In other respects, too, Xenophon's *Hellenika* looks back to Herodotus: there are echoes of phraseology, and a similar sense of décor, a characteristic generally absent from Thucydides' wartime narrative (2. 34, 3. 104, and 6. 32 are exceptional). And Xenophon's use of speeches and digressions puts him closer to Herodotus than to Thucydides. Some of these similarities were already noted in antiquity, by Dionysius of Halicarnassus.[59]

[56] C. J. Tuplin, *The Failings of Empire* (n. 25 above), corrects many standard views of its subject; see esp. ch. 1 against those who see Xenophon as trying and failing to be Thucydides. For the literary qualities of the *Hellenika* see V. J. Gray, *The Character of Xenophon's Hellenica* (London, 1989). On Xenophon and the other fourth-century historians see W. Will, 'Die griechische Geschichtschreibung des 4. Jahrhunderts: Eine Zusammenfassung', in J. M. Alonso-Nuñez (ed.), *Geschichstbild und Geschichtsdenken im Altertum* (Darmstadt, 1991), 113–35. And see *CAH* vi² (1994), ch. 1, where I also look at Xenophon's other writings. For instance the *Agesilaus* has importance as a contribution to biography; see Momigliano, *The Development of Greek Biography* (Harvard, 1971), 50f. And on the *Cyropaedia* see Deborah L. Gera, *Xenophon's Cyropaedia: Style, Genre, and Literary Technique* (Oxford, 1993).

[57] Tuplin refutes the simple ideas that Xenophon was biased in favour of Sparta, or Athens, or was Panhellenist (i.e. concerned to bring about Greek unity against Persia): Tuplin, *The Failings*, 163–8, cp. 60, 67, 121 (Panhellenism), 157–62 (Athens).

[58] For Thucydides see S. Hornblower, 'The Religious Dimension to the Peloponnesian War, Or, What Thucydides Does not Tell Us', *HSCP* 94 (1992), 169–97.

[59] For Xenophon's literary debt to Hdt. see Gray (above, n. 56); Tuplin (above, n. 56), 167, accepts some similarity but finds that 'the overall effect is not very like

Thematically, it is Xenophon's other major work, the *Anabasis*, which suggests Herodotus, in its vitality, its charm, and its wide geographical extension. The *Hellenika*, like Thucydides, is more parochially Greek. The *Anabasis*, which recounts the Persian expedition of Cyrus the Younger in 400–399 BC, is the first autobiographical attempt at history: Xenophon was himself one of the chief commanders. (Herodotus and Thucydides do feature in their own histories, but only just.[60]) The auto-biography in the *Anabasis* is a vehicle for personal apologetic: Xenophon, particularly in the last couple of 'books', tries to answer his critics, e.g. to defend himself against the suspicion that he had been the leader of a dangerous rabble (6. 4. 8).

The Oxyrhynchus Historian, on the whole a more arid—but more reliable—writer than the chatty, judgemental, and Herodotean Xeno-phon,[61] was another continuator of Thucydides in a rough sense. What survives is a section of a detailed political and military history of part of the 390s; we also have, from an earlier section, some scraps dealing with the late fifth century. Xenophon's *Hellenika* never mentions Thucydides, and we are left to infer what we like from his general handling and the way he allowed a small interval to elapse since Thucydides' terminal point, thus perhaps signalling his independence in a small way. By contrast the Oxyrhynchus Historian, intriguingly, *did* mention Thucy-dides (ch. 2, Bartoletti), but the text is so fragmentary at that point that we can only guess what he said about him. The context is a discussion of the Spartan Pedaritos, who had featured in Thucydides book 8; it seems possible that Oxyrhynchus Historian here digressed, with some back-reference to Thucydides, on the ambition and empire-building of various prominent Spartans.[62] Words like 'seems' are specially appropriate for

---

Herodotus'. T. S. Brown, 'Echoes from Herodotus in Xenophon's *Hellenica*', *Ancient World* 21 (1990), 97–101, interestingly compares Xen. *Hell.* 3. 1. 4 with Hdt. 3. 44, and Xen. *Hell.* 5. 4. 48 with Hdt. 3. 129–30, and suggests that the echoes are deliberate. Dionysius: 777–9 Usener–Radermacher, with Gray, 4f.

[60] Herodotus is hardly an agent in his own history, but 2. 143 is autobiographical after a fashion, and there are a large number of assertions in the first person singular throughout. Thucydides is an agent in his own work, see 4. 104. 4–106. 3 and below, p. 132, citing Genette.

[61] Xenophon judgemental: see e.g. Xen. *Hell.* 5. 1. 4, 6. 5. 51 on Teleutias and Iphi-krates, with Tuplin (above, n. 25), 37f., 73, 162. For a slight reaction against the usual view of the 'Thucydidean' reliability of the Oxyrhynchus Historian (HO) see Tuplin 12–13, who, however, cites only V. J. Gray, 'Two Different Approaches to the Battle of Sardis', *CSCA* 12 (1979), 183–200. But the battle of Sardis is a problematic topic, see *CAH* vi² (1994), ch. 3.

[62] See the interesting discussions by Andrewes, *HCT* v (1981), 84, citing unpublished suggestions by H. T. Wade-Gery. (HO ch. 2 has the words φι]λοτιμίας and δυ]ναστείαν.)

the Oxyrhynchus Historian, who has come down to us in such tatters. But what does survive, especially the opening on the origins of the Corinthian War, looks like serious Thucydidean grappling with ideas about causation (see Peter Derow, below, p. 80).

The authorship of this anonymous history was much discussed in the early years of the present century; inconclusively. The most obvious candidates would be historians, other than Xenophon, known to have begun their histories in 411. Names like Kratippos and Daimachos of Plataia[63] (*FGrHist* 64–5) take their plausibility, or such plausibility as they have, not from their starting dates (unknown), but from their interest in Boiotian affairs, which were a particular strength of the Oxyrhynchus Historian. A more substantial figure than these two, of each of whom only a handful of fragments remain, is Theopompus (*FGrHist* 115). His claim does rest partly on his known starting date of 411/410 (T 13). That was in his *Hellenika*. But Theopompus surely disqualifies himself as author of the *Oxyrhynchus History* by his waspishness and extreme views, as evidenced above all in another of his works, the *Philippika*, a history of Greece with Philip II of Macedon at the centre. The title of that work was one of the most important things about it: it denotes a single man, not a *polis* or area. It is, to that extent, an *Odyssey*, a poem about the man Odysseus, not an *Iliad*, a poem about Ilion or Troy.[64] Poems traditionally celebrated individuals and glorified their deeds (above, p. 7); it was a departure for history to put one man centre-stage in the way poetry had always done. In that sense Theopompus was innovative (though see below); but he schooled himself very tradition- ally, by doing an epitome of Herodotus (above, p. 21). Much that we think we know, or that we believe, about a *diadoche* or apostolic

---

[63] For Daimachos see F. Schwarz, 'Daimachos von Plataia: zum geistegeschichtlichen Hintergrund seiner Schriften', in R. Stiehl and R. Stier (eds.), *Beiträge zur alten Geschichte* (F. Altheim Festschrift), i (Berlin, 1969), 293–304. I have discussed Daimachos in 'Thucy- dides on Boiotia', forthcoming in the *Proceedings of the Second International Congress of Boiotian Studies*, where I argue that Jacoby was right to identify (and Schwarz was wrong to refuse to identify) the poliorcetic writer, i.e. writer about sieges, with the fourth-century not the Hellenistic Daimachos. Kratippos: H. Bloch, 'Studies in the Historical Literature of the Fourth Century BC', in *Athenian Studies Presented to W. S. Ferguson* (*HSCP* suppl. 1; 1940), 303–76 at 313: Kratippos unlikely to be the HO because Dionysius of Halicarnas- sus, who knew Kratippos, (wrongly) says that nobody after Th. used his summer-and- winter scheme; but HO does; therefore HO was not Kratippos.

[64] True, it has been said that the *Iliad* might just as well have been called the *Achilleia*: G. S. Kirk, *The Iliad: A Commentary*, i: *Books 1–4* (Cambridge, 1985), 52. But it is also the tragedy of Hector, so the title we have is at least even-handed in its refusal to single out any one man. See J. Redfield, *Nature and Culture in the Iliad: the Tragedy of Hector* (Chicago, 1975).

succession of historians is conjecture, but Theopompus' close attention to Herodotus is fact, whatever use he later made of the knowledge so acquired. Thucydidean influence on Theopompus is harder to establish, beyond the bare awareness implied by the opening date of the *Hellenika*; but maybe the Theopompus of the *Hellenika*, in his treatment of Philip and other charismatic individuals, took further Thucydides' building-up of the individual Alcibiades. By the time Theopompus' contemporary Aristotle wrote his *Poetics* (1451ᵇ7), history could be defined as 'what Alcibiades did and suffered'.[65]

Aristotle's own contribution to Greek historiography was indirect, but nevertheless profound. It took two forms. First, methodological. He defined what history was: it concerned itself with the particular ('what happened to Alcibiades'), by contrast with poetry, which deals with 'the kind of thing that might happen',[66] or as we might say 'the universal'.

Aristotle's second contribution was practical. He wrote nothing which is normally (i.e. by Aristotelian criteria!) categorized as history; that is, no diachronic account of a war or expedition or of the deeds of an individual. But by the more hospitable standards of a modern structuralist approach, the *Politics* of Aristotle, and that important but lost comparative treatise the *Laws* of his pupil Theophrastus (on whom see below, Ch. 6), should certainly count as history. The *Politics* is a synchronic analysis of civic organization, real and ideal, and draws on historical exempla for the construction of its models. So ambitious an edifice would not have been possible (and this is equally true of Theophrastus' *Laws*) without the laying of deep foundations in *historia*— using that word in its original Herodotean sense of 'enquiry' or 'research', the sense in which Herodotus had offered his work as a setting-forth of an enquiry. Of Aristotle's lost works, the list of victors at the Pythian (Delphic) games, drawn up in collaboration with his young relative Kallisthenes (see below) looks back to the Olympic researches of the fifth-century Hippias of Elis (Tod 187). It was presumably raw material for history, not history proper. But more sophisticated groundwork,

[65] For the idea that Thucydides became increasingly prone to stress the role of individuals see S. Hornblower, *Thucydides*, 146–7. Theopompus: see R. Lane Fox, 'Theopompus of Chios and the Greek World 411–322 BC', in J. Boardman and C. Vaphopoulou-Richardson (eds.), *Chios: A Conference Held at the Homereion in Chios 1984* (Oxford, 1986), 105–20; G. Shrimpton, *Theopompus the Historian* (Montreal—London—Buffalo, 1991). Alcibiades: see next n.

[66] Aristotle on history: G. E. M. de Ste Croix, 'Aristotle on History and Poetry (*Poetics* 9, 1451ᵃ36–ᵇ11)', in B. M. Levick (ed.), *The Ancient Historian and his Materials: Essays in Honour of C. E. Stevens on his Seventieth Birthday* (Farnborough, 1975), 45–58, repr. in A. Rorty (ed.), *Essays on Aristotle's Poetics* (Princeton, NJ, 1992), 23–32; S. Hornblower, *Thucydides*, 9f. and refs.

carried out by Aristotle or his pupils, was necessary for the *Politics*: no less than a city-by-city treatment of the Greek states, the so-called *Constitutions* or *Politeiai*. Theophrastus' *Laws* drew, it is thought, both on the *Politics* and on the individual *Constitutions*. One achievement of the Aristotelian project survives, the *Constitution of the Athenians* (*Athenaion Politeia*, here abbreviated as *Ath. Pol.*), discovered in 1890. This, to put it in modern language, blends the diachronic and the synchronic, the developmental and the structural: the first part is a chronological account of the development of the Athenian constitution. This is followed by an analysis of its structure.[67]

The *Ath. Pol.* was surely one of the longest, most complex, and important of the series of Aristotelian *Constitutions*. Its author's dependence on Herodotus is both demonstrable and at one point explicit (Herodotus is cited at 14. 4); his dependence on Thucydides is merely demonstrable.[68] But there agreement between modern scholars stops. In particular we do not know how much the *Ath. Pol.* owes to Hellanicus' fourth-century successors in the field of Atthidography (see above), men like Androtion.[69] ('Men' is the right word. There were few women historians after Homer's Helen, no doubt because history was so often concerned with war, and the general ancient feeling was, as Homer makes Hector say to his wife Andromache, that war should be a concern to men: *Iliad* 6. 492–3. The exceptions are very few. The female Alexander-historian Nikoboule (*FGrHist* 127) may be a man writing for some extraordinary reason under a pseudonym—the opposite of the 'George Eliot' phenomenon. And a woman called Hestiaia wrote in an apparently historical way about the topography of Homer's Troy (Strabo 13. 599). But we know very little about her. We have rather more fragments of Pamphile, a historian of the time of Nero (*FHG* 3. 520–2).

[67] For the *Ath. Pol.* see P. J. Rhodes, *A Commentary on the Aristotelian Athenaion Politeia* (Oxford, 1981); J. Keaney, *The Composition of Aristotle's Athenaion Politeia: Observation and Explanation* (Oxford and New York, 1992). On the *Politics*, especially in its relation to Theophrastus' *Laws*, see Keaney ch. 3 and Bloch (above, n. 63), 355–76. For Theophrastus' *Laws* see now S. C. Todd, *The Shape of Athenian Law* (Oxford, 1993), 39. Keaney, in M. Piérart (ed.), *Aristote et Athènes* (Paris, 1993), 267, seems disposed to deny Theophrastus' *Laws* the title of history, on the grounds that chronology is the backbone of history, whereas Theophrastus' treatment was not chronologically based. But we do not have to accept the narrow definition of history here implied, and Keaney (p. 270) acknowledges that Theophrastus' *Laws* was 'based on comparative material, reflecting historical research'. Note, however, Keaney's point, in both his 1992 and 1993 studies, that Theophrastus' work was closer in spirit and approach to Plato than to Aristotle.

[68] For Thucydides see Rhodes, *A Commentary* (above, n. 67), 363, and for the *Ath. Pol.*'s sources generally, Rhodes, 15–30.

[69] See Rhodes, *A Commentary*, 19, discussing the scepticism of Phillip Harding.

In view of the scholarly disagreement about the sources for the *Ath. Pol.*, which survives, it is hardly surprising that we are so totally in the dark about the sources drawn on by the Aristotelian school when it turned, in treatises now lost, to more distant *poleis*, for instance those in the *west* (Italy and Sicily: cp. e.g. *FGrHist* 566 Timaios F 51 for an item about Italian Siris which probably shows Timaean use of Aristotle).

In his paper on Aristotle's pupil and successor as teacher, Theophrastus, P. M. Fraser (below, pp. 182–8) shows that the Aristotelian school had some knowledge of the west, including even Corsica, and of the peculiarities of western Hellenism. But this knowledge was imperfect. There was no systematic study of the west, from the perspective of Old Greece, before Timaeus of Tauromenium (*FGrHist* 566) did his researches in Athens, *c*.300 BC;[70] though Hieronymus of Cardia, the historian of Alexander's Successors, sketched Rome's early history (*FGrHist* 154 F 13).[71]

Sicily, and in particular the rulers of Syracuse, produced or attracted historians at much earlier dates than Aristotle, let alone Timaeus. We have already noticed Thucydides' contemporary Antiochus of Syracuse. He seems to have ended his history with the conference of Gela in 424 BC, but there were other local historians for the biographer Plutarch (*c*.50–120 AD) to use when he wrote his *Life of Nikias*, above all Philistos (*FGrHist* 556). Plutarch was an eclectic reader,[72] and wanted alternatives and supplements to Thucydides' account, in books 6–7, of the Athenian disaster in Sicily (415–413). Philistos provided one.

Philistos was also a western source (Jacoby thought a 'main source', a *Hauptquelle*) used by a key figure in fourth-century Greek historiography, indeed Greek historiography generally: Ephorus (*FGrHist* 70).[73]

---

[70] Timaeus: T. S. Brown, *Timaeus of Tauromenium* (Berkeley–Los Angeles, 1958); P. M. Fraser, *Ptolemaic Alexandria* (Oxford, 1972), 763–74; A. Momigliano, *Essays in Ancient and Modern Historiography* (Oxford, 1977), 37–66; L. Pearson, *The Greek Historians of the West* (Atlanta, 1987); K. Meister, 'The Role of Timaeus in Greek Historiography', *SCI* 10 (1989–90), 55–65; F. W. Walbank, 'Timaeus' Views on the Past', *SCI* 10 (1989–90) 41–54, Ger. tr.=Xenia 29 (Konstanz, 1992), *Timaios und die westgriechische Sicht der Vergangenheit*.

[71] J. Hornblower, *Hieronymus of Cardia* (Oxford, 1981), 248–50.

[72] See further below, p. 63, citing Pelling (see esp. n. 158).

[73] Jacoby's comment is at *FGrHist* 556 Philistos, commentary, iii(b), 501. The classic account of Ephorus remains that of Ed. Schwartz, reprinted in his *Griechische Geschichtschreiber* (Leipzig, 1959), 3–26. G. L. Barber, *The Historian Ephorus* (Cambridge, 1935), exaggerates Ephorus' debt to Isocrates. A. Andrewes was working on a book about Ephorus at the time of his death; according to Lewis, *PBA* 1993, 230, there is 'surely matter to salvage here', but no real hope of a book. On Ephorus see also Sacks, below, pp. 217 ff. and R. Drews, 'Diodorus and his Sources', *AJP* 83 (1962), 383–92.

Ephorus certainly admired Philistos personally—see Plutarch, *Dion* 36.
Ephorus, who was active till about 330 BC, covered everything from
mythical times (the 'Return of the Heraklidai', i.e. roughly the time of the
Dorian Invasion) to 340 BC, four years before the death of Philip.
Ephorus' main concern was Greek affairs, but he wrote up Persian
history too, perhaps using Dinon of Kolophon (*FGrHist* 690), the father
of the Alexander-historian Kleitarchos. The relation between the writings of father and son is controversial, but some influence is surely
plausible. These two important figures, and we can add the well-
informed Heraclides of Kyme (*FGrHist* 689), are surely evidence that
Greek historiographic interest in Persia in the fourth century was not all
trivial—especially if Dinon ultimately (i.e. via Ephorus) lies behind
Diodorus' account (15. 90–2) on the Revolt of the Satraps in the 360s.
Even if those chapters exaggerate the scale of the affair, the information
is precious and surely not fabrication.[74] Another author of *Persika*,
Ctesias (*FGrHist* 688), cannot be ruled out as a source for Ephorus.

For Greece before the fourth century, Ephorus used the obvious
sources, Herodotus and Thucydides, supplementing them from authors
like Hellanicus. But Ephorus had his own way of doing things. Polybius
seems to imply that Ephorus recognized that some readers wanted stories
about city foundations and kinship ties (Pol. 9. 1), whereas others liked
to hear about genealogies; both these sorts of writing were distinct from
the third, which was political history in the strict sense. Ephorus supplied
all these demands, but particularly the second (though kinship ties are by
no means absent from Thucydides); and he also added plenty of moral
praise and censure.[75] He is usually thought to have been, together with
Timaeus, the direct source of the relevant parts of the *Library* or
universal history of Diodorus of Sicily, who wrote in the time of

[74] On Dinon see R. B. Stevenson, *Persica* (forthcoming), and provisionally 'Lies and
Inventions in Deinon's Persica', in Sancisi-Weerdenburg and Kuhrt (above, n. 13) 27–35.
A. Momigliano, *Alien Wisdom* (Cambridge, 1975), 134–5 and ch. 6 generally, took a
minimalist view of Greek historiographic interest in Persia in the 4th cent. BC. See however
Stevenson, forthcoming. Momigliano (p. 135) wrote that 'not one of Clitarchus' fragments,
nor any of the sections of Diodorus Book 17 which can be traced back to him, alludes to
Persian institutions'. But see *FGrHist* 137 F 5 on the wearing of the upright tiara, a sign of
rebellion if done by a subject. Dinon's known interest in Persia could have had some influence on his son Kleitarchos, who was presumably born in Kolophon when it was a subject
Persian city. For the Revolt of the Satraps see M. Weiskopf, *The So-called 'Great Satraps'
Revolt', 366–60 BC: Concerning Local Instability in the Achaemenid Far West* (Stuttgart,
1989), with *CR* 40 (1990), 363–5.

[75] For the Polybius passage, the interpretation of which is tricky, see Walbank's commentary, vol. 2 (Oxford, 1967); also K. S. Sacks, *Polybius on the Writing of History*
(Berkeley–Los Angeles–London, 1981), 178–9.

Augustus. Books 1–5 (mythology and early civilizations) and 11–20 of Diodorus survive complete. Of these, 11–16 are from Ephorus. They deal with mainly Greek history in the period from the early fifth century to 340 BC. (Books 17–20, for which see further below, covered the rest of the fourth century.) Books 6–10 survive only in fragments, but here too it is plausible to assume dependence on Ephorus. But Kenneth Sacks (below, pp. 217 ff.) shows that the identification of specifically Ephoran material in Diodorus is not straightforward, because Diodorus had independent views and did not reproduce Ephorus mechanically.

For the fourth century, the identification of the traditions lying behind 'Diodorus/Ephorus' (if that simple formula is permissible) becomes very much more important because of the absence of any surviving primary source of the reliability of Herodotus or Thucydides: Xenophon is not in the same class as those two. Ephorus made evident use of the Oxyrhynchus Historian, or (to express it more cautiously) the same tradition as the Oxyrhynchus Historian, for the late fifth and early fourth centuries. This produces an interesting and unusual situation because in this period we can for once place two surviving traditions side by side: Xenophon's *Hellenika*, and the alternative tradition, represented by Diodorus, which on the simplest view goes back via Ephorus to the Oxyrhynchus Historian. The result of such comparative operations is usually to the advantage of Diodorus/Ephorus, though Diodorus can introduce error which is unlikely to have featured in the Oxyrhynchus Historian, while Xenophon tends to cut a poor figure by comparison.[76]

At some point in his account, probably but not certainly at the watershed represented by the King's Peace of 386,[77] Ephorus, and so ultimately Diodorus, switched sources, presumably because the 'Oxyrhynchus Historian' tradition had dried up. The new source was Kallisthenes, whom we have already met as Aristotle's relative and collaborator. Kallisthenes' *Hellenika*, which was also a main source for Plutarch's important *Life of Pelopidas*[78], was as interested in Thebes as

---

[76] R. Littman, 'The Strategy of the Battle of Cyzicus', *TAPA* 99 (1968), 265–72; A. Andrewes, 'The Arginusai Trial', *Phoenix* 28 (1974), 112–22, and 'Notion and Kyzikos: The Sources Compared', *JHS* 102 (1982), 15–25. Note the convincing argument of H. D. Westlake, *Studies in Thucydides and Greek History* (Bristol, 1989), ch. 17, repr. from *Phoenix* 41 (1987), that the Oxyrhynchus Historian, mediated by Ephorus, lies behind Diodorus' account of Cyrus' expedition. But see above, n. 61, for doubts (Gray, Tuplin) about this historian's invariable superiority to Xenophon. But the essential case remains, I think, secure.

[77] See S. Hornblower, 'When was Megalopolis Founded?', *BSA* 85 (1990), 71–7 at 73 n. 6.

[78] H. D. Westlake, 'The Sources of Plutarch's *Pelopidas*', *CQ* 33 (1939), 11–22.

Xenophon was dismissive and hostile.[79] It has been said that not all literary losses from the ancient world are inconsolable, and there is something in this;[80] but the loss of the *Hellenika* of Kallisthenes is surely one of the more serious losses.

Kallisthenes' output spanned the half-century and more from the King's Peace to Alexander, because he wrote not only the *Deeds of Alexander*[81] but also a monograph on the Third Sacred War or war for control of the sanctuary of Delphi, fought between Thebes and Phokis in 355–346. This war, cf. Davies, p. 197 below, may have influenced the detailed tradition about the First Sacred War. The Third Sacred War was the episode which brought Philip into Greece, and Philip must have featured in it. But we have only one uninformative fragment (*FGrHist* 124 F 1), so generalization is not advisable.

Philip (as we have seen) was a main subject of Theopompus' historical work. And Diodorus' book 16 is crucial for modern students of Philip, though his sources for the period between 340 and 336 (in source terms, the period between the end of Ephorus and the beginning of Kleitarchos, Diodorus' source for Alexander in book 17) are an insoluble problem.[82]

Otherwise, it is a feature of the mid-fourth century, the age of the great Athenian political orators, and of Philip, that we do not derive our most important evidence for it from 'historians' in the usually accepted sense of that word. Naturally the orators are crucial: there is a good deal of history in Demosthenes' *On the Crown*, a long speech of retrospective political self-justification. Diodorus' source in the latter part of book 16 certainly knew and exploited this speech.[83] But there are other and more subtle traditions as well.

History is a way of looking at the past, and historiography, which is the work of fallible human beings with often strong political and other beliefs, cannot help but manipulate the past. Such manipulation was particularly intense in the fourth century, in ways discussed by John

[79] Note in particular *FGrHist* 124 F 11 and F 18: the battle of Tegyra (375 BC), omitted by Xenophon. It was important because it was a Theban victory over Sparta, and so anticipated Leuctra in 371.

[80] G. L. Cawkwell, *CR* 39 (1989), 245, reviewing Pearson (above, n. 70); Cawkwell was referring, very surprisingly, to Timaeus.

[81] L. Pearson, *Lost Histories of Alexander the Great* (New York, 1960); P. Pédech, *Historiens compagnons d'Alexandre* (Paris, 1984), ch. 1.

[82] A. B. Bosworth, 'Philip II and Upper Macedonia', *CQ* 21 (1971), 93–105 at 98, suggests a late Hellenistic writer, aware of e.g. Demosthenes 18, *De Corona*. Why not Diodorus himself? I suggested this at *CR* 34 (1984), 263.

[83] Diod. 16. 84–5 is based on Dem. 18. 169–74, the famous ἑσπέρα μὲν γὰρ ἦν narrative.

Davies in Chapter 7 below. It was during or soon after the Third Sacred War of 355–346 that there finally crystallized a tradition about a 'First' Sacred War, allegedly fought two and a half centuries earlier, in the early sixth century. Davies shows that though Herodotus and Thucydides do indeed ignore this war (this can be explained, see Section IV below), nevertheless there are problems about the radical sceptical position which treats the war as total fiction. But there is no doubt that one result of the increasing document-mindedness (as it has been called[84]) of the fourth century was a corresponding increase in the *invention* of documents.[85] In particular, there was, already in the fourth century, much creative working of the tradition about the Persian Wars. This is a topic explored, with special reference to the Hellenistic and Roman periods, by Antony Spawforth (below, Ch. 8). But it was in the fourth century that the so-called Themistocles Decree (ML 23 = Fornara 55) was concocted in the form it has come down to us, though the actual carving of the inscription was done in the third century. The decree purports to date from the Persian Wars (480 BC) and urges solidarity against 'the barbarian' (Xerxes? or Philip?) 'on behalf of the freedom of Greece'. This is only one of a proliferation of 'false documents'[86] and 'invented traditions'[87] from the fourth century, which Anthony Grafton has described as the beginning of the heyday of forgeries.[88] Another example is the tradition about Messenia and its wars of independence.[89] There certainly was an eighth-century struggle between Sparta and Messenia, and another in the seventh; but it was only after the re-establishment of Messenia as an independent state in the early 360s that the war of subjection was shaped into an epic with heroes to match: see *FGrHist* 265 Rhianos of Bene F 42 (early Hellenistic) for a startling comparison between the ?seventh-century Messenian hero Aristomenes and Achilles.[90] It is hardly controversial to say that the modern historian of the fourth century cannot neglect such evidence for perceptions and propaganda. The difficulties arise for the historian of the early fifth

[84] Thomas (above, n. 10), 42, cp. 38, drawing on a term of M. T. Clanchy, *From Memory to Written Record* (London, 1979).

[85] C. Habicht, 'Falsche Urkunden zur Geschichte Athens im Zeitalter der Perserkriege', *Hermes* 59 (1961), 1–35. [86] Habicht's term; see previous n.

[87] E. Hobsbawm and T. Ranger (eds.), *The Invention of Tradition* (Cambridge, 1983).

[88] A. Grafton, *Forgers and Critics* (Princeton, NJ, 1990), 11.

[89] L. Pearson, 'The Pseudo-History of Messenia and its Authors', *Historia* 11 (1962), 397–426.

[90] H. T. Wade-Gery, 'The "Rhianos-hypothesis"', in E. Badian (ed.), *Ancient Society and Institutions: Studies Presented to Victor Ehrenberg on his 75th Birthday* (Oxford, 1966), 289–302; Pearson (above, n. 89), 410–11, 417–26.

century, or the sixth, or seventh, or eighth—that is, the more remote and even shadowy periods which the documents and traditions purport to deal with.

Kallisthenes had a hand in this creative working-up of early Messenian history: *FGrHist* 124 F 23–4 (from the *Hellenika*), with which compare Diodorus 15. 66, surely Kallisthenes mediated via Ephorus. That was by no means Kallisthenes' only topical treatment of long-past events. He may himself have been responsible for a propaganda stroke of brilliance: the depiction of Alexander's invasion of the Persian empire as an act of reprisal for the Persian Wars of 150 years earlier; specifically, for the impiety of Xerxes' destruction of the temples of Athens.[91] Perhaps Kallisthenes got the idea of a Graeco-Macedonian jihad from his work on the history of Delphi—not so much the list of Pythian victors as the monograph on the ten-year Sacred War, which punished the impiety of the Phokians and ended in the year (346) when we first hear of serious Macedonian designs on Persia.[92]

Kallisthenes' most famous literary work was the *Deeds of Alexander*. It is to Alexander that we must now briefly turn. The historical tradition about Alexander is a very long one indeed. None of the surviving full-length accounts is less than 300 years later than Alexander's own time (Diodorus' book 17, which derives from Kleitarchos, was written in the time of Augustus); and Arrian and Plutarch are more than 400 years later. Quintus Curtius Rufus' dates are disputed but he probably flourished in the first century AD. Within this group two main sub-traditions have traditionally been discerned, the 'vulgate', which goes back to Kleitarchos the son of Dinon (this tradition is represented above all by Diodorus and Plutarch, also Justin); and the 'main source' tradition which goes back to Ptolemy, the later king of Egypt, and to Aristoboulos of Kassandreia (*FGrHist* 138–9). This second tradition is represented by Arrian, who in his preface notoriously expressed trust in Ptolemy on the grounds that as a king he could not have lied without special dishonour. Not every historian in antiquity shared this special sort of respect for rank: Pliny the Elder remarked that 'persons of high position, although not inclined to search for the truth, are ashamed of ignorance and consequently are not reluctant to tell falsehoods' (*NH* 5. 12). But though there have been attempts to convict Ptolemy of particular bias towards contemporaries and rivals, the better view is that his neglect of their achievements during Alexander's campaigns is due not to malice but to

---

[91] A. Momigliano, *Filippo il Macedone* (Florence, 1934), 165.
[92] On the revenge motif see further *CAH* vi[2] (1994), Epilogue.

epic highlighting of the central figure.[93] In any case, for the first few years of Alexander's reign, both the 'vulgate' and the 'main sources' must have used Kallisthenes, the official historian of the campaigns. So it is too simple to talk of Ptolemy's bias.

The most important vulgate source is Diodorus' book 17, in effect a monograph on Alexander. It was unusual both for its length (it is the second longest book of the *Library*, after book 1) and because it contains no Sicilian material, a characteristic it shares with book 18. It illustrates the more extravagant features of the 'Kleitarchan' vulgate. Kleitarchos' account must have contained many true and uncontroversial facts; but when condensing it for his own purposes,[94] Diodorus was probably on the look-out for colour, so that what we have may not represent the original specially fairly. He no longer had Ephorus as intermediary, so the selections are on the simplest view Diodorus' own and may therefore be unusually erratic: there is after all some very incompetent material in the latter part of book 16, after Ephorus dried up and Diodorus was perhaps forced to do the job himself.[95] In view of all this, the low reputation of the Kleitarchan 'vulgate', at least until very modern times, may therefore be to some degree unjustified. (This very modern tendency to upgrade the vulgate goes together with a more critical attitude to the previously revered 'main sources' and to Arrian.)[96] Equally unjustified are

---

[93] Ptolemy and lies: in the *Letter of Aristeas* (a treatise on kingship addressed to Ptolemy's son Ptolemy II Philadelphus) 206 it is said that lying brings special shame on kings. Pearson, *Lost Histories* (above, n. 81), 194 n. 27, suggests that the point may have been stressed in Ptolemaic theories of kingship (and perhaps in Ptolemy's own preface? see J. Hornblower, *Hieronymus*, 129 n. 102). For rulers and truth see also Brunt, 'Marcus Aurelius in his *Meditations*', *JRS* 64 (1974), 9, esp. the end of n. 50 for Arrian. Pliny the Elder: M. Beagon, *Roman Nature: The Thought of Pliny the Elder* (Oxford, 1992), 7. Bias in Ptolemy: R. M. Errington, 'Bias in Ptolemy's History of Alexander', *CQ* 19 (1969), 233–42, but see J. Roisman, 'Ptolemy and his Rivals in his History of Alexander', *CQ* 34 (1984), 373–85, and P. A. Brunt, *Arrian*, ii (Loeb; London, 1983), 510.

[94] P. A. Brunt, 'On Historical Fragments and Epitomes', *CQ* 30 (1980), 447–94 at 493 points out that Kleitarchus' work ran to more than 12 books (*FGrHist* 137 F 6), which Diodorus compressed into one. This article is reprinted in Alonso-Nuñez, *Geschichtsbild* (n. 56 above), 334–62.

[95] Above, n. 82.

[96] This more critical attitude is to be found above all in Brunt's revised Loeb edn. and in the writings of A. B. Bosworth. See his 'Arrian and the Alexander-vulgate', in E. Badian (ed.), *Alexandre le grand: image et réalité* (Entretiens Hardt 22; Geneva, 1976), 1–46; *A Historical Commentary on Arrian's History of Alexander*, i (Oxford, 1980); *From Arrian to Alexander* (Oxford, 1988). But behind both Bosworth and Brunt stands the figure of E. Badian, see Brunt, *Arrian* (Loeb), i, p. lxxxiii, and Bosworth, *Commentary*, i, p. vi. For a different but not preferable view of Arrian see N. G. L. Hammond, *Sources for Alexander the Great: Plutarch's Life and Arrian's Anabasis Alexandrou* (Cambridge, 1993).

those attempts made from time to time[97] to detect more than one strand and therefore more than one chief source in Diodorus 17. Inconsistencies of handling can be explained in other and better ways; and Diodorus' ordinary working method was to use one chief source at a time, occasionally inserting material from a supplementary source, but not routinely interweaving two parallel sources.

Plutarch drew on both of the above two sub-traditions; and Aristoboulos was also used by Strabo (Augustan period, see further below) in his eastern books, which dealt with countries visited by Alexander in the final phases of the campaign. But Strabo also used other authorities for India (*FGrHist* 715 Megasthenes), and on the Persian Gulf (*FGrHist* 711 Androsthenes of Thasos). Androsthenes, as P. M. Fraser shows below, was also a source for some exotic material in the botanical writings of Theophrastus (see p. 176). Of surviving writers, then, Strabo and Theophrastus are of interest for the later years of Alexander's expedition, as is Arrian's *Indike*, which exploited the Herodotean monograph of Nearchos of Crete (*FGrHist* 133).

Nearchos was not the only early Hellenistic writer to see the relevance of Herodotus to the new worlds exposed to Greek investigation by Alexander: Hecataeus of Abdera and Manetho, both of whom wrote about Egypt, and Berossos, who wrote about Babylon (*FGrHist* 264, 609, 680), all built on the old Herodotean ethnographic tradition. But they seem to have had a new and specifically Hellenistic motive. It has been suggested that they wanted, perhaps in rivalry with each other, to provide their Ptolemaic and Seleucid masters with an ideology of conquest, fitting them into the series of earlier native rulers. The theory in its simple form does not quite work for Megasthenes, who has to be regarded as giving the Seleucids a justification for *non*-conquest, by stressing Indian resources. But all this shows that Greek historiography could be twisted in new directions, as Hellenistic kingdoms swallowed up older cultures.[98]

---

[97] W. W. Tarn, *Alexander the Great*, ii (Cambridge, 1948), 63–87, thought that Diodorus used Kleitarchos and Aristoboulos. See however H. Strasburger, *Bibliotheca Orientalis* 9 (1952), 202–11. More recently N. G. L. Hammond, *Three Historians of Alexander the Great* (Cambridge, 1983), substituted for Aristoboulos the far more shadowy and therefore even more improbable Diyllus. See *CR* 34 (1984), 261–4.

[98] On Megasthenes see S. Sherwin-White and A. Kuhrt, *From Samarkhand to Sardis: New Perspectives on the Seleucid Empire* (London, 1993), 95–100. For the relevance of Herodotus see O. Murray, 'Herodotus and Hellenistic Culture', *CQ* 22 (1972), 200–13. For Hecataeus see O. Murray, 'Hecataeus and Pharaonic Kingship', *JEA* 56 (1970), 141–71. Compare S. West, *JHS* 1991, 154, speaking of a much earlier period of Egyptian

If these ingenious theories are right, they throw light on some Hellenistic royal policies, or the ways in which writers sought to influence those
policies. But it has to be acknowledged that as evidence goes it is all very
exotic and heavily 'coded'. The situation is much more transparent when
we turn to the straight political and military history of the first generation
of the Successors of Alexander, because we have a superlative and fully
surviving narrative preserved by Diodorus.

We saw above, when discussing book 17, that Diodorus' usual working method was to use one main source at a time. This proposition is
best exemplified by the last three complete books of Diodorus (18–20),
covering the period from Alexander's death to just before the battle of
Ipsus (323–302 BC). Those were full years of warfare between Alexander's former generals. Diodorus demonstrably followed one source
for these books (with the exception of the Sicilian narrative about
Agathocles in books 19–20, and some Rhodian and Ptolemaic inser-
tions, see below). That source was Hieronymus of Cardia,[99] a historian
of the first rank, who combined Thucydides' godless grasp of war and
politics (and his interest in causation, see Diod. 18. 8 on the Lamian
War[100]) with some of Herodotus' ethnographic perceptiveness, slanted
for political reasons.[101] Hieronymus spent his long life in the service
first of his older relative Eumenes of Cardia, then of Antigonus the
One-Eyed: a good grounding for a 'pragmatic' historian in the sense
used by Polybius in the second century BC.[102] Pragmatic history is
political and military history written from direct experience and
intended to be useful.

Diodorus did not however drink Hieronymus quite neat; there are
traces of an Alexandrian or pro-Ptolemaic source (this is not likely to
emanate from Hieronymus, the employee of Ptolemy's enemy Antigonus), and of a Rhodian source from which Diodorus derived the

history: 'the discovery of hitherto unsuspected antiquity is a well-attested response to
foreign domination and the loss of sovereignty'.

[99] See J. Hornblower, *Hieronymus*, ch. 2, 'Diodorus and Hieronymus'; 49–62 for the
Ptolemaic and Rhodian supplements. See also G. Lehmann, 'Der "Lamische Krieg" und die
"Freiheit der Hellenen": Überlegungen zur Hieronymianischen Tradition', *ZPE* 73 (1988),
121–49; 127f. on the Ptolemaic and Rhodian *Nebenquellen*.

[100] J. Hornblower, *Hieronymus*, 152 and n. 205 (Samos crucial to the outbreak of the
war); 108.

[101] For Hieronymus on the Nabataean Arabs (19. 94–7) see J. Hornblower (above,
n. 99), 144–53 (Hieronymus' 'sub-text' is a message about freedom, ἐλευθερία, see
below); G. W. Bowersock, *Roman Arabia* (Cambridge, Mass., and London, 1983), 12–16;
Sherwin-White and Kuhrt (above, n. 98), 97.

[102] Walbank, *Polybius*, 56f.

Rhodian flood in book 19 and the siege of Rhodes, by Antigonus' son
Demetrius, in book 20.[103]

With the year 302 the complete narrative of Diodorus ends: a decisive
date, from our point of view, in the history of Greek historiography. The
third century BC is much less clear to us, at least until the last two decades
of the century, when Polybius' detailed narrative begins. But it would be
wrong to think that, just because the Hellenistic period was an Age of
Kings, politically motivated history came to an end. Hieronymus himself
wrote about kings, but parts of his history seem to have had a 'subtext'
about liberty, a feature which has invited comparisons with Tacitus.[104]
Nor did partisan local history come to an end: the Atthidographer
Philochorus (*FGrHist* 328) was put to death by Antigonus Gonatas, son
of Demetrius and grandson of the One-Eyed Antigonus, for alleged
Ptolemaic sympathies.[105]

Plutarch in his *Lives* of third-century Greeks (Demetrius, Pyrrhus,
Aratus of Sicyon, the Spartan kings Agis and Cleomenes) made use of
lost historians like Duris of Samos and Phylarchus of Athens (*FGrHist*
76, 81; Duris was the earlier of these two writers and was himself a
source for Phylarchus). Plutarch also used Hieronymus for the earlier
part of the century, and Aratus himself, the leader of the Achaean Confederacy, for the later. Duris and Phylarchus are conventionally held to
exemplify what since Polybius has been known as 'tragic history':[106]
Polybius 2. 56, a famous passage, is the main text about Phylarchus. And
Agatharchides of Cnidus (*FGrHist* 86) is another who has a claim to be
categorized as 'tragic'.[107] But in 1994 'tragic history' is well on the way to
becoming what anthropologists call a 'disgraced concept'; that is, a
concept which, like 'fertility ritual', explains and includes too much.
After all, Thucydides (below, Ch. 5) had already made use of rhetorical
and tragic techniques;[108] and in Roman times even Plutarch, who

---

[103] J. Hornblower, *Hieronymus*, 118–63.   [104] J. Hornblower, *Hieronymus*, 178f.
[105] T 1, from the *Suda*. For the historical context of this event see W. S. Ferguson, *Hellenistic Athens* (London, 1911), 188.
[106] The classic discussion is by F. W. Walbank, 'History and Tragedy', in *Selected Papers*
(Cambridge, 1985), 224–41, repr. from *Historia* 9 (1960), 216–34; see also 'Tragic
History: A Reconsideration', *BICS* 2 (1955), 4–14.
[107] P. M. Fraser (above, n. 70), 786 n. 217, objecting to Walbank's 1955 (see previous
n.) elimination of peripatetic, i.e. Aristotelian, influence.
[108] C. W. Macleod, 'Thucydides and Tragedy', in *Collected Essays* (Oxford, 1983),
140–58; P. A. Brunt, 'Cicero and Historiography', in *Studies in Greek History and Thought*
(Oxford, 1993), 181–209 at 205 f. This paper has an importance wider than its title
suggests. For another view see A. J. Woodman, *Rhetoric in Classical Historiography*
(London, 1988), ch. 2.

affected to despise tragic history, turned his *Alexander* into a tragedy; or rather, both in the *Alexander* and the *Pyrrhus* he blended epic and tragic techniques: epic for the positive sides, tragic for the negative.[109] Perhaps we should stop talking of 'tragic history' as something distinct; but it is helpful to have a term which, like 'expressionism' in art, suggests a direct assault on the emotions.

With the *Aratus*, Plutarch was already touching on topics which overlap with the subject-matter of Polybius: historiographically, the clouds from the west are gathering (cp. Pol. 5. 104). With Polybius, we shall move from purely Greek to Graeco-Roman history: Polybius described how Rome became first a, then the leading, then the only, Mediterranean power, through war against and diplomacy with the confederacies and rulers of Greece and Macedon. In particular, book 6 described the Roman constitution and Roman military arrangements, with a (lost) 'Archaeology' dealing with Rome's very early history.

But before we pass to Polybius, Timaeus of Tauromenium needs a word, because already in the late fourth and early third centuries, Timaeus (perhaps himself drawing on Lycus of Rhegium, *FGrHist* 570) had carried out the first systematic work on the west in general and Rome in particular: above, p. 35. But J. Geffcken probably went too far when he said that 'Timaeus was to the west what the Alexander-historians were to the east'.[110]

Timaeus is the subject, or object, of Polybius' polemical digression on history-writing in book 12. Much modern generalization about Timaeus' failings derives from the hostility of Polybius. But Brunt is rightly cautious: 'rather too much respect is given to Polybius' denigration of his predecessors, who should certainly not be condemned in the absence of sufficient evidence by which we can control his judgment.'[111] In any case, Polybius owed to Timaeus an essential element in his own structural organization: the system of Olympiads.[112] We must again turn aside for a moment to consider chronology.

The periodization of history by reference to the Olympic Games was

---

[109] J. M. Mossman, 'Tragedy and Epic in Plutarch's *Alexander*', *JHS* 108 (1988), 83–93, esp. 85.

[110] J. Geffcken, *Timaios' Geographie des Westens* (*Philologische Untersuchungen* 13; Berlin, 1892), 177. Between Timaeus and Polybius, others had written about Rome in Greek: not just Hieronymus (*FGrHist* 154 FF 11 and 13 with J. Hornblower, *Hieronymus*, 248–50) but Diokles of Peparethos and Fabius Pictor, *FGrHist* 820 and 809.

[111] Brunt (above, n. 94), 480.

[112] Walbank, *Polybius*, 101.

(as we have seen) as old as the fifth century and Hippias of Elis; but it was Timaeus who first made the Olympiad the basic unit of chronological punctuation in a complex historical work. The new system had the obvious advantage that it avoided the parochialism against which Thucydides had warned; and it was, unlike a division by campaigning seasons, well suited to a vast Polybian narrative which took in the whole Mediterranean, the Near East, and even, occasionally, the Far East (Antiochus III's eastern expedition featured in Polybius' book 5): events in these areas might be related to each other only very loosely if at all. The disadvantage, as has often been pointed out[113] was that the Olympic Games, which happened in roughly August, fell in the middle of a campaigning season. This means that August is an unsuitable break-point for a conspicuously military narrative like that of Polybius. (The system may have suited Timaeus better, if he had less fighting and more ethnography.) Of later important figures who worked on chronography, Eratosthenes followed the Timaean system of Olympiads, but Apollo-dorus of Athens, who wrote in verse, preferred (if only for that reason) to arrange by Athenian archons (*FGrHist* 241; 244). Apollodorus also used a system known as the ἀκμή- or *floruit*-system, the middle-life date at which a cultural figure was presumed to have 'flourished'.

Polybius, as we saw above (p. 17), wrote forty books, of which the last ten were an afterthought. (The original idea was to cover *c.*220 BC, with flashbacks, down to 168 and the Roman defeat of Perseus of Macedon at the battle of Pydna. But Polybius then extended the terminal date to 146 and the destruction of Carthage and Corinth.) The complete work was enormous, some five times the length of Thucydides,[114] though we should remember that Thucydides broke off with seven years to go before his announced terminus of 404.[115]

As Peter Derow remarks below (p. 84), Polybius was a 'pro'. Derow is referring above all to the professionalism of the historian who is self-conscious about what he is doing. Of the Greek historians of antiquity, Thucydides and Polybius stand out for their methodological material. But Polybius (whether or not he was engaging in conscious argument with Thucydides, see Section IV below) goes further than Thucydides because he treats causation theoretically, and categorizes the types of cause (3. 6ff.). And at the beginning of book 3 he is much more

---

[113] Walbank, *Polybius*, 101.

[114] Ibid. 25; F. Millar, 'Polybius Between Greece and Rome', in J. T. A. Koumoulides (ed.), *Greek Connections: Essays on Culture and Diplomacy* (Notre Dame, Ind., 1987), 1–18 at 3.

[115] 2. 65. 12 and my comm.

prosaically explicit about his project, and how it developed, than Thucydides had been at 5. 26, the so-called 'second preface'. And we have seen that Polybius' chronological scheme was both more ambitious than that of Thucydides (Olympiads rather than campaigning seasons) and also in some respects more precise: spring for Polybius *does* have a fixed starting point.[116]

Neither this professionalism, nor his wordy hellenistic Greek,[117] prevented Polybius, any more than irritation with 'fairy-tales' had prevented Thucydides, from being a 'tragic historian' when he chose to be. Polybius' Philip V, in particular, is as tragic a creation as Plutarch's Alexander or Pyrrhus.[118]

For Polybius, Rome's rise to empire was a wonder. For Posidonius of Apamea (*FGrHist* 87), who started in 146 BC where Polybius left off, and ended with the 80s or perhaps 60s,[119] Rome's empire was an established fact. (For Posidonius' attitude see further Sacks, below, p. 221.) Posidonius resembles Hieronymus and Timaeus in that his importance is in inverse proportion to our detailed knowledge of his work.[120] The problems of identifying Posidonian material are however greater because at least there are fully surviving books of Diodorus which can be shown to derive largely from Hieronymus; and Timaeus is also present in surviving Diodoran books, although the exact relation of Timaeus to Diodorus' surviving Sicilian sections is more problematic than that of Hieronymus to Diodorus' books 18–20. By contrast, not only is Posidonius himself lost apart from fragments, but the relevant books of Diodorus also survive only in fragments. We badly need a proper investigation of the relation of Posidonius to these parts of Diodorus.

---

[116] P. S. Derow, 'Polybius', in J. Luce (ed.), *Ancient Writers* (New York, 1982), I 525–39 at 527: beginning of spring closely associated with the vernal equinox.

[117] See E. Norden, *Die antike Kunstprosa vom VI Jhdt. v. Chr. bis in die Zeit der Renaissance* (Leipzig, 1898), 152–5, for Polybius' 'chancellery style'.

[118] Walbank, *Selected Papers* (above, n. 106), ch. 14 'Φίλιππος τραγῳδούμενος: a Polybian experiment', originally *JHS* 58 (1938), 55–68.

[119] For Posidonius see now L. Edelstein and I. Kidd, *Posidonius*, i²: *The Fragments* (Cambridge, 1989; abbreviated E/K), and I. Kidd, *Posidonius*, ii (1) and (2): *The Commentary* (Cambridge, 1988). See also A. D. Nock, 'Posidonius', *JRS* 49 (1959), 1–15 = *Essays on Religion and the Ancient World*, ed. Z. Stewart (Oxford, 1972), 853–76. For the terminal date in the 80s see Kidd's commentary on E/K frag. 51; J. Malitz, *Die Historien des Poseidonius* (Zetemata 79; Munich, 1983), arguing for 86 BC; E. Ruschenbusch, 'Der Endpunkt der Historien des Poseidonius', *Hermes* 121 (1993), 70–6, pushing it back to 88. For the 60s possibility, H. Strasburger, *JRS* 55 (1965), 40–53 at 44; Ger. repr. in Strasburger's *Studien* (n. 39 above), 920–45 at 927.

[120] For the formulation about Hieronymus see F. Jacoby, *RE* viii, col. 1450 = *Griechische Historiker* (Stuttgart, 1956), 245. On the need for work on Posidonius in Diodorus see Kidd, *Posidonius*, ii (1), pp. ix–x and 295.

Modern scholars divide into 'pan-Posidoniasts' and 'oligo-Posidoniasts', that is, respectively, those who attribute as much uncertain material as they can to Posidonius, and those who attribute the minimum.[121]

It is not even certain that history was a primary activity for Posidonius: Athenaeus and Plutarch cite him as 'Posidonius the philosopher', and he regarded philosophy as without question the paramount science (frag. 284 E/K, from Seneca). He has even been called 'the first major professional philosopher of the ancient world who also wrote a large and important historical work',[122] a judgement which however needs modifying if the view of Aristotle offered above is correct. But Posidonius was hardly a historical dabbler: we know that his work was in fifty-two books (Polybius and even Diodorus restricted themselves to forty). He may like Timaeus and Hieronymus have included a sketch of early Rome. All this suggests that his scale and scope were wide and bold.

How far did Posidonius' (Stoic) philosophy colour his history? An extreme but not absurd view might be, 'Not at all'. After all, David Hume, it has been noted, managed to write a history which owed nothing to his philosophical beliefs, just as in our time Iris Murdoch claims that her novels owe little to her philosophy. But in fact Posidonius does seem to have fitted history into his system of beliefs: he was really interested in people, their habits and beliefs (T 80 E/K, from Athenaeus is basic here), rather than in history as a catalogue of events. History, or rather ethnography, was for Posidonius a kind of descriptive ethics.[123]

Much of the content of the fragments is skewed towards drinking habits and so forth, because Athenaeus preserved them, and those were his preoccupations. But that could be put another way: Athenaeus quarried Posidonius for that kind of thing because he offered so much to quarry; in any case the importance of the symposium in the culture of ancient Greece, and of the places to which that culture transplanted itself, is only now beginning to be recognized.[124] Strabo shows us Posidonius' more serious geographical side.

[121] Kidd, *Posidonius*, ii(1), 310 and elsewhere (see next n.), thinks that Diodorus preserved only the 'superficial broth' of Posidonius' 'potent historical brew'. Panposidoniasts: Kidd, ii(1), p. xviii. Oligo-posidoniasts: K. Algra, *CR* 41 (1991), 317.

[122] Kidd, 'Posidonius as Philosopher-Historian', in M. Griffin and J. Barnes, *Philosophia Togata: Essays on Philosophy and Roman Society* (Oxford, 1989), 38–50 at 38. Froth and brew (see also previous n.): 48.

[123] Kidd in *Philosophia Togata* (above, n. 122), 39 (on Hume, though see N. Phillipson, *Hume* (London, 1989) for a different view); also pp. 40, 49 (descriptive ethics); and in *OCD*³ (forthcoming), under 'Posidonius'.

[124] See O. Murray, 'The Symposium as Social Organisation', in R. Hägg (ed.), *The Greek Renaissance of the Eighth Century BC: Tradition and Innovation* (Stockholm, 1983),

A major question in any enquiry into Greek historians after 200 BC concerns their attitude towards Rome: they could not avoid the topic, just as for earlier historians it was impossible to avoid a definite position about Athens, Sparta, or Macedon.

Whether or not Posidonius shared Strabo's (and Polybius'?[125]) positive estimate of Roman imperialism is an important and much debated but insoluble question. The most authoritative modern view denies that a famous fragment of Posidonius about voluntary subjection (F 60 E/K = *FGrHist* 87 F 8) was offered as part of any kind of general critique or eulogy of Roman imperialism.[126]

It is more certain that Strabo, the geographer and historian of the time of Augustus, was an enthusiast for what he regarded as the civilizing activity of Rome, see for instance 2. 127 (but note 6. 253 *ad fin.*). Strabo's seventeen-book work on geography survives, but he also wrote an explicitly historical work which does not (*FGrHist* 91). This loss should not obscure the claim of the *Geography* to be regarded as a very considerable work of history, written by a man with an interesting mind of his own. For instance Strabo provides not only our main evidence for such social topics as the *hierodouloi* or sacred slaves attached to Anatolian temples, but our best, indeed our only, sequential account of his own part of the world, the kingdom of Cappadocia.[127] We have already seen that Strabo is a major source for our knowledge of some of the lost Alexander-historians (p. 42; he treated Alexander in his lost *History* as well (F 3), and must have drawn on them here too). He is also important as a transmitter of Posidonius (above).

The same is true of Strabo's contemporary Diodorus of Sicily, a man we have met frequently already (see esp. n. 99 above) as the prime medium through which Ephorus, Kleitarchus, Hieronymus, Posidonius, and other major figures of Greek historiography have come down to us. Diodorus' *Bibliotheke* or *Library*,[128] or what is left of its originally forty books (books 1–5 and 11–20 complete, the other books in fragments) is

195–9; 'The Greek Symposium in History', in E. Gabba (ed.), *Tria Corda: Scritti in onore di Arnaldo Momigliano* (Como, 1983), 257–72; and (ed.), *Sympotica: A Symposium on the Symposium* (Oxford, 1990).

[125] Pol. 36. 9 with Walbank's commentary; 'Political Morality and the Friends of Scipio', *Selected Papers* (above, n. 106), ch. 11, repr. from *JRS* 55 (1965), 1–16.

[126] Kidd, *Posidonius*, ii(1), 297; cf. Erskine, *Hellenistic Stoa* (1990), 200–4.

[127] *Hierodouloi*: F. Millar, 'The Problem of Hellenistic Syria', in A. Kuhrt and S. Sherwin-White, *Hellenism in the East* (London, 1987), 110–33 at 120. Cappadocia: Strabo, 12. 1–2. On Strabo see generally Bowersock, *Augustus* (1965), 126–35.

[128] For Diodorus' title see J. Hornblower, *Hieronymus*, 22–7. On Diodorus generally, Ed. Schwartz, *Griechische Geschichtschreiber*, 35–97, remains basic.

thus of crucial importance for what it transmits. But Diodorus has recently been fruitfully studied for himself as well as for what he transmits; the difficulty is that Diodorus is neither a mindless copyist on the one hand nor a Thucydides on the other. (By 'a Thucydides' I mean a penetrating intellect who certainly used sources, oral as well as written, but transmuted what he found and made it almost entirely his own.) It has recently been shown that Diodorus' diction reflects that of his sources, but it is also certain, and has long been noticed, that he has his own favourite turns of phrase.[129] Kenneth Sacks (below, Ch. 8) addresses the problem of Diodorus' views as well as his vocabulary, and argues that on important issues Diodorus' opinions parted company with those of his sources.

Sacks makes an intriguing case for Diodorus as cool or even negative towards Rome. There were precedents or parallels for such an attitude, above all in Timagenes (*FGrHist* 88), one of those whom the patriotic Livy (9. 18. 6 = Timagenes T 1) called 'frivolous Greeks'; and—more remarkably perhaps, given his date—in the second-century AD 'periegete' or guidebook-writer Pausanias (a major and learned figure who drew extensively on the work of historians without quite deserving the title himself). If this view of Diodorus is right he represents a nice contrast with Dionysius of Halicarnassus, and the two can be seen as the cool and the grateful provincial respectively.[130] (Herod's friend Nicolaus of

[129] Occasionally even Thucydides, whose narrative has so seamless an appearance, betrays the language of his sources. For the linguistic peculiarities of the Pausanias–Themistocles excursus (1. 128–38) see H. D. Westlake, 'Thucydides on Pausanias: A Written Source?', in *Studies* (above, n. 76), ch. 1 = *CQ* 27 (1977), 95–110. For verbal echoes of Antiochus in the *Sikelika* (Th. 6. 2–5) see Dover (above, n. 15). On Diodorus' stylistic relation to his sources see above all J. Hornblower, *Hieronymus*, ch. 2 and app. II. In addition to Sacks's contribution to the present volume see his *Diodorus Siculus and the First Century BC* (Princeton, NJ, 1990). F. Chamoux has now done a useful general introduction to Diodorus in vol. i of the Budé edn. (Paris, 1993). At his p. xxv n. 58 he discusses some improbable older views according to which, for instance, Diodorus knew Herodotus only via Ephorus/Agatharchides.

[130] Timagenes: G. W. Bowersock, *Augustus and the Greek World* (Oxford, 1965), 109f., esp. 109 n. 2, noting that Livy need not have been referring *only* to Timagenes. Cp. also Spawforth, below, p. 243. Dionysius: E. Gabba, *Dionysius and the History of Archaic Rome* (Berkeley and Los Angeles, 1991), esp. ch. 1. Pausanias: C. Habicht, *Pausanias' Guide to Ancient Greece* (Berkeley–Los Angeles–London, 1985), rescues Pausanias from the contempt of Wilamowitz. For P.'s attitude to Rome see Habicht ch. 5; on his knowledge of earlier historians see Habicht 97ff., concluding that he read widely and then wrote from memory. The degree of Pausanias' independence is tricky: at p. 111 Habicht attributes to P. himself a sentiment certainly lifted from Thucydides (3. 62. 3 with my commentary). On Pausanias' relation to Plutarch note C. J. Tuplin, 'Pausanias and Plutarch's *Epaminondas*', *CQ* 34 (1984), 346–58. See also F. Chamoux, 'Pausanias historien', in *Mélanges A. Tuilier* (Paris, 1988), 37, and *Mélanges P. Leveque* (Besançon, 1985), 85, for Pausanias as a sub-

Damascus, *FGrHist* 90, was probably closer in this respect to Dionysius than to Diodorus.[131]) Dionysius, unlike Diodorus, has left other, purely literary treatises for us, including the essay *On Thucydides*, which has great if secondary value for the historiographer. But his *Roman Antiquities* is a prime source for historians not only of Roman politics and war but also of Roman religious belief, an area where he was anxious to stress Greek influence.[132]

A comparable figure, in certain respects, is Josephus, the author of two long surviving works on *Jewish Antiquities* and the *Jewish War* (i.e. the war against Rome in the late 60s AD). Josephus perhaps modelled himself on Dionysius, and certainly applied Greek historical methods to a people—the Jews—who had started to write their own history very early indeed, but had then mysteriously abandoned historiography.[133] Josephus was not alone. There were other writers who applied Greek historical methods to past Jewish history. The most famous is the author of the Second Book of Maccabees, which goes back to a work by Jason of Cyrene. This was a rhetorically coloured history, hardly impartial stuff, but good evidence for Seleucid institutions (in particular, the documents it cites, e.g. at chapter 11, are nowadays accepted as usable evidence by

scriber to the 'great man view of history', brought out through description of tombs and statues. See also the very interesting study by J. Elsner, 'Pausanias: A Greek Pilgrim in the Roman World', *Past and Present* 135 (1992), 3–29, stressing the way Pausanias organized his narrative by a pattern of mythology: he was not just a traveller but a religious traveller, in fact a pilgrim. See also Arafat, *BSA* 87 (1992), 387–409.

[131] Valuable account of Nicolaus in E. Schürer, rev. G. Vermes and F. Millar, *History of the Jewish People in the Age of Jesus Christ (175 BC–AD 135)*, i (Edinburgh, 1973), 28–32. Nicolaus wrote a world history which drew on and invites comparison with Ephorus, but the work survives in fragmentary form and we have nothing later than the Achaemenid Persian period. (The fragments are nevertheless important as preserving the Ephoran picture of archaic tyrannies like the Cypselids of Corinth and the Orthagorids of Sikyon, frags. 57–61.) But despite this gap in our definite knowledge, the Roman parts of this large work are not likely to have been hostile to Rome because Nicolaus also wrote an encomiastic *Life* of Augustus. See the still useful edition, with tr. and comm., by C. M. Hall (Northampton, Mass., 1923)=Smith Classical Studies 4; more recent edns. are by J. Bellemore (Bristol, 1984) and B. Scardiglio (Florence, 1983). On Nicolaus' 'universalist ideology' and stress on Roman peace and stability see E. Gabba, 'The Historians and Augustus', in F. Millar and E. Segal (eds.), *Caesar Augustus: Seven Aspects* (Oxford, 1984), 61–88 at 61–3.

[132] Dionysius' treatise *On Thucydides* has been edited with commentary by W. K. Pritchett (Berkeley–Los Angeles–London, 1975); see the admiring review of G. W. Bowersock, *CP* 71 (1976), 361–4. For Dionysius and religion see Gabba, *Dionysius*, 118–32, particularly good on the differences between Strabo's and Dionysius' view of myth, something famously absent from Roman religion.

[133] A. Momigliano, *Classical Foundations* (above, n. 18), 20–4. For Josephus, T. Rajak, *Josephus* (London, 1983).

hard-headed Hellenistic historians[134]). But there were other Greeks who wrote about the Jews. We know about the most important of them (Demetrius, Eupolemus) through the use made of them by Alexander Polyhistor (*FGrHist* 273).[135] Finally Luke, or whoever wrote the Acts of the Apostles, can be seen as a Greek treatment of an up-to-date Jewish topic, with proper Greek respect for the evidence of eye-witnesses.[136]

With the second century AD Greek historians turned, with some nostalgia, to their remoter past. Arrian and Plutarch prefigure what is sometimes known as the Second Sophistic, an age of modest intellectual renaissance with an emphasis (hardly new in Greek culture but carried to extremes) on rhetoric and antiquarianism.[137] It is a controversial question how far this sort of archaizing represented a 'flight from the present', that is from the powerlessness of contemporary Greece before the fact of Roman domination. A more positive view of such 'cultural archaism' is possible: the past was, in fact, being exploited in the interests of an ideology of Greek participation in Roman ecumenicalism. The artificially constructed Hadrianic organization known as the Panhellenium fits into this picture. So does the strong imperial Roman emphasis on the Persian Wars, a topic discussed by Spawforth in Chapter 9 below, where he shows how pride in the Persian Wars could be entirely compatible with loyalty to Rome. It is true that Plutarch the biographer was a

---

[134] E. Schürer, rev. G. Vermes, F. Millar, and M. Goodman, *History of the Jewish People* (above, n. 131), iii(1) (Edinburgh, 1986), 531–7; Sherwin-White and Kuhrt (above, n. 94), 226–8; C. Habicht, 'Royal Documents in Maccabees II', *HSCP* 80 (1976), 1–18; R. Doran, '2 Maccabees and "Tragic History"', *Hebrew Union College Annual* 50 (1979), 107–14 (arguing that 2 Maccabees is not properly 'tragic history' but belongs to a nevertheless Greek genre of histories of epiphanic deliverance, i.e. deliverance due to direct divine intervention).

[135] See the classic book by A. Freudenthal, *Alexander Polyhistor und die von ihm erhaltenem Reste judäischer und samaritanischer Geschichtswerke* (Breslau, 1875)= *Hellenistischer Studien* 1–2; cp. also M. Stern, *Greek Authors on Jews and Judaism*, i (Jerusalem, 1974), 157–64; Schürer-Vermes-Millar-Goodman, iii(1) (above, n. 134), 509–58.

[136] C. Hemer, *The Book of Acts in the Setting of Hellenistic History* (Winona Lake, Ind., 1990), 63–100; Lane Fox, *The Unauthorized Version*, 304–10; J. T. Squires, *The Plan of God in Luke-Acts* (Cambridge, 1993): Luke–Acts resembles some Hellenistic historians in the central role it assigns to Providence.

[137] The term 'Second Sophistic' goes back to Philostratus, *Lives of the Sophists*, 481, see G. W. Bowersock, *Greek Sophists in the Roman Empire* (Oxford, 1969), 8–10, and 'Introduction to the Greek Renaissance', in Bowersock (ed.), *Approaches to the Second Sophistic* (Pennsylvania, 1974). But Philostratus actually says it began with Aischines in the 4th cent. BC, as opposed to Gorgias and the 5th; on this see Bowersock, *Greek Sophists*, 8–10. Brunt (*BICS* 39 (1994)) says that the Second Sophistic is a 'bubble' in that its intellectual significance has been overrated and many of its supposedly characteristic features are not new at all. For a full-length treatment see now G. Anderson, *The Second Sophistic: A Cultural Phenomenon in the Roman Empire* (London, 1993).

priest of the ancient sanctuary of Delphi, a man who devoted many of his *Lives* to much earlier Greek figures; it is not, however, obvious that this implies much dissatisfaction with the contemporary scene. But what of Arrian of Nicomedia, the politically fulfilled consul and Roman governor? It is surely even less plausible to suppose that Arrian had a psychological need to compensate, through his writings, for the unpleasant fact of Roman domination: he owed too much of his own worldly success to the Roman system.[138]

For his Alexander-history, Arrian drew on Ptolemy and Aristoboulos as we have seen; for his history of the Successors of Alexander (*FGrHist* 156), he used the excellent Hieronymus, and this alone should be enough to protect him from the more extreme modern detractors, however faulty his understanding of the institutions and history of the Macedon of nearly half a millennium earlier.

Appian of Alexandria, another Greek historian of the second century AD, was even more whole-hearted in his acceptance of the Roman empire: his *Romaike* chronicles the military and political extension of Roman power, shadowing Polybius part of the way, and providing incidentally unique material, not in Polybius, about Hellenistic history (for instance, the Seleucid history in the *Syriake*, which may derive from valuable, chronologically organized lists compiled in the Hellenistic period).[139]

Cassius Dio, in the early third century, took the process further, supplying a Greek narrative account of the whole of Roman history but with a strong interpretative bias of his own, above all a preoccupation with the link between monarchy and stable government. With later developments, such as the classicizing fifth-century historians Priscus (who

[138] For the influential theory that Greek archaizing literary tendencies in this period represent a 'flight from the present' see the important paper of E. L. Bowie, 'Greeks and their Past in the Second Sophistic', in M. I. Finley (ed.), *Studies in Ancient Society* (London, 1974), 166–209, quotation from p. 194. But for the different and more positive view of cultural archaism given in my text see A. J. S. Spawforth, *JRS* 76 (1986), 328, review of O. Andrei, *A. Claudius Charax di Pergamo*, and generally Ch. 9 below. Panhellenium: A. Spawforth and S. Walker, *JRS* 75 (1985), 78–104; 76 (1986), 88–105; and Spawforth, p. 239 below. Plutarch: C. P. Jones, *Plutarch and Rome* (Oxford, 1971), 120 and n. 74, distances himself from Bowie's view, as far as Plutarch goes. For Arrian, see P. Stadter, *Arrian of Nicomedia* (Chapel Hill, NC, 1980) with *CR* 31 (1981), 12: again, a slightly different perspective from Bowie.

[139] K. Brodersen, *Appians Abriss der Seleukidengeschichte: Syriake 45.232–70.369. Text und Kommentar* (Munich, 1989), and *Appians Antiochike: Syriake 1.1–44.232* (Munich, 1991) = *Münchener Arbeiten zur alten Geschichte*, i and ii. See also Brodersen, 'Appian und sein Werk', *ANRW* 2. 34. 2 (Berlin and New York, 1993), 339–63, and in *OCD*[3] (forthcoming).

certainly imitated Herodotus and Thucydides), Olympiodorus, Procopius in the sixth century, and after them the whole tradition of
Byzantine historiography, we look beyond the scope of the present
volume. But even so late a writer as Procopius definitely imitated Thucydides, as had Cassius Dio—a fashionable thing to do, on the evidence of
the second-century AD Lucian.[140] On this note we may turn to the general
problems raised by intertextuality.

IV. INTERTEXTUALITY AND THE GREEK HISTORIANS

As I have already said (above, p. 1), I think that one of the most important questions raised by the papers in this book is the question of implied
or explicit interrelation, what modern literary critics call intertextuality,
between the Greek historians of antiquity and each other; and between
the historians and writers in other genres like epic, tragedy, and oratory.
Thus Kenneth Sacks's paper, which is about Diodorus' sources, and my
own paper, which discusses Thucydides' use of poetic, specifically
Homeric, techniques, are most obviously concerned with the relation
between writers and their predecessors (although Sacks actually argues
that Diodorus showed a fair degree of independence). Peter Derow also
deals with the way the historians of ancient Greece seem to have tried to
improve on each other's ideas about causal explanation. (We shall see,
p. 61 below, that the writers of antiquity are more likely to have been
aware of each other's views on methodological topics of this sort, than of
the detailed factual content of each other's works.)

Again, John Gould's chapter deals at more than one point with the
debts owed, or not owed, by Herodotus to his literary precedessors. Thus
he argues that Herodotus' acknowledgement of what one might call the
'uncertainty principle' in Greek religion need not be an allusion to
Xenophanes or Alcmaeon, or be based on 'specific historiographic
principles' at all, but rather expresses something basic about Greek
religion. On the other hand Gould also shows how much Herodotus
owed to Homer and Hesiod for his ideas about Greek religion. (See
pp. 94, 104 ff.) P. M. Fraser asks how Theophrastus knew what he knew
about the lands opened up by Alexander. Was it from returning veterans,
or from the writings of men like Androsthenes of Thasos? And how far

---

[140] F. Millar, *A Study of Cassius Dio* (Oxford, 1964); J. W. Rich, *Cassius Dio: The
Augustan Settlement (Roman History 53–59.9)* (Warminster, 1990); J. Edmondson, *Dio:
The Julio-Claudians* (London, 1992), 40 and n. 86, for Dio and other writers keen to
imitate Thucydides, with refs. to Lucian, *How to Write History*, 15, 25–6, 42.

did Theophrastus' knowledge of the west derive from the corpus of knowledge accumulated by the Aristotelian school? Antony Spawforth and John Davies are each concerned with the literary development of a tradition, in each case a tradition about a war. But Spawforth's war, the Persian War, was a real war: ultimately it was Herodotus' war. The problem with Davies' war, the First Sacred War, is that an exuberant literary tradition developed about it in the fourth century, but its literary roots are (unlike the Persian Wars) hard to see because Herodotus and Thucydides do not mention it. Finally, Ernst Badian's paper can be seen as a deconstruction of a particular tradition. The general framework is again (compare above on Spawforth) the Persian Wars; the particular tradition, in effect the premiss from which Badian begins, is the tradition that Athens and Macedon were the historical enemies of Persia. Badian shows that Herodotus gently and elegantly masks an earlier and very different truth, namely the sixth-century medizing or pro-Persian policies of both states. The need for Herodotus' care and tact in this matter arose from the later, fifth-century, position when Herodotus was writing or reciting. By this time, the playwrights (like Aeschylus in the *Persae*), the architects, and the panegyrists, had invented a patriotic tradition which was not easily compatible with the truth as Herodotus knew or suspected it to be. So here too we have an ancient writer who is wrestling with an existing, and above all a literary, tradition.

Handbooks with titles like 'Ancient Greek Historians' are many and excellent,[141] and they naturally acknowledge the existence of influence, argument, polemic, and so forth. But they suffer from a major defect. They tend (by presenting the sequence of historians as a *diadoche* or succession) to take for granted, what is in fact not always self-evident, that writer *B* was reasonably thoroughly acquainted with the works of earlier writer *A*. In the previous section I have of course used the inevitable *diadoche* arrangement myself, but I have tried to avoid assuming too much direct and automatic influence, and have symbolically called it the 'story' not the 'development' of Greek historiography.

---

[141] Starting with J. B. Bury's excellent book of that title (London, 1909). More recently there are S. Usher, *The Historians of Greece and Rome* (London, 1969) and T. S. Brown, *The Greek Historians* (Lexington–Toronto–London, 1973). A useful German handbook is K. Meister, *Die griechische Geschichtsschreibung* (Cologne, 1990). J. M. Alonso-Nuñez (above, n. 56) is a chronologically arranged set of surveys and papers by different authors. Some methodological problems are addressed in Averil Cameron (ed.), *History as Text: The Writing of Ancient History* (London, 1989); see esp. ch. 2 by M. Wheeldon, '"True stories": the reception of historiography in antiquity'.

This defect presumably occurs because there is not the space, in a handbook, for a proper analysis of the way in which a tradition develops. Such an analysis (and the present section is merely a set of suggestions) must try to answer two separate but related questions: (i) what is the debt owed by a historian to earlier or other kinds of writing including poetic? (Cp. Ch. 4 below, for poetic techniques in Thucydides.) And (ii) what is the force or weakness of arguments from silence? What conclusions are we entitled to draw from statements of the form 'writer X does not mention item Y which *we* know from historian or source Z'? (Where Z might be an inscription.) I return to this question at pp. 57f. below.

To put the issue behind both these questions another way, should Greek historiography be seen as an organically developing coral reef rather than as a set of pigeon-holes?

I begin with question (i) above, on the face of it a purely factual question, and perhaps a paradoxical one, given the polemical tone of, for instance, Polybius book 12 (*Polemic against* implies, or ought to imply, *knowledge of*). How sure can we be that one historian was aware, or unaware, of the writings of a predecessor? This is a large issue: it begins, as we saw (p. 16), very early on indeed—with the problem, recently posed, of Herodotus' detailed awareness of Hecataeus. It must be narrowed down. I shall, after some introductory illustrative points, glance (again) at the question, did Thucydides know Herodotus' work well? Then I shall spend rather longer discussing whether Polybius knew Thucydides' work well.

But first, some illustrated warnings, taken from other writers.

There is absolutely no doubt that Strabo was acquainted with Thucydides. At one point (8. 374) he even notes a manuscript variant in the texts of Thucydides, 'Methone' for 'Methana'; and (to mention only one example) at 8. 359 he inaccurately quotes Thucydides (4. 3. 2) for the 400-stade distance from the Messenian harbour-town Pylos to Sparta. (Thucydides had said *about* 400 stades, and he does not quite say that Pylos was the harbour of the Messenians.) So it is a surprise to find that at 6. 266 Strabo cites Ephorus, not Thucydides for the circumference of Sicily. Ephorus said the *periplous* or circumnavigation took five days and nights. Thucydides (6. 1. 2) had said it was eight days for a merchant ship. One can think of various reasons for Strabo's preference for Ephorus here, such as the desire silently to correct Thucydides; but the information in Thucydides and Ephorus is not glaringly irreconcilable, so it does not look like a firmly rooted preference (in the context, Strabo merely wants an example of a writer who gives a distance in a more general way than by miles or stades). But Ephorus recurs at 6. 267, so perhaps he was in Strabo's mind;[142] whereas Thucydides is not cited before Strabo's book 7 and the Greek books proper.

A harder and more remarkable example is at 14. 635: Strabo cites *Kallisthenes* for the fact that the tragic poet Phrynichus was fined 1,000 drachmai for his play about the fall of Miletus. This information differs not at all from that given in a celebrated chapter of Herodotus (6. 21). Now Herodotus is an author whom Strabo often quotes or shows apparent knowledge of (not all of it, surely, derived at second hand from Eratosthenes, though some of it may have been[143]); and it is likely, as Jacoby assumed in his commentary on the fragment in question (*FGrHist* 124 F 30), that Kallisthenes' source was Herodotus. Why then does Strabo here cite Kallisthenes not Herodotus? No very respectable motive comes to mind. Perhaps he just forgot that Herodotus had mentioned it. Perhaps on the relevant roll of Kallisthenes' *Deeds of Alexander* was on Strabo's desk for those west Anatolian places; or perhaps Strabo had looked at him very recently. (He had last cited him at 13. 627, only half a dozen chapters earlier, for an item not in Herodotus.) But it would be facile and wrong to draw any conclusions from all this about Strabo's 'knowledge of Herodotus'.

For this sort of reason we should not look too ingeniously for answers to such problems as, why Arrian cites the obscure Aristos and Asklepiades for the Roman embassy to Alexander, rather than the more famous Kleitarchos, who we know mentioned this embassy.[144] Ancient historians did not feel themselves bound by modern conventions about citing or preferring the best or earliest source. And sometimes citation of one source may conceal more general and extensive dependence on another; thus Thucydides' citation of Hellanicus at 1. 97 masks a greater dependence on Herodotus. Or a source *not* generally followed may be cited out of pride at unearthing an item not given by more obvious authorities. An example is Tacitus' citation of the memoirs of Agrippina.[145]

We mentioned above the possibility of 'silent correction'. A related

[142] On Strabo's use of Ephorus see the interesting remarks of Jacoby in the introduction to his Ephorus commentary, *FGrHist* ii(c), 32 f.: Strabo tends to cite Ephorus, like other of his subsidiary authorities, at the *end* of a section. Sicily: see Kidd on F 249 E/K.

[143] Cp. O. Murray, 'Herodotus and Hellenistic Culture' (above, n. 98), 210 n. 3; K. A. Riemann, *Das herodoteische Geschichtswerk in der Antike* (Diss.; Munich, 1967), 47–55.

[144] Arr. *Anab.* 7. 15. 5; *FGrHist* 143 Aristos F 2 and 144 Asklepiades F 1. For full discussion see Bosworth, *From Arrian to Alexander* (above, n. 96), 83–93.

[145] Tac. *Ann.* 4. 53, with the admirable comments of P. Fabia, *Les Sources de Tacite dans les histoires et les annales* (Paris, 1893), 330–4.

phenomenon is silent preference. This arises when an author confessedly operates with just two main sources; the classic example is Arrian's use of Ptolemy and Aristoboulos. Several times in the early parts of the *Anabasis*[146] Arrian gives a fact or a statistic without indicating a variant. If we had no control on him, we would assume that Ptolemy and Aristoboulos agreed (cp. ταὐτά at pr. para. 1). Actually, however, there are several points where we can prove that Aristoboulos' version differed from Ptolemy's (usually we know this because Plutarch happens to tell us what Aristoboulos said). Here, then, Arrian has 'silently preferred' Ptolemy.

Enough has been said to show the difficulties which await the modern scholar who wants straightforwardly to use ancient citations of earlier writers. (This has a bearing on the controversial issue of Herodotus' source-citations, see p. 18 above.) I leave undiscussed the more general problems of drawing inferences from fragments and epitomes of ancient historians, because Brunt (above, n. 94) has treated this topic so well.

What, though, of non-citation, where an author gives a fact in a form which leads us to suspect that he has used some earlier writer, not mentioned? Much will depend on how generally learned and systematic the later writer is. When the scholarly but reticent Theophrastus refers (see Fraser, below, p. 171) to the burning down of the sacred olive-tree on the Athenian acropolis, it seems reasonable to assume knowledge on his part of Herodotus 8. 55. Again, when Thucydides describes Chaironeia as 'the last city of Boiotia' (4. 76. 3), this suggests awareness of Hecataeus of Miletus, who had said (*FGrHist* 1 F 116) that Chaironeia was the *first* city of Boiotia, sc. if you are coming from Phokis. (There is an interesting issue of focalization here, cp. pp. 134f. below for this word; both Hecataeus and Thucydides imply a viewpoint. Thucydides' imagined focalizer/traveller comes from Athens, Hecataeus' comes from Phokis. There is in fact an implied second person singular here: 'if *you* are coming from . . .'.) But (it may be said) the geographical position of Chaironeia was uncontroversial; Thucydides could have noticed it for himself. Or he could slip into ethnographic 'mode' and put the matter in a Hecataean way without owing a conscious debt to Hecataeus.

It is time to pass on to Thucydides' use of Herodotus. I have argued for a *factual* dependence more fully elsewhere (see n. 41), and have suggested that it takes a particular form, being more in evidence in

---

[146] Arr. *Anab.* 1. 10. 4 (cp. Plut. *Demosthenes* 23), 1. 11. 3 (cp. Plut. *Alex.* 15), 1. 16. 4 (cp. Plut. *Alex.* 16. 7). See Brunt's notes on the Arrian passages, in his outstanding Loeb edn., vol. i (1976).

Thucydides' speeches than in his narrative. Thucydides' 'treatment' of the two Sacred Wars he knew about or could have known about, both fought for possession of Delphi, is illuminating both for Thucydides' attitude to Herodotus and also perhaps to Hellanicus.

The First Sacred War of the early sixth century BC is not mentioned by either Herodotus or Thucydides, hence the inverted commas (above) round the word 'treatment'. But this silence is, in its Thucydidean aspect, really a 'secondary silence' as we may call it. If we accept Davies's tentative hypothesis (below, Ch. 7) that the war was historical, we need to explain its absence from both Herodotus and Thucydides (an absence which has led some scholars to say the war never happened). Herodotus' silence is not troublesome. He was not a systematic writer, and did not offer a linear account of the archaic age. Herodotus' failure to mention something cannot be taken as evidence that it did not happen. *Thucydides* does not mention the First Sacred War in the polemical *Archaeology*, because Herodotus had not done so. Thucydides therefore had no need to say, as he said or implied of the Trojan, Persian, and Lelantine Wars, that 'they were insignificant in scale compared to *my* war, the Peloponnesian'.

The Second Sacred War was fought in the middle of the fifth century BC; Thucydides describes it at 1. 112, where he uses a curious expression about it, 'the so-called Sacred War', or 'the war called the sacred'. As I show below (pp. 162f.), this is an unusual expression for Thucydides to use about something other than a place. In the present context, I should like to draw attention to one aspect of the focalization implied by the word καλούμενον. Just *who* called it a sacred war? The immediate subject is the Spartans, so perhaps it was they who called it sacred (and the word would be a respectful or cynical Thucydidean allusion to Spartan piety). Or (a second possibility) did Hellanicus perhaps call it sacred? If so, this would be a distancing device of a special polemical sort. The remaining possibility is that Thucydides is here expressing his own attitude to religion, and it is this possibility which I chiefly consider in Chapter 5 below. But the others cannot be ruled out. Is it significant that Thucydides calls it just 'the so-called Sacred War' as if there were no other? This would have a bearing on the question, whether he was aware of a First Sacred War (see above). But the point cannot be pressed. Thucydides' ways of referring to wars are capricious, so for instance at 5. 32. 2 the six words to the effect that 'the Phokians and Lokrians began a war' may conceal a 'sacred war' (the preceding sentence mentions Delphi). Again,

Thucydides calls the Persian War a Greek war at 1. 128. 3,[147] and never calls the First Peloponnesian War by any name at all.

From Thucydides' knowledge of Herodotus, I pass to the question, did Polybius know Thucydides' work, and if so, how well? With one trivial exceptional Polybius does not quote or refer to Thucydides; the only mention is 8. 11. 3, but this merely says that *Theopompus* began where Thucydides left off. However, the same sort of consideration would 'prove', wrongly in my view, that Thucydides did not know Herodotus (since he never mentions him); in any case we should always remember that we do not have Polybius anything like complete. What of close Polybian echoes of Thucydides? The most specific text is the methodological sentence at Polybius 3. 31. 12, where the distinction between ἀγώνισμα and μάθημα (a virtuoso prize composition and a lesson), and between usefulness and pleasure, has often been thought to echo Thucydides 1. 22. 4.[148] And Polybius' views about, for instance, causation (see especially 3. 6–7, with its distinction between causes, pretexts, and first actions) are usually taken[149] to be an attempt at a refinement on Thucydides' simpler 'cause-and-pretext' model: see Derow's paper below. J. B. Bury, however, believed[150] that 'Thucydides exercised no great influence on him [Polybius], and the extant parts of his work indicate that he was not one of the historians with whom he was familiar'. That is very strong. But Bury is not alone: it has often been pointed out[151] that there is an oddity in Polybius' treatment of an episode which we know was reported by Thucydides. In book 12, Polybius (12. 25k = *FGrHist* 566 F 22) criticizes Timaeus' version of Hermokrates' speech at Gela (Th. 4. 59– 64). But he does not use the (to us) obvious following arguments: (i) Timaeus' speech was evidently very different from Thucydides'; and therefore (ii) Timaeus' report cannot be right unless, (iii) very improbably, Timaeus had access to a different and superior tradition from that

[147] A. Andrewes, 'Spartan Imperialism?', in Garnsey and Whittaker (eds.), *Imperialism in the Ancient World* (Cambridge, 1978), 303 n. 11.

[148] H. G. Strebel, *Wertung und Wirkung des Thukydideischen Geschichtswerk in der griechisch-römischen Literatur* (Diss.; Munich, 1935), 23 and n. 72. See also Derow, below, p. 85 on Pol. 38. 4. 8 as an echo of Th.

[149] Strebel (above, n. 148), 23; Walbank ad loc.; K. Ziegler, 'Polybios', *RE* xxi(2) (1952), col. 1503. See Sacks, *Ath* 1986, 394f.

[150] Bury, *The Ancient Greek Historians*, 210; cp. above, n. 141.

[151] See Walbank's n. in his *Historical Commentary on Polybius*, vol. ii (Oxford, 1967), and G. Lehmann, 'The "Ancient" Greek History in Polybios' *Historiae*: Tendencies and Political Objectives', *SCI* 10 (1989–90), 66–77 at 72f. *FGrHist* 556 Timaios T 18 (=Plut. *Nik.* 1) says that Timaeus tried to outdo Thucydides in δεινότης, but this may just represent Plutarch's own literary judgement and not be based on anything explicitly said by Timaeus about Thucydides. Cf. my forthcoming paper on Th.'s reception.

used by Thucydides. Even if so, it does not remove the main oddity: Polybius' failure to mention Thucydides in this context at all. Even allowing for Momigliano's point that Polybius' remarks about speeches (in 12. 25) were aimed at the 'obviously invented speeches of Thucydides',[152] it was very high-handed of Polybius to disregard a particular Thucydidean speech completely.

What are we to conclude? The reference to ἀγώνισμα seems decisive evidence that Polybius knew Thucydides; the Hermokrates example seems decisive that he did not. The way out may be to accept that Polybius' knowledge was uneven. The early, methodological, chapters of Thucydides were striking and memorable, part of a post-classical prose writer's mental furniture (cp. *FGrHist* 124 Kallisthenes F 44); the detailed narrative, including even pivotal episodes like the Gela conference, was perhaps something to read once and no more. Ignorance of the total kind postulated by Bury seems improbable; it may be more helpful to think in terms of Polybius' mental horizons, rather than putting particular passages or phrases under the microscope. He seems to have had a clear vision of Greek history as far back as 480, but hardly to have entered into the thought-world of Herodotus at all[153] (the name 'Herodotus' appears in our texts of Polybius 12. 2, a fragmentary passage on the lotus, but it is really Athenaeus, not Polybius, who mentions him). If that is right, Polybius must—though even this has been doubted—have had a working knowledge of the fifth century, that is, the Thucydidean period; but if so, it was perhaps mediated through Ephorus, a more palatable drink than neat Thucydides. Thucydides was a difficult writer, and direct evidence that he was imitated by Hellenistic historians is rare; stylistic taste had changed.[154] Agatharchides was an 'imitator', ζηλωτής, of Thucydides, according to Photius (*FGrHist* 86 T 2); but this may just be a vivid way of saying that Photius himself saw resemblances.[155]

---

[152] Momigliano, *Classical Foundations* (above, n. 18), 47.

[153] For Polybius' knowledge of Thucydidean methodology (but not necessarily detailed facts): Lehmann (above, n. 151), 73. Polybius' 'clear view' of later Greek history: it is not in dispute that Polybius was thoroughly familiar with Kallisthenes and Hieronymus of Cardia (J. Hornblower, *Hieronymus*, 236, and Lehmann, 69 n. 7, and 74). Polybius' ignorance (or neglect) of Herodotus: Millar (above, n. 114), 13, but see Lehmann, 68. There is a possible echo of Hdt. at Pol. 11. 5. 9, cp. Hdt. 5. 97. 3, ἀρχὴ κακῶν. But as Walbank says the phrase may be proverbial; it is certainly Homeric, *Iliad* 5. 63.

[154] Polybius and the 5th cent.: Lehmann (above, n. 151), 72f. and the exchange between Momigliano, Pédech, and Walbank in E. Gabba (ed.), *Polybe* (Entretiens Hardt 20; Geneva, 1974), 62f. Changes in stylistic taste: Strebel (above, n. 148), 26.

[155] *Ibid.* 24. Thucydides, then, was difficult. But more transparent and immediately

General points now arise about the awareness of one writer of another writer's work: points to do with access to libraries, availability of texts, and ease of consultation; nor (as Momigliano insisted) should we over-look access to *lectures and public readings* by historians in every period of ancient history.[156] As far as libraries are concerned, Timaeus' half-century of access to Athens' libraries was surely unusual, though this must have changed over time: even allowing for the persistently oral character of ancient society, it is surely uncontroversial that Plutarch's culture was more bookish than that of Thucydides. Ancient books were clumsy and it would not be (as it would for a modern scholar with shelves full of Teubner and Oxford texts or Loeb editions) the work of a moment for Polybius to turn up the Gela conference in a copy of Thucydides—always assuming he possessed one, or had borrowed one from Scipio Aemilianus (for exchanges of books between these two see Pol. 31. 23. 4). Or take Theophrastus. How did the scientific results of Alexander's conquest get distilled into written scientific research? (Cp. Fraser, pp. 174f. below, on the danger of anachronistically assuming that Alex-ander's staff were like Napoleon's scientists in Egypt.)

Diodorus' working methods (compare Kenneth Sacks's paper, Ch. 8 below) have a close bearing on this problem. Diodorus' method was to use one basic source at a time. This is surely to be explained, at least in part, by the simple principle that an ancient writer tends to work with what is on the desk in front of him (hence Strabo's preference for Ephorus or Kallisthenes where we might have expected Thucydides or Herodotus).

attractive writers naturally left more obvious traces on their successors. For Herodotus see Murray, 1972, and Riemann (above, nn. 98 and 143), and n. 59 above for Xenophon's debt to Herodotus, charmers both. For Xenophon's own influence in antiquity see K. Münscher, *Xenophon in der griechisch-römischen Literatur* (*Philologus* suppl. 13(2); Leipzig, 1920), and, for real-life influence exercised on military practice by Xenophon's writings, J. Hornblower, *Hieronymus*, 205–11. Arrian was famously an admirer of Xenophon, see for instance Bosworth's commentary on the *Anabasis* (above, n. 96), 1, 6–7. Whatever the motives for the cultural archaizing of the Second Sophistic (above, n. 138 for this problem), there is no doubting the 'intertextual' facts in this area: writers like Arrian and Pausanias (see n. 130) were well read and showed it off.

[156] On availability of books see E. Kenney, *Cambridge History of Classical Literature*, ii (1982), 20. Plutarch, *Demosthenes* 2, is good on the need for a historian, about to embark on work requiring out-of-the-way reading, to go to a city well-stocked with books. (But note that Plutarch adds that this has the additional advantage that you may also pick things up by hearsay; very relevant to R. Thomas's view, see *Literacy and Orality*, 158–70, that Graeco-Roman culture continued to have an oral aspect even in its later phases. Plutarch amusingly adds that he belongs to a small city—Chaironeia in Boiotia—and chooses to go on living there so that it does not get any smaller.) For readings by historians see the import-ant paper by A. Momigliano, 'The Historians of the Classical World and their Audiences: Some Suggestions', *Sesto contributo* (Rome, 1980), 361–76.

This principle does not preclude more sophisticated techniques like that of Plutarch: the memorizing, over time, of numbers of different writers and then the drawing up of detailed notes, followed finally by the end draft.[157] That is essentially Christopher Pelling's plausible picture. But Pelling himself acknowledges that the clumsiness of ancient papyrus rolls must have made it inconvenient, though possible (cp. Strabo 17. 790), systematically to compare two accounts. And in *Lives* where Plutarch was following where Thucydides had once trodden, it is clear that even the sophisticated Plutarch follows Thucydides, the source 'open in front of him', closely and at first hand; though he supplements him with new facts, or strikes out with different and far from worthless political insights.[158] Returning from Plutarch to the less complex Diodorus, we may note finally that here too (compare above for Polybius' access to Scipio's books) we can prove that he could have had good access to historical works. A second-century BC catalogue of historical authors, collected in the library of the town gymnasium of Tauromenium (Timaeus' home town, only eighty kilometres from Diodorus' native Agyrrhium) includes Kallisthenes, Philistos, Fabius Pictor, and (?) Paraballon of Elis (cp. *FGrHist* 416 F 1): *SEG* xxvi. 1123. But Diodorus seems to have used Ephorus (who in turn used Kallisthenes), rather than using Kallisthenes direct. In that way he did not need to chop around between sources.

The whole notion of 'sources' is not however straightforward (see n. 156). Further questions must be put, if not answered, questions to do with orality and literacy—and something in between: the extent to which an author is known from a written text, and to which a written text is cited or checked with bookish precision. How much did authors like Plutarch or Polybius have by heart? More of Homer than of a difficult prose writer like Thucydides, surely. And what can we infer from 'accurate' as opposed to 'inaccurate' quotation from earlier authors whose texts we have? It has been shown that the second-century AD

[157] C. Pelling, 'Plutarch's Method of Work in the Roman Lives', *JHS* 99 (1979), 74–96 at 93; note also p. 96, for valuable warnings against simplistically talking of the 'sources' of a writer like Plutarch, who may have read something recently, or had it dictated to him, or remembered it from a while ago. See also C. Pelling, 'Plutarch's Adaptation of his Source-Material', *JHS* 100 (1980), 127–40. See also J. P. Hershbell, 'Plutarch and Herodotus: The Beetle in the Rose', *Rh. Mus.* 136 (1993), 143–63: Plutarch sometimes gives Hdt. passages from memory; his knowledge is not just anecdotal but based on an intelligent grasp of Hdt.'s narrative as a whole. Clumsiness of papyrus rolls: Pelling, 1979, 92f. Cp. Habicht (above, n. 130) on Pausanias' procedures.

[158] C. Pelling, 'Plutarch and Thucydides', in P. A. Stadter, *Plutarch and the Historical Tradition* (London, 1991), 10–40 at 24.

writer Athenaeus was a particularly reliable transmitter of earlier writers. And where we can compare Arrian's use of sources with Strabo's use of the same sources, it seems that Strabo abbreviates more heavily but gives the gist efficiently; Arrian takes more artistic liberties but avoids the unclarity of Strabo's more ruthless précis technique. But even this sort of comparative operation can be dangerous. When we speak of reliable transmission of earlier writers, we should be careful not to assume that the texts of those earlier writers were stable. A good example is quotation of Homer. How 'accurate' was Plutarch's use of Homer? In *Moralia* 26 he cites four lines of *Iliad* 9 (lines 458–61) which are found in modern texts. But Plutarch is our *only* authority for those lines! They occur in no manuscripts or Ptolemaic papyri and seem to have been unknown to the scholiasts. Nevertheless, several editors since Wolf have promoted them to the text. Whatever decision one comes to about the 'authenticity' of the lines, Plutarch must have got them from somewhere, they are not just the result of faulty memory.[159]

Let us take further this issue of historians' knowledge of epic. Thucydides' similarities to Homer (see my Narratology paper, Ch. 5 below) seem to provide a clear case of a writer definitely aware of a particular literary tradition; but it is necessary to answer the sceptic who pleads more general influence, or says 'these are standard story-telling devices; show me a chapter of Thucydides' wartime narrative which could not have been written if he'd never heard of Homer'. I think this *can* be answered. True, it is not enough merely to show that Thucydides was aware of Homer in detail. For that limited purpose it is enough to point to 3. 104, with its extensive quotations from the Homeric *Hymn to Apollo*. And there are a few other passages. Generally, he is sparing with specific references to Homer; apart from the '*Homeric Hymn* chapter', there are just three mentions by name early in the *Archaeology*: 1. 3, 1. 9 (including a whole line quotation at para. 4), and 1. 10; and a slighting remark by Pericles in the *Funeral Oration*: 2. 41. (Other chapters in the *Archaeology* deal with Homer without naming him, above all 1. 11.) And there are other probable allusions, like the adjectives 'wealthy',

---

[159] Athenaeus: Brunt (above, n. 94), 480f.; Arrian and Strabo: see Bosworth, *From Arrian to Alexander*, ch. 3. For the *Iliad* lines see Hainsworth (above, n. 30), 123. (Note that Plutarch was aware of the existence of a problem because he says the lines were excised by Aristarchus, probably just a guess, see Hainsworth ad loc., also R. Janko, *The Iliad: A Commentary*, iv: *Books 13–16* (Cambridge, 1992), 28.) Compare, for Plato's use of Homer, A. Lesky, *RE* Suppbd. xi (1968), cols. 836–7, discussing J. Labarbe, *L'Homère de Platon* (Liège, 1949): Plato's variants may be due in part to genuinely alternative traditions.

applied to Corinth, and 'Minyan', applied to Boiotian Orchomenos: 1. 13. 5 and 4. 76. 3. Both these 'come from' the *Catalogue of Ships*, *Iliad* 2. 570 and 511. But the inverted commas round 'come from' are necessary because the adjectives were no doubt traditional, i.e. pre-Homeric; and at 1. 13. 5 Thucydides actually says that 'the old poets', in the plural, called Corinth 'wealthy', surely not just a rhetorical plural.

How intense a study does all this imply? The expressions 'in the Handing down of the Sceptre' at Thuc. 1. 9. 4 and 'in the *Catalogue of Ships*' at 1. 10. 4 imply a rather modern and academic study of Homer; and the casual and surprising mention of the straits of Messina as a candidate for Charybdis (4. 24. 5) is evidence that the game of locating Odysseus' wanderings had already begun by Thucydides' day. In the same way the allusion to the Phaiakians at 1. 25. 4 and to the sacred precinct of Alkinoos at 3.70 is evidence that Thucydides was aware of the equation between Homeric Phaiakia and historical Corcyra (see above, p. 9). All that is proof of a pretty thorough knowledge of Homer; but it is still not strict proof of Homeric influence on the wartime narrative, or speeches. Such strict proof is perhaps not to be had. It is true, but too vague, to assert that Thucydides' concern with war and suffering is epic; for one thing Herodotus shared that concern in a general way, but made something different of the topic.[160]

For the better understanding of this problem it will be necessary to approach it indirectly; in fact, to return to Herodotus, against whom Thucydides reacted. Homeric influence on Herodotus is more transparent than Homeric influence on Thucydides. The number of actual mentions is not conspicuously greater—ten to Thucydides' six; and six of Herodotus' mentions are from the Egyptian book 2, where he is rationalizing Homer, and two others concern Libya (4. 29).[161]

But it is hard to avoid the impression that Homer is closer to the surface of Herodotus. Speakers in Herodotus drop into Homeric or quasi-Homeric discourse very readily; and Herodotus in his own person unselfconsciously uses Homeric devices like rhetorical questions,[162] and

---

[160] H. Strasburger, 'Homer und die Geschichtschreibung', in *Studien* (above, n. 39), 1058–97; C. W. Macleod (above, n. 108). J. Cobet, 'Herodotus and Thucydides on War', in I. S. Moxon *et al.* (eds.), *Past Perspectives*, 1–18, suggests that Thucydides thought war atrocious but basically inevitable, whereas Herodotus thought that 'war should not happen at all' (p. 17). Protesilaus' temple is a closure for both Th. (8. 102) *and* Hdt. (9. 116).

[161] Hdt. 2. 23, 53 (twice), 116, 117 (twice); 4. 29, 32 (twice); 7. 161. There are Ὁμήρεια ἐπέα at 5. 67. Th. 1. 3. 3, 9. 4, 10. 3; 2. 41. 4; 3. 104. 4 and 6. Note also Hdt. 5. 94. 2 and the other passages in Richardson (n. 25 above), 27.

[162] Rhetorical questions: below, p. 149 n. 48, and M. Lang, 'How Could Herodotus

such resonant epic phrases as 'the cities of mortals' (1. 5. 3, cp. *Odyssey* 1. 3). The opening sentence of Herodotus, with its declaration of desire not to let the deeds of men become 'fameless', ἀκλεᾶ, recalls the Homeric phrase with which we began this Introduction: κλέα ἀνδρῶν, the 'glorious deeds of men' at *Iliad* 9. 189. The important difference is that Herodotus celebrates men of his own time, or rather the recent past.[163] A particularly interesting Homeric echo in a speech occurs at 7. 159. The Spartan Syagros loses patience with the exorbitant demands of the Sicilian Gelon, who has just demanded the supreme leadership of the war against the invading Persians. 'Agamemnon descendant of Pelops would groan aloud', he says, if he learnt of such a thing, ἦ κε μέγ' οἰμώξειε ὁ Πελοπίδης Ἀγαμέμνων. The line is almost a perfect Homeric hexameter, and recalls ἦ κε μέγ' οἰμώξειε γέρων ἱππηλάτα Πηλεύς, 'old horse-driving Peleus would groan aloud...' (*Iliad* 7. 125).

Herodotus' line *almost* scans, but not quite: the first vowel of Πελοπίδης would have to be lengthened to Πηλοπίδης (and the optional letter ν added to the end of the verb). Why did Herodotus avoid making the line a hexameter? There is surely no doubt that he did deliberately avoid it: a simple change of adjective was all that was needed, not Πελοπίδης but Ταντάλιδης, 'descendant of Tantalus'. (For this word see Aesch. *Ag.* 1469.) There are four possible reasons why Herodotus used Πελοπίδης rather than Ταντάλιδης. First, 'Tantalus' was more ill-omened than 'Pelops' (not that Pelops had an altogether reassuring career). Second, 'Pelops' suggested Spartan lordship of the Peloponnese and the acquisition of the bones of Agamemnon's son Orestes (cp. Hdt. 1. 67–8 and Xen. *Hell.* 3. 4. 3). Third, Πελοπίδης may have been suggested to Herodotus' conscious or unconscious mind by the similar-sounding Πηλεύς of Homer's line. Fourth and last, Herodotus did not *want* a perfect hexameter; it is a noticeable feature of such echoes that they often avoid perfect metricality (see further below).

Another example occurs in Herodotus' account of the final phase of the Ionian revolt, a solemn moment (6. 11. 2). Dionysius of Phokaia tells

Imitate Homer?', in *Herodotean Narrative and Discourse* (Cambridge, Mass. and London, 1984), ch. 3.

[163] For this difference between Homer and Herodotus see P. A. Brunt, *Studies in Greek History* (above, n. 108), 400; and cp. p. 14 and n. 22 above, citing R. Thomas, for the breakthrough represented by Herodotus' attention to the *recent* past. Simonides, it now appears, made an explicit Persian/Trojan War comparison, see *POxy.* 3965 and commentary by P. J. Parsons; also M. L. West, 'Simonides Redivivus', *ZPE* 98 (1993), 1–14. See also E. L. Bowie, 'Early Greek Elegy, Symposium and Public Festival', *JHS* 106 (1986), 13–38. For the opening of Herodotus see also G. Nagy, 'Herodotus the Logios', in *Arethusa* 20 (above, n. 18), 175–201.

the Greeks that things are on the razor's edge, ἐπὶ ξυροῦ γὰρ ἀκμῆς
ἔχεται τὰ πράγματα. Compare *Iliad* 10. 173, ἐπὶ ξυροῦ ἵσταται ἀκμῆς.
(But note again that the insertion of γάρ breaks up the metre.) This
expression may sound merely proverbial; but one feature of the introductory narrative suggests that the Homeric echo was definite and
deliberate. 'Others spoke, and also Dionysius . . .' The rather pompous
word for 'spoke', ἠγορόωντο, is used here only in Herodotus. It is in fact
Homeric, see for instance *Iliad* 4. 1.[164]

Thucydides does not work quite like this. Nikias at 7. 77. 7 tells his
men there is no safe place for cowards to fly to. If this recalls Ajax, as I
think it does (*Iliad* 15. 733–41), the echo is not verbal but an echo of a
thought. So too the uncharacteristic rhetorical question at 7. 44 ('in a
night battle, who could know anything for certain?') suggests the poet's
despair at *Iliad* 12. 176, a line there is no good reason to suspect: 'it
would take a god to describe all the fighting'. (Note the density of
Homeric echoes in the epic Sicilian books 6 and 7, and that the Homeric-
looking rhetorical question occurs in one of these books.) And we have
seen (above, p. 9) that Thucydides alludes to the genealogically framed,
i.e. Homeric, appeals of Nikias in the same book, and here the word
πατρόθεν ('by their fathers' names') *is* a verbal echo. Similarly, Hermokrates' language about Sicily at 4. 64. 3 recalls *Odyssey* 19. 173 on
Crete, especially the word περίρρυτος ('sea-girt'), used of both islands.
Such exact verbal echoes are rare. More normal is the description of
Pericles as 'good at speaking and doing' (1. 139. 4). This recalls the
thought, but not the language, of *Iliad* 9. 443, Phoenix's definition of a
good leader. Similarly Thucydides' statement (6. 59. 3) that Hippias, an
Athenian, gave his daughter in paradoxical marriage to a Lampsacene is

---

[164] On Tantalides in Aeschylus see E. Fraenkel in his commentary (Oxford, 1950), 695,
noting the possibility that Aeschylus intended the name to sound ominous, but in the end he
doubts whether it was meant to 'recall the idea of a family curse going back to the founder
of the family'. That is, the word is fairly neutral. For Homeric reminiscences in Herodotus
see Jacoby, *Griechische Historiker*, 502f., pointing out that Hdt. often adapts them
slightly: he is 'far from abolishing the boundaries between poetry and prose' (for Th. 4. 64/
Od. 19. 173 see Jacoby, p. 503); also W. Aly, *Volksmärchen, Sage und Novelle bei Herodot
und seine Zeitgenossen*² (Göttingen, 1969), 263–77 (noting at pp. 267 and 272 that at
Hdt. 7. 159 κε is Aiolic not Ionic, and that the colouring of ἠγορόωντο at 6. 11 is epic).
L. Hüber, 'Herodots Homerverständnis', in H. Flashar and K. Gaiser (eds.), *Synusia: Festgabe für W. Schadewaldt*, 29–52, discusses (see esp. pp. 32, 35) the kind of solemn occasion on which Herodotus is most obviously Homeric, and has good remarks on some of the
more complex echoes, e.g. (p. 31) the phrase χειμάρρῳ ποταμῷ at Hdt. 3. 81. 2, cp. *Iliad*
5. 87–8, 13. 138, etc. See also W. Schmid and O. Stählin, *Geschichte der griechischen
Literatur*, i(2) (Munich, 1934), 553 n. 3. Finally, there is much relevant material in Gould's
Herodotus (1989), see e.g. 76f. See also A. Saugé, *De l'épopée à l'histoire: Fondement de la
notion de l'historié* (Frankfurt am Main–Berne–New York–Paris, 1992).

expressed in language chosen to 'mark the singularity of the occasion', Ἀθηναῖος ὢν Λαμψακηνῷ. It is however more than that: the emphatic juxtaposition (but not the actual choice of words) is Homeric; compare *Iliad* 16. 176 for the phrase γυνὴ θεῷ εὐνηθεῖσα, 'a mortal woman sleeping with a god'.

But there *are* some subtle verbal echoes of Homer in Thucydides. I believe that there is one concealed in Pericles' description (2. 60. 5) of himself as good at 'devising and explaining a sound policy', γνῶναί τε τὰ δέοντα καὶ ἑρμηνεῦσαι ταῦτα. This suggests Homer's description of Halitherses the diviner as 'good at understanding bird-lore and interpreting omens', ὄρνιθας γνῶναι καὶ ἐναίσιμα μυθήσασθαι (*Odyssey* 2. 159). (Note the similarity of subject-matter, the divination of the seer and of the politician. Euripides was to say that the best seer was the one who guessed correctly: frag. 793 Nauck.) But like Herodotus, Thucydides avoids a perfect hexameter; for that, the τε would have to be lengthened.

It should not incidentally be doubted that great literary artists tend to avoid jingles such as hexameter endings. The first paragraph of Jane Austen's *Emma* ends with the words 'very little to distress or vex her'. Transpose the verbs and you get the uglier 'very little to vex or distress her', the second half of a hexameter. When composing the opening paragraph of a masterpiece, Jane Austen surely considered the alternatives and rejected them. By contrast, the first sentence of Daphne du Maurier's *Rebecca* is unsatisfactory because too perfectly rhythmical, actually an alexandrine: 'Last night I dreamt I went to Manderley again.' This recalls Edwardian romantic poetry like the opening of Ernest Dowson's *Cynara*: 'Last night, ah yesternight, betwixt her lips and mine . . .'. The Authorized Version of the Bible gets away with lines like 'how art thou fallen from heaven, O Lucifer son of the morning', or 'I cannot dig, to beg I am ashamed', a hexameter and an iambic line respectively (Isaiah 14: 12; Luke 16: 3). But even in antiquity such accidental metricality was disapproved of, see Cicero, *Orator* 189.[165] All this makes the perfect but

---

[165] See the commentary by W. Kroll in his 1913 edn. (repr. Berlin, 1958). Note D. L. Page's collection of accidental hexameters in Demosthenes: *History and the Homeric Iliad* (Berkeley and Los Angeles, 1963), 211f. Dem. 21. 179 (ταῦτ᾽ ἔλεγεν μὲν ἐκεῖνος, ἐχειροτονήσατε δ᾽ ὑμεῖς) is, as Page says, a very neat verse. On natural verse rhythms in English, see H. Gardner, *The Art of T. S. Eliot*, 34 n. 1. Relevant also is Max Beerbohm, *A Peep into the Past*, ed. R. Hart-Davis (London, 1972), 35, on the 'saliently iambic ending' which mars the section on Botticelli's Madonnas in Pater's *The Renaissance*. The offending words are 'and fair white linen on their sunburnt throats', W. Pater, *The Renaissance* (London, 1910; repr. Chicago, 1977), 57. I have discussed verse snatches in Thucydides in my *Commentary* on the following passages: 1. 13. 6 (cp. 3. 104. 2), 38. 2, 129. 3; 2. 36. 4,

little noticed iambic line at the end of Thucydides' 'tragic' Sicilian expedition the more remarkable, 'and most miserable for the defeated', καὶ τοῖς διαφθαρεῖσι δυστυχέστατον (7. 87. 5). Again, the exceptional instance comes from a Sicilian book.

But it is not possible to show that when Thucydides uses such devices as 'presentation through negation' or delayed introduction of a key actor (see Ch. 5 below), the influence of Homer was direct and conscious. All we can do is point to the specific passages which prove that Thucydides, and still more obtrusively Herodotus, knew their Homer, and to the Homeric-looking devices in their works; and then draw the inference, legitimate but not watertight, that the poet influenced the narrative technique of the historians. Thucydides may well have had Homer in mind at 1. 24. 1 ('Epidamnus is a city . . .', cp. *Iliad* 6. 152, 'There is a city Ephyre'), especially since Ephyre was an old name for Corinth, and Epidamnus was a colony of a colony of Corinth. But for naturalness this particular story-telling device almost rivals 'once upon a time'.

Finally, arguments from silence. (Here I return to question (ii) posed at p. 56 above.) It is as important to look at what historians do not say as what they do say. John Gould raises the problem of what Herodotus *does not* say about religion, and Ernst Badian the problem of what he *does not* say about Macedon. The problem about religious silences or selectivity may be extended to include Thucydides. But the reasons for the religious silences of the two authors are different. Herodotus' silences are part of a reticence which is often made explicit (see especially 2. 65. 2) and which by no means indicates contempt or lack of interest.[166] Thucydides also

61. 2. Note that Thucydides never allows the formula 'so ended the nth year which Thucydides recorded' to become a hexameter, though the ending is one (ὃν Θουκυδίδης ξυνέγραψεν). The topic of Thucydides' *detailed* debt to Homer deserves a monograph; on 7. 87. 5 and other verse rhythms see Lamb, *Clio Enthroned* ch. viii. There have been some excellent individual suggestions, e.g. F. Cairns, 'Cleon and Pericles: A Suggestion', *JHS* 102 (1982), 203f. (comparison between the Thucydidean Kleon's echoes of the Thucydidean Pericles and Thersites' echoes of Achilles); and note Macleod, *Essays*, 157—remarks which I should dearly have liked to see amplified. My quotation in the text about the daughter of Hippias is from Wade-Gery, *Essays in Greek History* (Oxford, 1958), 160 n. 2. For *Iliad* 16. 176 see the good n. in R. Janko's commentary (above, n. 159). See also Richardson (above, n. 25) on *Iliad* 22. 75: Homer in the *Iliad* tends to avoid αἰδοῖα ('genitals'). Here Hdt. is actually *less* like Homer than is Th. Another insufficiently remarked poetic debt in Thucydides is to Solon, see my *Commentary*, p. 346: the thought of 2. 65. 9 is Solonian and so is the language of para. 8 (with κατεῖχε τὸ πλῆθος compare Solon F 37. 7 West, οὐκ ἂν κατέσχε δῆμον; the phrase at 2. 65. 8 is picked up at 8. 86. 5 where it is applied to Alcibiades). For Th. and Euripides see J. H. Finley, *Three Essays on Thucydides* (Cambridge, Mass., 1967), ch. 1.

[166] See Gould, Ch. 3 below.

has his silences. He says nothing about the Delphic amphiktiony, the international organization which ran the affairs of the sanctuary. It would be rash and wrong to infer from this silence that the amphiktiony did not function or matter in the fifth century, as it undoubtedly did in the adjoining centuries.[167] Thucydides often neglects religion, for reasons which reflect his basically secular outlook.[168] The silences of Xenophon, by contrast, are deafening; but they do not include religion. In this area, as in others (above, p. 30), he returns to Herodotus' priorities.

Again, P. M. Fraser raises questions about gaps in Greek historiographic awareness of the west before Timaeus. Thucydides knows about Sicily, but the South Italian dimension to the great Athenian expedition of 415 is curiously understated, only passages like 7. 33 offering a hint (Athenian diplomacy with Artas of Messapia, and with Metapontum, see below, Ch. 5, p. 147 n. 43. And 6. 88. 7 is revealing: some Syracusan envoys say that Athens has designs on Italy as much as on Sicily.) Even Herodotus, despite having settled in Thurii in South Italy for a time, had curiously little detailed knowledge about South Italy, despite its plethora of Greek communities; the Tarentine massacre at 7. 170 is an exception (on this see below, p. 151). This may be an intermittent blind spot in Greek historiography, and helps to explain a conspicuous cluster of omissions in Herodotus' history, a proper treatment of Phoenicia, Carthage, and Etruria. (Hecataeus had perhaps done better: his fragments 80–9 give the names of a number of South Italian Greek and native communities, several absent from Herodotus. And Antiochus, whose work Thucydides knew, certainly talked about Italy as well as his native Sicily—see p. 22 above for his possible aim to improve on Herodotus here.) Kenneth Sacks's paper about Diodorus on Rome (below, Ch. 8) is relevant here too. One of his general arguments for Diodorus' negative attitude towards Rome proceeds from the absence of the obvious comparison between Rome and other successful empires. But even in Diodorus there is (as in the other writers we have been considering) a 'western', in fact a Sicilian silence, which is at first sight surprising given that Diodorus is himself a Sicilian historian: Aeneas' connection with Sicily is (Sacks argues) minimized. This, Sacks ingeniously suggests (p. 224), is a use of the Aeneas legend as part of 'resistance historiography'.

A classic case of silence or omission, though of a different kind, is the

---

[167] See my paper in *HSCP* 94 (1992), 169–97. For the earlier position see C. Morgan's important *Athletes and Oracles* (Cambridge, 1990), 135 f.

[168] Ibid.

subject of John Davies's paper, the First Sacred War, which was 'omitted' (assuming that it happened at all) by all the big historians. For this see above on the 'omission' of the war from Thucydides' Archaeology.

Trickier is a tradition about a war which *did* happen, but acquired gross accretions, viz. the Persian—the subject of Antony Spawforth's paper. It would be wrong to treat this as a set of facts merely transmitted or garbled by more or less competent historians. The theme had a propagandist life of its own.

But arguments from silence are, like many of the arguments we have been considering, treacherous in that they often assume a perfectly conscientious (rather than lazy or slapdash) historian whose aim was the correct recording of fact (rather than the achieving of a literary creation). Such arguments can, for instance, be abused to 'prove' dates of composition. If a historian, especially one like Arrian with a dated official career in a given part of the world, shows ignorance of that part of the world in some section of his writings, then that section ought in a perfect world to pre-date the relevant part of his career. But in fact historians do not always show knowledge of areas they had visited or in which they had carried out official duties. Syme pointed that out in connection with Tacitus and Frontinus; but it holds for Arrian too.[169] When operating in 'literary mode', a writer might simply fail to draw on his own experience, or to update the obsolete information that he found in his sources.

The thesis or theses of this section can now be summed up. In a sentence, the historians of antiquity should be treated as products of their age, not as if they themselves were carrying out *Quellenforschung*, the scientific study of sources. The word 'sources' is itself problematic, and covers a range from recent inspection of a text to an inaccurate memory of a snatch of a poem. The motives of ancient historians may be 'literary' rather than or as well as historical; and this permitted, for instance, the retention of anachronisms in such 'sources', and the preferring of easier and more charming authorities to harder if earlier and better ones. Given the nature of ancient book-production, even collation was hard work, as the younger Pliny said ('onerosa collatio', 5. 8. 12). Ancient historians can often be shown to have been aware of each other's writings, though there was (I have suggested) a tendency to be more aware of the methodological sections of a predecessor's work than of his routine

---

[169] R. Syme, *Tacitus* (Oxford, 1958), 779 (Frontinus), 127–9 (Tacitus' *Germania*). For Arrian see *CR* 31 (1981), 14.

narrative.[170] But even the most polemical of them—and polemic does not necessarily imply thorough and fair acquaintance—may show what may to us seem surprising gaps in such knowledge of predecessors. Or else they may cite what we would regard as the 'wrong' authority: so Strabo cites not the older, primary source Herodotus but the younger, secondary Kallisthenes, for a fact given in identical terms by both. Finally, not all puzzles in this area are soluble by making rational or charitable assumptions. Oversights happened,[171] and were not so easily corrected on one of a large number of rolls of papyrus as on the disk of a 1990s word-processor. Some of these points (though not I hope all) may seem elementary and obvious, but the speed with which our own techniques of information storage and retrieval are advancing perhaps makes it worth insisting on them.

[170] An exception should be made for a specialist military writer like the 4th-cent. Aeneas Tacticus, who took facts from Th.'s wartime narrative, see Th. 2. 2–6 (siege of Plataia) and Aen. Tact. 2. 3, with D. Whitehead, *Aineias the Tactician* (Oxford, 1990), 102. Aen. Tact. 38. 2 = Th. 5. 8. 9—a *speech*. Note also that in his essay *On the Malice of Herodotus*, Plutarch makes sophisticated use of a non-methodological section of Thucydides to attack Herodotus, *Mor.* 870d.

[171] Thus the sheer awkwardness of making corrections to an ancient text should always be remembered. This, rather than any subtler explanation, may lie behind such Thucydidean repetitions as the Aiginetan material at 2. 27 = 4. 56.

2

Historical Explanation: Polybius and his Predecessors[1]

PETER DEROW

The title of this paper is not intended to imply that developments in Greek historiography are best studied back to front. But I do feel obliged to take a gentle ramble through some of Herodotus and Thucydides *en route* to Polybius. Without this it would be hard to see how much changed along the way and, in what may be the most essential respects, how little.

That said, the story must begin with Hecataeus. He was the first to identify, if I may put it so, the past as a field of critical study. The past in this case is what the ridiculous tales of the Greeks are about, and it is the truth of what happened that he seeks.

[1] These 'predecessors' of Polybius are, as will emerge very quickly, Herodotus and Thucydides, with a brief but necessary glance at Hecataeus. There were, of course, others, and amongst them some whose attitudes towards historical explanation might be discussed with profit. But none of these lends himself to a discussion of anything like the depth or security that is possible with Herodotus and Thucydides. Many potentially promising candidates are impeded partly by fragmentary preservation and even more by the fact that what we have of them comes to us refracted through other minds and intentions. This last is of particular concern when matters of thought and language are so centrally at issue. For reasons such as these this version retains the agenda and dramatis personae of the original. In other respects, too, it remains essentially the same, save for the incorporation into the text of passages that seemed (and seem) to me importantly illustrative. In these, Greek is retained when at least some of the words require to be focused upon. The translations provided are mostly based upon Kirk and Raven (for Anaximander and Heraclitus), Grene (for Herodotus), Crawley (for Thucydides), and Shuckburgh and Paton (for Polybius): see General Bibliography for full references. Treatments of these authors are limitless, citations of these treatments here the opposite. With some of what is said about Herodotus compare Donald Lateiner, *The Historical Method of Herodotus* (*Phoenix* suppl. 23; Toronto, 1989), chapters 6 and 9; my direction is different and more limited, my analysis different, too, I think, and more simple. On the sections of Thucydides book 1 at issue see Simon Hornblower, *A Commentary on Thucydides*, i (Oxford, 1991), not only for citation of (especially recent) work but also for his own observations. Some of what is said about Polybius refers more or less obliquely to my 'Polybius, Rome and the East', *JRS* 69 (1979), 1–15, and on Polybius there is above all Frank Walbank (*Historical Commentary on Polybius*, 3 vols (Oxford, 1957–79) and *Polybius* (Berkeley, 1972)). If my debt to his work and to him were thought to be visible here, I should be pleased.

Ἑκαταῖος Μιλήσιος ὧδε μυθεῖται· τάδε γράφω, ὥς μοι δοκεῖ ἀληθέα εἶναι· οἱ γὰρ Ἑλλήνων λόγοι πολλοί τε καὶ γελοῖοι, ὡς ἐμοὶ φαίνονται, εἰσίν.

Hecataeus the Milesian speaks so: I write the things that follow as they seem to me to be true. For the stories of the Greeks are both many and, as they appear to me, ridiculous.   (*FGrHist* 1 F 1)

The nature of his critical method soon becomes clear.

ὁ δὲ Αἴγυπτος αὐτὸς μὲν οὐκ ἦλθεν εἰς Ἄργος, παῖδες δέ, ⟨ἐόντες⟩, ὡς μὲν Ἡσίοδος ἐποίησε, πεντήκοντα, ὡς ἐγὼ δέ, οὐδὲ εἴκοσι.

Aegyptus did not himself go to Argos, but his sons did—fifty of them in Hesiod's story, but as I reckon not even twenty.   (*FGrHist* 1 F 19)

He objects to the idea that fifty sons of Aegyptus came to Argos, apparently on the grounds that fifty was simply too many, and suggests something more moderate, less than twenty. He had analogous difficulties with the tales of Cerberus at the gates of the underworld, and again he has an answer.

ἀλλὰ Ἑκαταῖος μὲν Μιλήσιος λόγον εὗρεν εἰκότα, ὄφιν φήσας ἐπὶ Ταινάρῳ τραφῆναι δεινόν, κληθῆναι δὲ Ἅιδου κύνα, ὅτι ἔδει τὸν δηχθέντα τεθνάναι παραυτίκα ὑπὸ τοῦ ἰοῦ· καὶ τοῦτον ἔφη τὸν ὄφιν ὑπὸ Ἡρακλέους ἀχθῆναι παρ' Εὐρυσθέα.

Hecataeus the Milesian has found a likely account, saying that a terrible serpent grew up at Taenarum, and that it was called the dog of Hades because anyone bitten by it was killed immediately by the venom. And it was this serpent, he says, that was brought by Heracles to Eurystheus.   (*FGrHist* 1 F 27)

It wasn't an impossible dog at Taenarum, but a particularly poisonous serpent. This was something that reason and nature could countenance, and reason and nature were the twin criteria of his rigorously rationalizing method. In the same passage he may be thought to reveal the limits of his aims, for having reasonably got rid of the impossible dog, he leaves us with Heracles presenting a snake to Eurystheus.

   The process has begun, but it needed Herodotus to define the field of inquiry as he defined it and to ask the question that he asked before it could continue productively. He does both these things in the proem.

Ἡροδότου Ἁλικαρνασσέος ἱστορίης ἀπόδεξις ἥδε, ὡς μήτε τὰ γενόμενα ἐξ ἀνθρώπων τῷ χρόνῳ ἐξίτηλα γένηται, μήτε ἔργα μεγάλα τε καὶ θωμαστά, τὰ μὲν Ἕλλησι, τὰ δὲ βαρβάροισι ἀποδεχθέντα, ἀκλεᾶ γένηται, τά τε ἄλλα καὶ δι' ἣν αἰτίην ἐπολέμησαν ἀλλήλοισι.

Herodotus of Halicarnassus here sets forth the result of his inquiry, that the doings of men might not be forgotten with time, and that great and wonderful works and deeds—wrought by both Greeks and barbarians—might not be uncelebrated, and together with all this the reason why they warred with one another.

The field is τὰ γενόμενα ἐξ ἀνθρώπων, what people have done, human history; and the question is 'why?', δι' ἣν αἰτίην, 'for what reason?'. Here at the beginning, it is 'why did the Greeks and barbarians war with one another?'; but the question recurs in various contexts throughout Herodotus and was evidently an essential element of his inquiry. It would seem to be an essential part of any decent historical inquiry, although we have to wait for Polybius to insist upon this explicitly.

In general, Herodotus' answers to the question 'why?' might seem to operate at two levels, of which one would appear to involve some kind of notion of 'fate'. I would rather not insist on that precise English word, for I do not think that Herodotus had a Greek one to match it (ἡ πεπρωμένη in 1. 91. 3 approaches but is perhaps too verbal quite to get there). What I have in mind are the statements with ἔδει or χρῆν or the like.[2] The words χρῆν γὰρ Κανδαύλη γενέσθαι κακῶς mean that Candaules 'had to' or 'was bound to' come to an evil end, just as the words at 5. 33. 2 mean that it was not on the cards for the Naxians to be destroyed by this expeditionary force (another one still to come would achieve what this one did not), and those at 5. 92δ. 1 that it had to happen that misfortunes for Corinth would come from the offspring of Eetion. Some form of inevitability and, not least, of predetermination is at issue here, as I think becomes clear in the sequence of dreams early in book 7 and particularly in the words of the dream that visited Artabanus, clad in Xerxes' night-dress and asleep in the royal bed.

Σὺ δὴ εἶς ὁ ἀποσπεύδων Ξέρξην στρατεύεσθαι ἐπὶ τὴν Ἑλλάδα ὡς δὴ κηδόμενος αὐτοῦ; ἀλλ' οὔτε ἐς τὸ μετέπειτα οὔτε ἐς τὸ παραυτίκα νῦν καταπροίξεαι ἀποτρέπων τὸ χρεὸν γενέσθαι, Ξέρξην δὲ τὰ δεῖ ἀνηκουστέοντα παθεῖν, αὐτῷ ἐκείνῳ δεδήλωται.

Is it you then who would discourage Xerxes from marching against Greece with your claim to be concerned for him? Not for the future nor for the present will you get away with seeking to avert what must happen. What Xerxes must suffer if he disobeys has been shown to him directly.   (7. 17. 2)

---

[2] 1. 8. 2 χρῆν γὰρ Κανδαύλη γενέσθαι κακῶς; 5. 33. 2 οὐ γὰρ ἔδεε τούτῳ τῷ στόλῳ Ναξίους ἀπολέσθαι; 5. 92δ. 1 ἔδει δὲ ἐκ τοῦ Ἡετίωνος γόνου Κορίνθῳ κακὰ ἀναβλαστεῖν (and cf. 4. 79. 1, 6. 64, 6. 135. 3, 9. 109. 2).

Or in the story of the dream that came to Croesus and 'revealed to him the truth of the evils that were going to happen in connection with his son' (ὄνειρος, ὅς οἱ τὴν ἀληθείην ἔφαινε τῶν μελλόντων γενέσθαι κακῶν κατὰ τὸν παῖδα, 1. 34). For Herodotus the future was predetermined and therefore knowable, or knowable and therefore predetermined; either will do. This is not surprising, for such must be part of the mind-set of one who believes in true prophecy, whether through the medium of oracles, dreams, seers, or even through an individual like Croesus who returned from his glimpse of death with preternatural knowledge. Plainly, there could be no greater aid to planning one's actions than knowing what the future holds, but of course nobody took proper notice when they were told. Cassandra ruled OK.

Now, predetermination—some version of fate, or thereabouts—may be a fact, but it is not an explanation, and Herodotus knew this. The explanation of human affairs has to be done at the human level. The Persian expedition against Greece was, as we have seen, in the category of τὸ χρεὸν γενέσθαι, what had to happen, but it came about, in Herodotus, for very human reasons indeed. Recall the scene in 3. 134. Darius and Atossa are in bed one evening, having a discussion about conquering. She suggests that a king so young, rich, and powerful as he ought to be extending Persian dominions. It's interesting you should say that, he replies, for I have in mind to conquer the Scythians. Very well, she says, but why not leave the Scythian campaign for later. For now, won't you please put on a campaign against Greece for me. I've asked round and I should like to have some Laconian and Argive and Attic and Corinthian handmaidens. All right, says Darius, but we had better send some spies first. And off goes Doctor Democedes, exactly as planned. And there are the beginnings of the Persian attack upon Greece. Darius does in the end deal first with the Scythians. 'He wanted to take vengeance upon them because they had begun the injustice (ὑπῆρξαν ἀδικίης) by initially invading Median territory and defeating in battle those who came to oppose them' (4. 1. 1). He came to want similar vengeance upon the Athenians for their action during the Ionian revolt, and he was still wanting τιμωρήσασθαι Ἀθηναίους when he died (7. 4). Xerxes felt the same, and as well as wishing to bring retribution upon the Athenians he, like Darius, wanted to acquire additional territory for the Persians (7. 8α. 2–β. 1). Desire for revenge and desire to have more—call it greed—are pretty basic in human psychology. So they are in Herodotus' aetiology because that is wholly rooted in human psychology.

Basic they are to most of Herodotus' explanations, but he is of course capable of more subtlety, as he shows in his twin explanations of Croesus' attack upon Cappadocia.

(1) Κροῖσος δὲ ἐπὶ δύο ἔτεα ἐν πένθεϊ μεγάλῳ κατῆστο τοῦ παιδὸς ἐστερημένος· μετὰ δὲ ἡ Ἀστυάγεος τοῦ Κυαξάρεω ἡγεμονίη καταιρεθεῖσα ὑπὸ Κύρου τοῦ Καμβύσεω καὶ τὰ τῶν Περσέων πρήγματα αὐξανόμενα πένθεος μὲν Κροῖσον ἀπέπαυσε, ἐνέβησε δὲ ἐς φροντίδα, εἴ κως δύναιτο, πρὶν μεγάλους γενέσθαι τοὺς Πέρσας, καταλαβεῖν αὐτῶν αὐξανομένην τὴν δύναμιν.

Croesus, reft of his son, spent two years in great grief. Then the destruction of the rule of Astyages son of Cyaxares by Cyrus son of Cambyses and the increase of the Persian state stopped Croesus from his grief, and he began to consider whether he might be able to put a stop to the power of the Persians while it was still growing and before they became great.   (1. 46. 1)

(2) Ἐστρατεύετο δὲ ὁ Κροῖσος ἐπὶ τὴν Καππαδοκίην τῶνδε εἵνεκα, καὶ γῆς ἱμέρῳ προσκτήσασθαι πρὸς τὴν ἑωυτοῦ μοῖραν βουλόμενος, καὶ μάλιστα τῷ χρηστηρίῳ πίσυνος ἐὼν καὶ τείσασθαι θέλων ὑπὲρ Ἀστυάγεος Κῦρον.

Croesus led his army against Cappadocia for these reasons: because he wished to add to his own portion out of a desire for land, and especially because he was relying upon the oracle, and because he wished to take vengeance upon Cyrus on behalf of Astyages.   (1. 73. 1)

The explanation of 1. 73 provides the motives of greed and vengeance, but 1. 46 provides something else, namely the element of calculated response to circumstances. It is an interesting notion, being worried by the growth of someone else's power, and it recalls another passage in Herodotus. In 5. 90 Cleomenes returns from Athens to Sparta with oracles foretelling the injury the Spartans are to suffer at the hands of the Athenians. He continues (5. 91):

When the Lacedaemonians got these oracles and saw that the Athenians were gaining in power and not at all ready to be their subordinates, and when they took cognizance that the Attic race, in its freedom, would be the equal of themselves but, if controlled by a despotism, would be weak and disposed to subjection—when they understood all this, they sent for Hippias, the son of Pisistratus, from Sigeum on the Hellespont.

(The oracles, by the way, which foretold *inter alia* that the Spartans along with the other Dorians would be driven from the Peloponnesus by the Medes and the Athenians, were recalled again twenty-five years on: 8. 141.) And both passages bear a measure of resemblance to a reasonably well-known passage of Thucydides (viz. 1. 23. 6, on which see further below).

But that, for the moment at least, is by the way. Before leaving Herodotus one must ask how he answered the question at the beginning: why did they war with one another? At 1. 5 the humour of the first four chapters comes decidedly to an end.

ἐγὼ δὲ περὶ μὲν τούτων οὐκ ἔρχομαι ἐρέων ὡς οὕτως ἢ ἄλλως κως ταῦτα ἐγένετο, τὸν δὲ οἶδα αὐτὸς πρῶτον ὑπάρξαντα ἀδίκων ἔργων ἐς τοὺς Ἕλληνας, τοῦτον σημήνας προβήσομαι ἐς τὸ πρόσω τοῦ λόγου, ὁμοίως σμικρὰ καὶ μεγάλα ἄστεα ἀνθρώπων ἐπεξιών. (4) τὰ γὰρ τὸ πάλαι μεγάλα ἦν, τὰ πολλὰ αὐτῶν σμικρὰ γέγονε, τὰ δὲ ἐπ᾽ ἐμεῦ ἦν μεγάλα, πρότερον ἦν σμικρά. τὴν ἀνθρωπηίην ὦν ἐπιστάμενος εὐδαιμονίην οὐδαμὰ ἐν τὠυτῷ μένουσαν ἐπιμνήσομαι ἀμφοτέρων ὁμοίως.

About these things I am not going to say that they happened in this or in some other way, but I myself know who was the first to begin unjust acts against the Greeks and having signalled him I shall proceed with my account, treating alike of the small and the great cities of men. (4) For of the cities that were great in former times most are become small, and those that were great in my time were small before. Knowing therefore that human good fortune never stays in the same place I shall make mention of both alike.   (1. 5. 3–4)

'I myself know who was the first to begin unjust deeds against the Greeks. . . .' That seems to do it. And on that two observations. First, Herodotus' insistence here at the beginning and throughout his work upon the idea of vengeance and upon wrongs and the righting of wrongs (and remember that greed involves transgression) suggests to me that he subscribed to what might be called the conflict, or retributive, theory of world order and justice that had been developed particularly by Anaximander and, with typically paradoxical formulation, Heraclitus.

ἐξ ὧν δὲ ἡ γένεσίς ἐστι τοῖς οὖσι, καὶ τὴν φθορὰν εἰς ταῦτα γίνεσθαι "κατὰ τὸ χρεών· διδόναι γὰρ αὐτὰ δίκην καὶ τίσιν ἀλλήλοις τῆς ἀδικίας κατὰ τὴν τοῦ χρόνου τάξιν," ποιητικωτέροις οὕτως ὀνόμασιν αὐτὰ λέγων.

And the source of coming-to-be for existing things is that into which destruction, too, happens 'according to necessity; for they pay penalty and retribution to each other for their injustice according to the assessment of Time', as he describes it in these rather poetical terms.   (Anaximander, D–K 12A9)

εἰδέναι δὲ χρὴ τὸν πόλεμον ἐόντα ξυνόν, καὶ δίκην ἔριν, καὶ γινόμενα πάντα κατ᾽ ἔριν καὶ χρεών.

It is necessary to know that war is common and right is strife and that all things happen by strife and necessity.   (Heraclitus, D–K 22B80; cf. also B53)

Further, I think it can safely be said that, for Herodotus, the explanation of why something happened reduces to the explanation of why someone did something, and that this is achieved by the imputation of what are fundamentally personal motives. I say 'imputation' because I do not suppose that Herodotus knew what transpired in bed between Atossa and Darius or what was in the minds of Xerxes or Croesus or most, if not all, of the other people, let alone peoples, about whom he wrote. Nor does he seem to see this as a problem.

But it is time to move on in the direction of Polybius.

For Thucydides, predetermination is not an issue. He is resolutely anchored in the realm of humanity, and so alive to human suffering that it is in terms of it that he calculates the magnitude of his war. It is, moreover, his conviction that people will always respond similarly to similar circumstances that enables him to claim timeless value for his work. At least, this is what I take him to be saying in 1. 22. 4.

καὶ ἐς μὲν ἀκρόασιν ἴσως τὸ μὴ μυθῶδες αὐτῶν ἀτερπέστερον φανεῖται· ὅσοι δὲ βούλονται τῶν τε γενομένων τὸ σαφὲς σκοπεῖν καὶ τῶν μελλόντων ποτὲ αὖθις κατὰ τὸ ἀνθρώπινον τοιούτων καὶ παραπλησίων ἔσεσθαι, ὠφέλιμα κρίνειν αὐτὰ ἀρκούντως ἕξει. κτῆμά τε ἐς αἰεὶ μᾶλλον ἢ ἀγώνισμα ἐς τὸ παρα-χρῆμα ἀκούειν σύγκειται.

It may be that the unromantic character of my work will not make for pleasant hearing. But it will be enough if it is judged useful by those who wish to have a clear and accurate picture of what happened and of the similar and analogous things that will, given the condition of humanity, happen again at some point in the future: it is intended as something that will retain its value for all time, and not as a competitive utterance for the moment.

And this is also why he so insists upon the importance of τὸ σαφές there, the importance of clear and accurate description. He has just written of the great pains he has gone to in arriving at as accurate (ἀκρίβεια is the word in 1. 22. 1 and 2) a portrayal as possible of what was said and done. Words and deeds, ὅσα μὲν λόγῳ εἶπον and τὰ δ' ἔργα τῶν πραχθέντων, are equally historical facts, and the same kind of care must be taken with both. He has left us in no doubt in chapters 20 and 21 that errors and misconceptions are abroad, and both there and elsewhere he is ready to offer correction.

In all this the emphasis is most firmly and explicitly upon accuracy of narrative and care with facts. He is, of course, also concerned with explanation, but I think to a much lesser extent. He addresses the matter of his own war in 1. 23. 'The Athenians and Peloponnesians began it when they broke the thirty years' treaty they made after the capture of Euboea. As to why they broke the treaty, I have given first an account of the claims of either side and their differences, in order that no one should ever have to inquire from what (ἐξ ὅτου) so great a war came about for the Greeks.' (1. 23. 4–5) This refers to the businesses of Corcyra and Potidea and promises a definitive account of the circumstances and the dealings between Athenians and Peloponnesians out of which the war arose. And he goes on, famously, in 1. 23. 6 to say that he considers the truest explanation, although the least often heard, to be that the Athenians by growing great and causing fear amongst the Lacedaemonians forced them to go to war (τὴν μὲν γὰρ ἀληθεστάτην πρόφασιν, ἀφανεστάτην δὲ λόγῳ, τοὺς Ἀθηναίους ἡγοῦμαι μεγάλους γιγνομένους καὶ φόβον παρέχοντας τοῖς Λακεδαιμονίοις ἀναγκάσαι ἐς τὸ πολεμεῖν). Different things are at issue here. But to my mind we are not looking at a contrast between proximate or superficial causes on the one hand and an underlying cause on the other. Rather, there is on the one hand the context or set of circumstances out of which came the war (the ἐξ ὅτου, as it were), and on the other the historian's explanation (Thucydides' judgement, that is) of why it came about—two very different *kinds* of thing.

But that is still not quite right as it stands. Thucydides' explanation is not of anything so general as why the war came about. It is an explanation of why the Spartans began it, why they acted as they did and broke the treaty. (That both Thucydides and they saw it this way is clear from 1. 118. 2 and 7. 18. 2, respectively.) The explanation has two parts. One involves a response to circumstances, of the kind ascribed by Herodotus to his Croesus and even to his Lacedaemonians. (Thucydides 1. 118. 2 is in kind really very much like Herodotus 1. 46 and 5. 91.) The other is, if anything, even more Herodotean: the imputation (at the national level) of a fundamentally personal human motive, in this case fear.

Not Herodotean is the length and care to which Thucydides goes to describe the circumstances which so affected the Lacedaemonians and to which they responded as they did. He does this in chapters 89–118 of book 1. Rather, what he does is to describe what might be called the public facts of the growth of Athenian power and dominion and to attach to these from time to time very much less public, in fact worryingly private and frankly teleoscopic, things about the Spartans.

The trouble begins when the Athenians want to rebuild their walls.

Perceiving what was going to happen (τὸ μέλλον), the Spartans sent an embassy to Athens. They would themselves have preferred to see neither Athens nor any other city in possession of a wall; though here they acted principally under pressure from their allies, who were alarmed at the strength of their navy, which they had not had before, and at the daring they had displayed in the war against the Medes. They asked that they not only refrain from building walls for themselves but also that they join with them in pulling down any walls that survived of cities outside the Peloponnesus. In doing so they did not reveal the real intention of their advice or their suspicion of the Athenians. (1. 90. 1–2)

But Themistocles' trickery triumphed; the walls achieved defensible height unbeknown to the Lacedaemonians. And their response?

The Spartans when they heard this did not reveal their anger to the Athenians (they had after all not sent the embassy with the expressed purpose of prevention but rather to look like advice for the public good, and they were at that juncture on the best of terms with them on account of the spirit they had shown against the Mede). Yet, frustrated of their wish, they were secretly vexed (ἀδήλως ἤχθοντο). (1. 92)

No more is heard on this score for several chapters. Not until 1. 101. Thasos is being besieged by the Athenians, and the Thasians call upon the Spartans to help them by launching an invasion of Attica. Their response?

The Spartans undertook so to help them, unbeknown to the Athenians; and they were going to do so, but they were prevented by the occurrence of the earthquake, at which the Helots, and of the perioikoi the Thouriatai and Aithaieis, withdrew in rebellion to mount Ithome. (1. 101. 2)

But for a seismological accident war would have begun then, on Thucydides' account. And it could have begun that early, on Thucydides' account, because the explanation for it has been there from the beginning, in the form of private feelings imputed to the Lacedaemonians— concealed suspicion, unrevealed anger—and a measure of Peloponnesian fear. The story, of course, continues consistently, through the incident at mount Ithome when the difference, the διαφορά, first came out into the open (1. 102. 3), to the other differences and claims from which the war itself arose. This is interpretative narrative of high order indeed, although one might object that the narrative is informed by an interpretation that is supplied at the beginning and reasserted throughout. However compelling the interpretation is—and I think it very compelling—it is not analysis.

I hope it is clear that I do not by any of this mean to say that Thucydides was Herodotus, or a more closely-focused, intense, and streamlined version of him. For Thucydides was capable of analytical explanation in a way that Herodotus was not, or chose not to reveal. They both speak of Athens' rise to power and dominion, but they do it in very different ways. Partly, no doubt, because this phenomenon bore differently upon their purposes, but not only for that reason. For Herodotus, this rise of Athens was, along with its consequences for the rest of the Greeks, something foreordained. I have already mentioned the oracles that Cleomenes brought back to Sparta from the acropolis at Athens. And we can further recall the reply of Hippias to Socles, the Corinthian, who did most to dissuade the Peloponnesians from the Spartan project of restoring the Athenian tyrant:

That is what Socles said, the delegate from Corinth. Hippias answered him, invoking the very same gods against him: 'Verily,' he said, 'the Corinthians more than any other people will yet long for the Pisistratids when the appointed days are accomplished and they are sorely vexed by the Athenians.' That was the answer of Hippias, for he knew the oracles more accurately than any other man.

(5. 93)

In keeping with this is the Cyrus-like behaviour of the children of the Athenian women on Lemnos:

These women had children in great numbers, and they taught the children the Attic speech and Athenian ways. Their children would have nothing to do with the children born of the Pelasgian women, and, if one of them was struck by a Pelasgian child, all the others came to his assistance and so succoured one another. And the Athenian children absolutely claimed to rule the others and were far more authoritative. The Pelasgians took note of this and considered. In their consideration a strange and terrible thought overcame them: if these Attic-born children even now were making such a distinction, by coming to the help of their fellows against the more lawfully born, and were trying outright to rule them, what would they do when they grew up? So they determined to kill the children of the Attic women.   (6. 138. 2–4)

But in Herodotus there is never in the end any mileage in seeking to avert τὸ χρεὸν γενέθαι. These Attic-born children did not grow up, but others did, and Lemnos later fell to the Athenians and was ruled by them.

At the more usual level of human action, he gives us Aristagoras success-fully playing upon Athenian greed and desire for self-aggrandizement in order to secure support in his Ionian venture (5. 97). Miltiades after Marathon does much the same:

He asked the Athenians for seventy ships and an army and money, without saying against what country these would be used—only that they would grow rich if they followed him. For he would lead them to a country where they might easily win an abundance of gold. With these claims he asked for the ships. The Athenians were excited by his words and gave them to him. (6. 132)

Later on Themistocles emerges as the greediest person in the book (οὐ γὰρ ἐπαύετο πλεονεκτέων is Herodotus' comment at 8. 112. 1), and it is with him that collective and individual Athenian greed merge in 8. 111–12 and the Andrian Dialogue. Against this background, and maintaining the imputed motivation, Herodotus looks ahead to the events of 478/7:

There had been talk at the beginning, before ever they sent to Sicily about the alliance, that one ought to trust the fleet to the Athenians. When the allies objected, the Athenians gave way; they thought that what mattered most was the survival of Greece and knew very well that if there was a dispute about the leadership, Greece would perish—and that thought was correct, for strife within the nation is as much a greater evil than a united war effort as war itself is more evil than peace. So because the Athenians knew this, they put up no resistance, but yielded, but only so long as they had urgent need of the others, as they later proved. For as soon as they had driven out the Persian and were fighting for *his* territory rather than their own, the Athenians stripped the Lacedaemonians of their primacy (though nominally this was because of the arrogance of Pausanias). (8. 3)

Of this last episode Thucydides in 1. 95 gives a very clear and remarkably neutral account of what actually happened. It is enough on its own to reveal Herodotus' unfriendly bias. The assumption by Athens of confederate hegemony develops straightforwardly from the circumstances that precede it. And Thucydides' account of the growth of Athenian dominion over other Greeks is equally clear and neutral. He speaks in 1. 99 about the allies of Athens who rebelled from time to time:

Of all the causes of defection, that connected with arrears of tribute and vessels, and with failure of service, was the chief; for the Athenians were very severe and exacting, and made themselves offensive by applying the screw of necessity to men who were not used to and in fact not disposed for any continuous labour. In some other respects the Athenians were not the old popular rulers they had been at the start; and if they had more than their fair share of service, it was correspondingly easy for them to reduce any who tried to leave the confederacy. For this the allies had themselves to blame; the wish to get off service making most of them arrange to pay their share of the expense in money instead of in ships, and so to avoid having to leave their homes. Thus while Athens was increasing her navy with the funds which they contributed, a revolt always found them without resources or experience for war.

Taken as a statement about the developing dynamic imbalance between
the Athenians and their allies, about the relationship between neat power
and dominion, κράτος and ἀρχή, this is about as good as you can get.
Later on Thucydides lends his perception to the Athenians at Melos, who
say that they believe it of the gods and know it of men that as a matter of
necessity they exercise dominion wherever they have power (οὗ ἂν
κρατῇ, ἄρχειν, 5. 105. 2). Not 'that they rule wherever they can'. That
would make it a statement about some human need to seek dominion,
which would be something quite else; but that is not what the words
mean. Explanation can happen by analysis of circumstances and without
imputation of motive, and that is a serious advance.

Well then, what about Polybius? Two simple observations to begin
with. Polybius started writing his history some two and a half centuries
after Thucydides' breaks off, and Polybius was a pro.

One thing this passage of time means is that Polybius was a historian
writing for people who were accustomed to read histories, to use the
word in a general sense. He had to know and to talk about many of his
predecessors, and he did so. His bibliography, as it were, lists some
dozens of authors, which is pretty good for the ancient world. He was
aware of his place in a historiographical tradition, and very and explicitly
aware of the obligation that he as a historian had to his readers. This is
perhaps clearest in a passage towards the end of the work.

συγγραφέα δὲ κοινῶν πράξεων οὐδ᾽ ὅλως ἀποδεκτέον τὸν ἄλλο τι περὶ πλείο-
νος ποιούμενον τῆς ἀληθείας. (6) ὅσῳ γὰρ εἰς πλείους διατείνει καὶ ἐπὶ πλείω
χρόνον ἡ ⟨διὰ⟩ τῶν ὑπομνημάτων παράδοσις τῶν πρὸς καιρὸν λεγομένων,
τοσούτῳ χρὴ μᾶλλον καὶ τὸν γράφοντα περὶ πλείστου ποιεῖσθαι τὴν ἀλήθειαν
καὶ τοὺς ἀκούοντας ἀποδέχεσθαι τὴν τοιαύτην αἵρεσιν. (7) κατὰ μὲν τοὺς τῶν
περιστάσεων καιροὺς καθήκει βοηθεῖν τοὺς Ἕλληνας ὄντας τοῖς Ἕλλησι κατὰ
πάντα τρόπον, τὰ μὲν ἀμύνοντας, τὰ δὲ περιστέλλοντας, τὰ δὲ παραιτουμένους
τὴν τῶν κρατούντων ὀργήν· ὅπερ ἡμεῖς ἐπ᾽ αὐτῶν τῶν πραγμάτων ἐποιήσαμεν
ἀληθινῶς· (8) τὴν ⟨δ᾽⟩ ὑπὲρ τῶν γεγονότων τοῖς ἐπιγινομένοις διὰ τῶν
ὑπομνημάτων παράδοσιν ἀμιγῆ παντὸς ψεύδους ἀπολείπεσθαι χάριν τοῦ μὴ
ταῖς ἀκοαῖς τέρπεσθαι κατὰ τὸ παρὸν τοὺς ἀναγινώσκοντας, ἀλλὰ ταῖς ψυχαῖς
διορθοῦσθαι πρὸς τὸ μὴ πλεονάκις ἐν τοῖς αὐτοῖς διασφάλλεσθαι.

But a writer of public history above all deserves no indulgence whatever, who
regards anything of superior importance to truth. (6) For in proportion as written
history reaches larger numbers, and survives for longer time, than words spoken
to suit an occasion, both the writer ought to be still more particular about truth,
and his readers [lit.: listeners] ought to admit his authority only so far as he
adheres to this principle. (7) At the actual hour of danger it is only right that

Greeks should help Greeks in every possible way, by protecting them, veiling their errors or deprecating the wrath of the sovereign people,—and this I genuinely did for my part at the actual time: (8) but it is also right, in regard to the record of events to be transmitted to posterity, to leave them unmixed with any falsehood: so that readers should not be merely gratified for the moment by a pleasant tale, but should receive in their souls a lesson which will prevent a repetition of similar errors in the future. (38. 4. 5 ff.)

What he says is clearly reminiscent of Thucydides, especially when he warns his readers against the dangers of tales that are pleasant to hear but tinged with patriotic falsehood. Truth is an absolutely necessary consideration, and for Polybius as for Thucydides this applies to the accurate reporting of what was said as well as what was done. 12. 25b is good on this.

ὅτι τῆς ἱστορίας ἰδίωμα τοῦτ' ἐστὶ τὸ πρῶτον μὲν αὐτοὺς τοὺς κατ' ἀλήθειαν εἰρημένους, οἷοί ποτ' ἂν ὦσι, γνῶναι λόγους, δεύτερον τὴν αἰτίαν πυνθάνε- σθαι, παρ' ἣν ἢ διέπεσεν ἢ κατωρθώθη τὸ πραχθὲν ἢ ῥηθέν· (2) ἐπεὶ ψιλῶς λεγόμενον αὐτὸ τὸ γεγονὸς ψυχαγωγεῖ μέν, ὠφελεῖ δ' οὐδέν· προστεθείσης δὲ τῆς αἰτίας ἔγκαρπος ἡ τῆς ἱστορίας γίνεται χρῆσις. (3) ἐκ γὰρ τῶν ὁμοίων ἐπὶ τοὺς οἰκείους μεταφερομένων καιροὺς ἀφορμαὶ γίνονται καὶ προλήψεις εἰς τὸ προϊδέσθαι τὸ μέλλον, καὶ ποτὲ μὲν εὐλαβηθῆναι, ποτὲ δὲ μιμού- μενον τὰ προγεγονότα θαρραλεώτερον ἐγχειρεῖν τοῖς ἐπιφερομένοις· (4) ὁ δὲ καὶ τοὺς ῥηθέντας λόγους καὶ τὴν αἰτίαν παρασιωπῶν, ψευδῆ δ' ἀντὶ τούτων ἐπιχειρήματα καὶ διεξοδικοὺς λέγων λόγους, ἀναιρεῖ τὸ τῆς ἱστορίας ἴδιον.

The special province of history is first of all to ascertain the words actually spoken, whatever they may have been, and then to inquire after the reason why something that was said or done either failed or succeeded. (2) The simple report of what happened touches the fancy, but it is without use. When, however, the reason is added, the study of history becomes fruitful. (3) For it is the drawing of analogies between similar circumstances and our own that gives us the means of forming presentiments about what the future holds, and enables us at some times to act with caution and at others, by imitating what happened previously, to deal more boldly with what confronts us. (4) But a writer who passes over in silence the words that were spoken and the reason for what happened and gives us instead rhetorical exercises and discursive speeches destroys the peculiar virtue of history.

But truth is not a sufficient consideration. This emerges clearly from 12. 25b. 2 ff., as it does from 11. 19a in a complementary way.

τί γὰρ ὄφελός ἐστι τοῖς ἀναγινώσκουσι διεξιέναι πολέμους καὶ μάχας καὶ
πόλεων ἐξανδραποδισμοὺς καὶ πολιορκίας, εἰ μὴ τὰς αἰτίας ἐπιγνώσονται,
παρ' ἃς ἐν ἑκάστοις οἱ μὲν κατώρθωσαν, οἱ δ' ἐσφάλησαν; (2) τὰ γὰρ τέλη τῶν
πράξεων ψυχαγωγεῖ μόνον τοὺς ἀκούοντας, αἱ δὲ πρόσθεν διαλήψεις τῶν ἐπι-
βαλλομένων ἐξεταζόμεναι δεόντως ὠφελοῦσι τοὺς φιλομαθοῦντας. (3) μάλιστα δὲ πάντων ὁ κατὰ μέρος χειρισμὸς ἑκάστων ἐπιδεικνύμενος ἐπανορθοῖ
τοὺς συνεφιστάνοντας.

What benefit is it to readers to describe wars and battles and the stormings and
enslavements of cities, if they are to know nothing of the reasons for which some
succeeded and others failed on particular occasions? (2) The results of actions
merely touch the fancy; it is the proper investigation of the previous judgements
of those responsible that is of benefit to students. (3) And it is above all the
exposition of the detailed management of individual episodes that improves the
understanding of attentive readers.

It is even more necessary for the historian constantly to address the
question why? Why did something happen, why did a certain action or
policy or plan succeed or fail, and so on. Related is the emphasis in
11. 19a. 3 on the question how? The insistence upon establishing
accurately the who, what, where, and when he shares with Thucydides—for Polybius ἀλήθεια, for Thucydides ἀκρίβεια conducing to τὸ
σαφές. But the explicit insistence upon the paramount importance of
the how and above all the why is Polybius' own; it is new and, for some
while at least thereafter, unique. Polybius has defined the historian's
task as explanation.

That this task will inform the whole work is apparent early on. In book
1, chapter 1 he asks, 'could anyone be so indifferent or idle as not to wish
to understand how, and by what kind of state, almost the whole of the
known world was overpowered and fell under the single dominion of the
Romans in a space of not quite fifty-three years, something that never
happened before?' A fair question, about this, the Polybian, *pente-
kontaetea*. The answer to the question 'how and by what sort of state
. . .' occupied some twenty-six and a half of the forty books of his history.

As part of the larger task he asks the question 'why?' about episode
upon episode, not least about the wars that led to Rome's dominion.
Realizing, as we have seen, the importance of that question, he
formulated a method for answering it. Another first for Polybius. In the
context of the start of the war with Hannibal he feels it necessary, not
unreasonably, to distinguish the beginning, or first action, of a war (or
anything else) from the reason or reasons for that first action. This leads
to the methodological pronouncement of 3. 6. 7:

ἐγὼ δὲ παντὸς ἀρχὰς μὲν εἶναί φημι τὰς πρώτας ἐπιβολὰς καὶ πράξεις τῶν ἤδη
κεκριμένων, αἰτίας δὲ τὰς προκαθηγουμένας τῶν κρίσεων καὶ διαλήψεων·
λέγω δ' ἐπινοίας καὶ διαθέσεις καὶ τοὺς περὶ ταῦτα συλλογισμοὺς καὶ δι' ὧν
ἐπὶ τὸ κρῖναί τι καὶ προθέσθαι παραγινόμεθα.

I maintain that the beginnings of anything are the first attempts and actions of
those who have already taken decisions, but that the reasons are what lead up to
the decisions and judgements; I refer here to ideas and states of mind and reckon-
ings to do with these and the things through which we come to take decisions and
to form projects.

Beginnings are actions. Actions are preceded by decisions to act. And
decisions to act are processes involving various elements. A proper
explanation must delineate these processes and identify the various
elements. A tall order, it must be said, if a good one. He goes on in the
next chapter to offer a specimen explanation of Rome's war with
Antiochus.

καὶ μὴν τοῦ κατ' Ἀντίοχον καὶ Ῥωμαίους [*sc.* πολέμου] δῆλον ὡς αἰτίαν μὲν
τὴν Αἰτωλῶν ὀργὴν θετέον. (2) ἐκεῖνοι γὰρ δόξαντες ὑπὸ Ῥωμαίων ὠλιγωρη-
σθαι κατὰ πολλὰ περὶ τὴν ἔκβασιν τὴν ἐκ τοῦ Φιλίππου πολέμου, καθάπερ
ἐπάνω προεῖπον, οὐ μόνον Ἀντίοχον ἐπεσπάσαντο, πᾶν δὲ καὶ πρᾶξαι καὶ
παθεῖν ὑπέστησαν διὰ τὴν ἐπιγενομένην ὀργὴν ἐκ τῶν προειρημένων καιρῶν.
(3) πρόφασιν δ' ἡγητέον τὴν τῶν Ἑλλήνων ἐλευθέρωσιν, ἣν ἐκεῖνοι περι-
πορευόμενοι μετ' Ἀντιόχου τὰς πόλεις ἀλόγως καὶ ψευδῶς κατήγγελον, ἀρχὴν
δὲ τοῦ πολέμου τὸν Ἀντιόχου κατάπλουν εἰς Δημητριάδα.

And it is clear that the reason for the war between Antiochus and the Romans
must be taken to be the anger of the Aetolians. (2) Considering themselves to
have been slighted by the Romans in many matters to do with the conclusion of
the war with Philip, as I indicated earlier, they not only dragged in Antiochus but
were ready to do and to suffer anything because of the anger that developed from
the aforementioned circumstances. (3) As pretext one must reckon the liberation
of the Greeks, which the Aetolians went round the cities with Antiochus
proclaiming, without regard to reason or to truth, and as beginning the entry of
Antiochus into the harbour at Demetrias.   (3. 7. 1–3)

The landing of Antiochus at Demetrias was not the reason for the war; it
was the beginning and thereby the proper focus of explanation. (I do not
think he is right about the war with Antiochus, but I have no quarrel at all
with the method.) The example he gives suggests a risk of the method
leading to schematism, to a too rigorous separation of αἰτία, πρόφασις,
and ἀρχή. At the same time it reveals its inherent flexibility. The account
of the reasons can involve circumstances and people's response to them,
actions and reactions. It is this above all that shows Polybius' concern

was with explanation and not with the assignment of responsibility, or blame. The method does not work for the latter, as emerges even better from his account of the outbreak of the Hannibalic war. The most important reason for that conflict (the μεγίστη αἰτία: 3. 10. 4, and compare the notorious 3. 30) was the Roman seizure of Sardinia and the imposition of additional reparation a few years after the first Punic war, things which angered the Carthaginians. But it was still the Carthaginians who began, or performed the first actions of, the war itself, and that not at all unreasonably as Polybius has it. Where, on this kind of reckoning, would responsibility reside?

Yet there is, or may be, a problem inherent in the method. For what it means that Polybius is explaining is nothing so general as why a war broke out, but more precisely why whoever began it began it. In the case of the Hannibalic war it was Polybius' Carthaginians, just as in the case of the Peloponnesian war it was Thucydides' Lacedaemonians. Herodotus, of course, is just the same, in a still less complex way. A risk of one-sidedness, then, in the focus of the explanation. In a sense this is inevitable, at least in the case of wars, which are after all a collection of actions, and one of them has got to be the first. It is inevitable also in a more profound sense, if we agree with Polybius that the historian's task is to explain why people did things. The risk can, of course, be avoided, or at least minimized, by making proper use of the flexibility that we have seen also to be inherent in the method. And this Polybius can be seen to do pretty well where we have his aetiologies intact, which is too little of the time. Had more of books 22–5 survived, I reckon that we would have to spend less time trying to understand his explanation of the third Macedonian war, and the role therein of Philip V. I sometimes think I see what he means, but sometimes it looks downright silly.

Like his predecessors, Polybius includes basic human motives in his explanations, itself inevitable if we are to be dealing with people. We have met the anger (ὀργή) of the Aetolians in connection with the Antiochus war and the anger of the Carthaginians engendered by the Roman seizure of Sardinia. There is also the wrath of Hamilcar Barca, as his θυμός is often called, which figures in Polybius' account of the start of the Hannibalic war. But Polybius, I think, is more careful about these. The ascription of motive comes much more from the description and analysis of circumstances than from imputation. Polybius, that is, explains why Hamilcar Barca was upset after the end of the first Punic war in 241, why the Carthaginians a few years later joined him in his anger, why the behaviour of Flamininus and the Romans in 197 and after made the Aetolians cross, and in the latter part of book 18 we see some cross Aetolians in action. This is different from Thucydides. We sort of know (from 1. 118) what Lacedaemonian fear was about in the late 430s, but not what it was about in 465 or why they would have begun war then.

And Polybius is specially careful about the biggest motive, or intention, of them all, the aim of universal dominion he ascribes to the Romans, ἡ τῶν ὅλων ἐπιβολή. No further expatiation on that theme here. A look at book 1, chapter 3 (one of a number of passages of its kind) will begin to exemplify:

I therefore concluded that it was necessary to prefix this and the next book to my history. I was anxious that no one, when fairly embarked upon my actual narrative, should feel at a loss, and have to ask what were the designs entertained by the Romans, or the forces and means at their disposal, that they entered upon these undertakings, which did in fact lead to their becoming masters of land and sea everywhere in our part of the world. I wished, on the contrary, that these books of mine, and the prefatory sketch which they contained, might make it clear that the resources they started with justified their original idea, and sufficiently explained their final success in grasping universal empire and dominion.

The breadth of what Polybius looked for in his explanations is indicated nicely, and with similar reference, in 3. 2, where he announces book 6, on the Roman πολιτεία:

At this point I shall pause in my narrative to introduce a disquisition upon the Roman Constitution, in which I shall show that its peculiar character contributed largely to their success, not only in reducing all Italy to their authority, and in acquiring supremacy over the Iberians and Gauls besides, but also at last, after their conquest of Carthage, to their conceiving the idea of universal dominion.

And we will do well to remember that the 'πολιτεία' of book 6 is about more than 'constitution' commonly connotes: it comprises Roman military, religious, and political institutions.

It is probably as well to insist that for Polybius human behaviour requires human explanation, and this is what he manages to provide, save once. Even better to let Polybius insist upon this himself, for insist upon it he does, as he does so engagingly upon so much else.

Those things of which it is impossible for a mere man to ascertain the causes, such as a continuous fall of rains and unreasonable wet, or, on the contrary, droughts and frosts, one may reasonably impute to God and Fortune [ὁ θεός and ἡ τύχη], in default of any other explanation; and from them come destruction of fruits, as well as long-continued epidemics, and other similar things, of which it is not easy to find the cause. . . . But those things, of which it is possible to find the origin and cause of their occurrence, I do not think we should refer to the gods. I mean such a thing as the following. In our time all Greece was visited by a dearth of children and generally a decay of population, owing to which the cities were denuded of inhabitants, and a failure of productiveness resulted, though there were no long-continued wars or serious pestilences among us. If, then, anyone had advised our sending to ask the gods in regard to this, what we were to do or say in order to become more numerous and better fill our cities,—would he not have seemed a futile person, when the cause was manifest and the cure in our own hands? For this evil grew upon us rapidly, and without attracting attention, by our men becoming perverted to a passion for show and money and the pleasures of an idle life, and accordingly either not marrying at all, or if they did marry, refusing to rear the children that were born, or at most one or two out of a great number, for the sake of leaving them well off or bringing them up in extravagant luxury. For when there are only one or two sons, it is evident that, if war or pestilence carries off one, the houses must be left heirless: and like swarms of bees, little by little the cities become sparsely inhabited and weak. On this subject there is no need to ask the gods how we are to be relieved from such a curse: for any one in the world will tell you that it is by the men themselves if possible changing their objects of ambition; or, if that cannot be done, by passing laws for the preservation of infants. On this subject there is no need of seers or prodigies. And the same holds good of all similar things.   (36. 17)

Plus ça change...

And with this depressing picture of Polybius' Greece under Rome's dominion I conclude.

3

Herodotus and Religion

JOHN GOULD

'Herodotus and religion' is a big subject and one that has provoked much discussion.[1] That is hardly surprising: religion bulks large in Herodotus, both in what he records of other cultures and in his narrative of the events that concern him. In his accounts of each of the major cultures that fall within his chosen historical horizon (those of the Persians, the Egyptians, the Scythians, and the Thracians) he offers an extended and comprehensive description of their religious traditions and practices and even in his cultural references to populations of more fleeting importance to his narrative, peoples such as the Taurians, the Libyans, the Babylonians, the Geloni, or the Issedones, there are, more often than not, passing references at least to matters of religion. Furthermore, in his primary narrative, from first to last, there are very frequent discussions of divine or supernatural strands in the causation or the prefiguring of the events narrated. Religion, it would appear beyond doubt, was important in Herodotus' perception of things.

And yet there is constant and recurring unhappiness over the part played by religion in Herodotus' text. This unhappiness is of two, rather different kinds. One kind is a sort of generalized unhappiness which finds it difficult, even impossible, to accept that Herodotus really gave such weight as superficially he appears to do to the presence of divinity in human affairs—a sort of incredulity that things present themselves to him as they seem to do, a feeling that he 'must have been' more sceptical than he seems; and a determination to make him so. The other is somewhat more specific, a sort of disappointment with an author otherwise admired, a sense that Herodotus fails to display his usual sharpness of observation, the analytical clarity and sensitivity of response that we expect of him, when it comes to issues of religion, above all to the phenomena of other religious traditions than his own. In this paper I set

[1] Most recently, in Walter Burkert's essay, 'Herodot als Historiker fremder Religionen', in G. Nenci and O. Reverdin (eds.), *Hérodote et les peuples non-grecs* (Entretiens Hardt 35; Vandœuvres–Geneva, 1990), 1–32, with the following discussion on pp. 33–9. See his p. 2 n. 5 for further bibliography.

out to address both kinds of unhappiness and to suggest ways in which they are related.

The first kind of discomfort is well represented in Donald Lateiner's recent book, *The Historical Method of Herodotus*.[2] In his discussion of Herodotus' marked 'reticence' in matters of religion, most evident and most explicit in book 2, Lateiner ascribes the inhibition at work not to 'religious belief or fear' but to an overriding 'historiographical principle': 'Those things', he writes, 'which Herodotus cannot prove or disprove, topics that afford the historian no suitable, "down to earth" evidence, do not present material for *historie* [inquiry] as he understood it ... Herodotus shies away from stories without evidence' (pp. 65–6). The principle, which, as Lateiner acknowledges, is due to Ivan Linforth's seminal articles of almost seventy years ago,[3] is clearly one that has considerable force in understanding Herodotus' treatment of his data. But it cannot simply be used to sweep away all grounds for attributing to Herodotus a recognizably religious feeling of inhibition in the face of at least some manifestations of divinity. Indeed Lateiner is forced into some significant distortions of the passages under discussion in order to make his interpretation seem more plausible. Herodotus repeatedly uses terms such as 'impious' or 'unbecoming' to explain his silences. For Lateiner to describe this as 'an elegant excuse for avoiding an excursus into the irrelevant' (p. 65) is surely special pleading of an egregious kind. Nor is it likely that Herodotus' somewhat coy reference to the Samothracian mysteries in 2. 51, which reveals nothing to anyone who is not an initiate, is to be explained on the grounds that 'arcane religious doctrines and rituals do not illuminate his proper subject': I take it, by the way, that Herodotus' form of words ('whoever is an initiate in the rites of the Kabeiroi ... knows what I am saying') does indeed imply that he was himself an initiate.

Lateiner lays stress on Linforth's assertion that 'personal preference rather than religious scruple' dictates Herodotus' exclusions. He cites 2. 3. 2 as primary evidence for this view of the issue: 'I am not anxious [πρόθυμος] to expound the divine matters in the accounts that I heard [from the Egyptian priests] (apart from the [divine] names that they used), since I believe that all men have an equal sense [ἴσον ... ἐπίστασθαι: the Greek word implies something less than the certainty of

    [2] *Phoenix* suppl. 23 (Toronto–Buffalo–London, 1989).
    [3] 'Herodotus' Avowal of Silence in his Account of Egypt', *UCPCPh* 7 (1924), 269–92; 'Greek Gods and Foreign Gods in Herodotus', *UCPCPh* 9 (1926), 1–25; 'Named and Unnamed Gods in Herodotus', *UCPCPh* 9 (1928), 201–43.

knowledge] of them'. Lateiner paraphrases this highly significant sentence as: 'all men have beliefs and rituals which satisfy them, and they are inaccessible to testing for objective truth.' This last may very well be part of Herodotus' thinking on matters of religion but it is very far from obvious that it is implied by the argument of 2. 3. 2. More relevant to that argument, I think, may be another well-known comment on religious traditions, 3. 38. 1–2: Herodotus' evidence for his belief that Cambyses went mad in Egypt. This is what he writes:

It is absolutely clear to me that Cambyses suffered a major onset of madness. Otherwise he would not have attempted to mock things which were both sacred and traditional. For if one were to offer any community of men a choice and ask them to pick the most admirable traditions of all, each such group, after consideration, would pick their own traditions: this is the extent to which all communities of men regard their own traditions as the most admirable. So it is not plausible that anyone other than a man out of his mind would make such things an object of mockery.

This is not perhaps 'religious scruple' but an argument for respecting the traditions (which include the religious traditions) of all cultures, including the demand for religious silence on some matters of cult. So, too, is the argument of 2. 3. 2: all men have an equal (if necessarily limited) access to divine matters, so that to adopt a position of cultural superiority and regard oneself as exempt from the religious exclusions imposed by the traditions of another culture is unreasonable and unjustifiable.

It does not, of course, follow from this that Herodotus himself accepted these exclusions on any other grounds than respect for the cultural traditions of others. Lateiner indeed points to Herodotus' statement that he is 'compelled' to make some references to divine matters (2. 3. 2; 2. 65. 2) in order to reinforce his sense of Herodotus' reluctance, on specifically historiographical grounds, to introduce religious explanations into his narrative. Here, I have to confess, I find it hard to enter into the argument in the proper spirit of open-mindedness. For it seems to me self-evident that Herodotus took the possibility of supernatural causation in human experience quite as seriously as he took the involvement of human causation. His text is pervaded with the acknowledgement, both implicit and explicit, that the events of the historical past (even the immediate past) may display the presence of non-human powers at work.[4]

---

[4] See 9. 100. 2 for further evidence of Herodotus' acceptance of what he calls τὰ θεῖα τῶν πρηγμάτων as an element in human experience.

Of course, he is cautious in admitting such causation and still more so in identifying its source and rationale. But such caution, which at times may even involve the admission of authorial uncertainty and a verdict of *non liquet*, cannot be taken as evidence of a peculiar methodological scepticism in the face of the apparent activity of supernatural powers, still less of religious disbelief. It is due, as I have argued elsewhere,[5] to the built-in 'uncertainty' principle' which is a necessary part of any phenomenological religion; in such a religious system, the action of divinity is not revealed: it can only be inferred from the outward signs of that activity and these signs are almost never so unambiguous as to allow the inference to be certain. It is a further aspect of the same 'uncertainty principle' that the identity of the divine power at work and the motivation for divine action are still further removed from the possibility of certainty.

Thus I would argue that Herodotus' expressions of hesitation and uncertainty in questions of divine action in human experience are no more than the expression of a universal (and among ancient Greeks universally accepted) implicit acknowledgement of the limitations of human knowledge in such matters. Lateiner's citation of Xenophanes' and Alcmaeon's statements of this necessary uncertainty in matters concerning the gods is misleading precisely in so far as it suggests that only the more penetrating and sceptical minds of philosophical thinkers were capable of such insight and that such statements imply a positivist tenor of thought which sets the thinkers involved apart from the mainstream of ancient Greek ideas of divinity.[6] In fact, their statements in themselves (apart, that is, from the inferences that Xenophanes and Alcmaeon draw from them) are no more than generalized formulations of what all ancient Greeks implicitly took for granted in their response to the possibility of divine incursions into their experience.[7] Herodotus' acknowledgements of the same necessary uncertainty are not based on specific 'historiographical principle' but on the nature of Greek religion.

---

[5] See my essay, 'On Making Sense of Greek Religion', in P. E. Easterling and J. V. Muir (eds.), Greek Religion and Society (Cambridge, 1985), 1–33 (with notes on pp. 219–21), esp. pp. 9–14.

[6] Burkert ('Herodot als Historiker . . .', 26) points in addition to the similarity in expression between Herodotus 2. 53. 1 and Protagoras fr. 4 D–K.

[7] Edward Hussey, in an unpublished essay on 'The Religious Opinions of Herodotus', which he has been kind enough to let me see, argues that Herodotus is a follower of Xenophanes in his expressions of religious scepticism and indeed that he was probably, like Xenophanes, a monotheist. I still feel, however, that the case I argue above is a more likely interpretation of the tenor of Herodotus' comments on his religious data.

To reach a better understanding of the character of Herodotus' caution in matters of religion, we must look at some examples. One of the earliest instances of Herodotus' invoking a supernatural explanation for a historical event is in the question of the prolonged sickness of the Lydian king, Alyattes, which followed on the accidental destruction by his troops of the temple of Athena Assesia at Miletus (I. 19. 2–20. 1). Herodotus reports the account that he had received of Alyattes' consultation of Delphi in order to find an explanation for his illness. The reply involves an instruction from the Pythia to rebuild the temple before Apollo would give an oracular response. Herodotus comments that this account (which by implication was the account he received in Miletus) was confirmed by what he heard from the Delphians. Already the supernatural character of the event seems to call for some extra weight of evidence before Herodotus takes it into the fabric of his narrative.

A little later comes the famous comment on Croesus' loss of his son, Atys: 'After Solon's departure a great anger from a god overtook Croesus, because, to offer a guess,[8] he had supposed himself to be the most blessed of men' (I. 34. 1). The explicit admission that the identification of the motive for the (unidentified) god's anger is a 'guess' or probable inference, as well as the refusal to identify the divine power concerned, is another revealing example of Herodotus' caution in matters of divine action.

The story of Pactyes, the Lydian to whom Cyrus had entrusted the gold that he had taken as booty from Croesus and the Lydians, offers us a third example, different in character again from the first two (I. 153. 2–160). Herodotus tells the story of Pactyes' seizure of the gold, his taking refuge at Kyme from his Persian pursuers and the Kymaeans' consultation˙ of Apollo's oracle at Branchidae in unmarked, direct-speech narrative, indistinguishable from what precedes it. But when he reaches their second consultation of the oracle, with a new group of θεοπρόποι headed by Aristodikos, the son of Herakleides (the first having reported that the reply of the oracle was to hand over the suppliant), the narrative suddenly brings in the tell-tale 'it is said' and continues for three sentences in *oratio obliqua*, before going back into direct speech with the account of the Kymaeans' response to the story reported by Aristodikos on his return to Kyme. The three sentences cover precisely that part of

---

[8] More properly perhaps, 'to follow probability'. The Greek word εἰκάζειν, which Herodotus uses here, perhaps implies reference to notions of the probable to which he frequently appeals elsewhere in his discussions of supernatural strands in the causation of experience. (I owe this point to Edward Hussey.)

Aristodikos' report (no more and no less) which related that he had heard the voice of Apollo himself coming from the adyton and telling him that the first oracle was intended to bring about the destruction of Kyme for the impiety of even asking whether to surrender a suppliant.

These three examples between them tell us that, faced with the possible incursion of divinity into the events that he is narrating, Herodotus sometimes feels that he requires additional weight of evidence, that it is or may be impossible to identify the power concerned or be certain of its motive, and that such reports may be better distanced from the rest of the surrounding narrative by being given in *oratio obliqua*. The *oratio obliqua* is sometimes attributed, sometimes not; sometimes indeed introduced without even the use of any governing verb such as 'it is said', a stylistic usage which Lateiner, following G. L. Cooper,[9] calls 'intrusive oblique'. A further cautionary mode of narrative that Herodotus sometimes notoriously adopts is to offer several altern- ative explanations for a historical event, sometimes concluding with a decisive expression of preference for one explanation over another, sometimes leaving the question open. This strategy is conspicuously adopted when there are supernatural explanations among those offered by his informants. Thus in the case of the bizarre and macabre suicide of the Spartan king, Cleomenes, Herodotus reports that he had heard four explanations for this event; three of them, attributed to 'most Greeks', the Athenians, and the Argives respectively, all different but agreeing in this one respect, involved supernatural punishment for action against divinity; the fourth, offered by the Spartans themselves, was a 'common- sense', non-supernatural explanation (6. 75–84). Herodotus accepts ('my own view is . . .') the view of 'most Greeks', namely that Cleomenes was paying for his suborning of the Pythia and for his deposition of Demaratos on the basis of the Pythia's false reply. On that view, the agent was presumably, though Herodotus does not say so, Apollo him- self, avenging the insult to his divine τιμή ('worth' or 'status').

It should not surprise us that Herodotus gives his authorial approval to such an explanation, involving the anger of divinity. He does so again in the following book, in a case which makes the supernatural act even more uncanny in its execution, the case of the two Spartan heralds put to death by the Athenians, 'many years after Xerxes' invasion of Greece, during the war between the Peloponnesians and the Athenians', as Herodotus twice records. Their death was due, Herodotus asserts in his

---

[9] 'Intrusive Oblique Infinitives in Herodotus', *TAPA* 104 (1974), 23–76.

own person, to the 'anger' (μῆνις) of the hero Talthybius, Agamemnon's herald, who had a ἱρόν at Sparta (7. 134. 1); 'this action', he writes, 'was clearly the action of divinity (θεῖον . . . τὸ πρῆγμα)' (7. 137. 2). Talthybius was avenging the killing by the Spartans, a generation earlier, of two Persian heralds, sent by Dareius to demand earth and water. What makes this episode uncanny, and Herodotus' adoption of supernatural anger as explanation more striking even than in the case of Kleomenes, is that Talthybius' victims were not involved in the death of the Persian heralds but came to pay for it, as Herodotus sees the episode, by being the sons of two Spartan heralds who had been sent, years before, by the Spartans to Xerxes as volunteers, willing to atone by their deaths for those of the Persian heralds. The Spartans had decided to make atonement after 'a long period' during which their sacrifices had failed to give favourable omens. But Xerxes had refused the offer of atonement and Talthybius' anger had had to wait a generation to be assuaged. Herodotus' explanation of this event, which had occurred in what, at the point of writing, must have been the very recent past, is not based on Lateiner's principle of '"down-to-earth" evidence' but quite explicitly on the uncanniness of the coincidence that the heralds put to death by the Athenians were the sons of those not put to death by Xerxes.[10]

A special category of historical event which Herodotus repeatedly attributes to the action of a divine power is the event which prefigures another, usually catastrophic, event. Many of these instances involve occurrences which are themselves uncanny, such as ominous dreams or prodigies, but by no means always. In 6. 27. 1 Herodotus gives it as his own view that Histiaeus' massacre of 'large numbers of Chiots' and his conquest of Chios in the course of the Ionian revolt had been prefigured in two earlier disasters: the death through plague of ninety-eight members of a chorus of a hundred Chiot youths sent to Delphi, and the death in Chios town, 'at the same time, shortly before the naval battle [of Lade]', of a hundred and twenty boys, killed when the roof of their school collapsed on them as they were being taught to write. Herodotus cites these occurrences as evidence for a general proposition, which he puts forward in his own, authorial voice, that 'it is common for there to be indications beforehand whenever great disasters are going to occur to a city or a people'. The verbal phrase in this sentence (φιλέει . . . προση-μαίνειν), which I have had to paraphrase rather than translate, has no explicit subject but the implied subject is beyond question divinity.

---

[10] See further, on these last two examples, 'On Making Sense' (n. 5 above), 11–13.

There are very many other examples which I could cite but it is surely already clear that Lateiner's assertion that Herodotus 'generally omits the gods from his own explanations of historical events' (p. 67) is a massive over-simplification of this aspect of Herodotus' dealings with religion. Instead of looking at further examples, I will turn now to the second of the unhappinesses with the topic of 'Herodotus and religion' that I promised at the outset to tackle. This is a sense of disappointment with which I have much more sympathy and indeed, in some moods, come to share. It is, to recapitulate, a feeling that Herodotus lets us down, in terms of perspicacity and sharpness of thought and observation, when it comes to the interpretation of religion, above all the religion of other cultures.

But first we need to focus more precisely on what exactly it is that Herodotus gives us in his accounts of alien religions. The answer (and it is highly significant, as I shall argue) is almost exclusively: ritual.[11] Hence it should be no surprise that, when Herodotus wishes to support his assertion that the Egyptians are 'exceptionally religious' ($\theta\epsilon o\sigma\epsilon\beta\acute{\epsilon}\epsilon s$ ... $\pi\epsilon\rho\iota\sigma\sigma\hat{\omega}s$: 2. 37. 1), his evidence is drawn exclusively from ritual: washing and purifying ritual vessels and clothing, circumcision, shaving the entire body every other day, dressing in linen clothing only and wearing sandals of papyrus ($\beta\acute{v}\beta\lambda os$)—that is, not using animal products for clothing or footwear. Furthermore the great majority of his descriptions of other religions involve some account of sacrificial ritual practice, often an extremely detailed one and almost always defined in terms of its departure from the ritual of Greek sacrifice. So I hope to be forgiven if I now turn to the ritual procedures of sacrifice and look fairly closely at the use Herodotus makes of them in his accounts of alien religions.[12]

Thus Persian sacrifice involves no use of altars and no fire is lit beforehand for roasting the sacrificial meat; there is no libation,[13] no flute music, no garlands, no grains of barley: nothing, that is, of the essential preliminary rituals of Greek sacrifice and nothing corresponding to the

---

[11] Burkert ('Herodot als Historiker...', 4–5), supported by Giuseppe Nenci in the subsequent discussion, also underlines the central place of ritual in Herodotus' perception of religion. Perhaps we should recognize the beginning of a shift of attention from 'theology' to ritual in our approach to ancient Greek religion. If so, that is all to the good, though I have the feeling the point is still far from being taken with its full significance by most students of the subject.

[12] See J. Rudhardt and O. Reverdin (eds.), *Le Sacrifice dans l'Antiquité* (Entretiens Hardt 27; Geneva, 1981).

[13] An assertion which, as Burkert ('Herodot als Historiker...', 13 f. with n. 49) points out, is contradicted by Xerxes' ritual offering of libations at his crossing of the Hellespont (7. 54. 2).

procedure known as κατάρχεσθαι. Herodotus has an acute (and acutely Greek) sense of the religious importance of these ritual procedures; it is significant that he notes the refusal of the Chiots to use barley grown in the territory of Atarneus for sacrificial κατάρχεσθαι or for the making of sacrificial cakes to be offered to the gods (1. 160. 5) as an index of their sense of sacrilege on returning the suppliant Pactyes to his Persian pursuers in return for Atarneus. And we have another instance of Herodotus' interest in κατάρχεσθαι in his telling of the story (which he rejects as a false tale) of the Egyptian attempt to sacrifice Heracles (2. 45. 1: note also the garlands and the πομπή).

In Persian sacrifice, as Herodotus describes it, the animal is simply slaughtered in any convenient open space and its flesh is boiled, not roasted; it is placed on a bed of soft grass or (better still) of clover and a 'theogony' is chanted over it by a magus (an essential, but quite un-Greek, part of Persian sacrificial procedures, as Herodotus explicitly remarks), following which the meat is taken away, after a suitable pause, to be 'disposed of' by the sacrificer (1. 132. 1–3). 'Taking away' (ἀπο-φέρεσθαι) is another, but optional, part of Greek sacrifice, regularly subject to ritual regulation; for Herodotus' interest in ἀποφέρεσθαι, compare also his account of Egyptian pig sacrifices to Dionysus (2. 48. 1).

Interestingly, the one moment of Persian sacrifice that Herodotus' account does not give us is the actual moment of slaughter. He is concerned with what leads (or does not lead) up to the act of slaughtering and with the preparation thereafter of the flesh for consumption but he avoids the death act itself. In this his account is exactly similar to the vase iconography of Greek sacrifice, which is much concerned with processions, altars, and spits, and with the spitting of the entrails and flesh for roasting but seemingly avoids the cutting of the victim's throat.[14]

This is clearly not because the subject is in itself taboo for Herodotus, since elsewhere he records that among the Scythians the victim is garrotted, not slaughtered, once more without the lighting of a fire and without libation, or any form of κατάρχεσθαι, and that its flesh is boiled (4. 60–1). He records, too, that the Egyptians do use fire and libations, and do slaughter their victims, but that they then cut off the head (which is cursed and ejected from the ritual community). Moreover, he notes

---

[14] Burkert (Herodot als Historiker . . .', 18f.) draws attention to Karl Meuli's interpretation of Greek animal sacrifice as ritual slaughter of meat for eating and to the attendant notion of a 'comedy of innocence' surrounding the act of taking animal life; see further his *Homo Necans* (Berkeley, 1983), 1–12.

that before sacrifice the victim, if it is a bull, is minutely inspected for ritual purity by a priest whose especial task this is and then marked as 'clean' by having its horns bound with papyrus, daubed with clay, and sealed with the priest's ring: only such victims may be sacrificed, upon pain of death for anyone sacrificing an unmarked victim (2. 38–9). Here again the detail of Herodotus' account has as much to do with the alienness of these procedures from those of Greek sacrifice as it has to do with their importance in Egyptian religious culture.

He reports that the Libyan tribes cut off the ear of their sacrificial victims and throw it over the house in the ritual of καταρχεσθαι (4. 188); that the Taurians club their victims to death by clubbing their heads, after a rite of καταρχεσθαι (4. 103); that the Massagetae sacrifice horses to the sun (a practice for which he offers, unusually, an explanation: the swiftest of gods is given the swiftest of mortal creatures) and practice also ritual slaughter of their elders, along with animal victims, and eat the flesh of both after boiling it (1. 216. 2 and 4); and that the Babylonians sacrifice only suckling animals on the golden altar of 'Zeus' (Baal), while they make annual offerings of a thousand talents of frankincense, as well as sheep, on the much larger altar which stands next to it (1. 183. 1–2). All these accounts of sacrificial ritual practice, where they are not attributed to other informants (as, for example, with the report of Babylonian ritual), give every impression of being precisely observed and recorded as essential elements in the definition of alien religions. Even more importantly perhaps, they seem, as we have seen, to be recorded by an observer accustomed to registering the significant detail of ritual behaviour, as defined in Greek cult practice.[15] They are not random titbits of colourful information and they are in line with Herodotus' equally precise descriptions of other ritual procedures.

He offers accounts of pig sacrifices to 'the Moon' in Egypt—with careful attention to what is done with 'the tip of the tail, the spleen, and the caul', which are wrapped in pork belly fat and burnt (2. 47. 3)—and of the ways in which Egyptian processions in honour of 'Dionysus' (Osiris)

---

[15] Burkert ('Herodot als Historiker . . .', 21) argues that Herodotus' description of Persian sacrificial ritual, though presented as that of an eyewitness, is in fact 'a reconstruction, not field notes' and that his object is to offer a personal criticism of aspects of Greek religious practice and belief, attributed to the Persians. The point made in the text about the 'technical' appearance of Herodotus' comments on rituals of καταρχεσθαι and the disposal of sacrificial meat seems to me to tell against this reading of Herodotus' text. Granted that Herodotus' account is not the 'field notes' of an ethnographer, the relevant comparison may rather be with the landscape painting of Constable, based on observation 'recollected in tranquillity', as contrasted with that of Monet or Cézanne, painted *in situ*.

are like and unlike Greek phallic processions in honour of Dionysus: the procession is preceded by a flute and the women sing hymns, but the Egyptians do not carry separate phalli in procession but rather puppet figures, roughly a cubit high, with a movable phallus, the size of the puppet itself (2. 48. 2–3). Herodotus' interest extends to certain specific and memorable sacrifices, such as Croesus' colossal offering to Apollo of three thousand victims, as well as his stunningly lavish burnt offerings of gilded and silvered furnishings and of purple-dyed fabrics (1. 50. 1), or the Delphians' sacrifice to the winds before Artemisium (7. 178. 1–2; the beginning of a tradition which had continued to Herodotus' own day). It extends to the ritual privileges of the Spartan kings in the disposal of sacrificial victims (6. 56–7. 1); to the origin of Athenian sacrifices to Pan (a result of Phillipides' reported encounter with the god: 6. 105. 3); and to the fact that only the Carians sacrifice to Zeus Stratios (5. 119. 2), while the Scythians sacrifice sheep, horses, and human prisoners to Ares (4. 62. 2–3, again with details of ritual practice).

It is no surprise, then, to find Herodotus listing sacrificial rituals, along with temples of the gods ($\theta\epsilon\hat{\omega}\nu$ $i\delta\rho\acute{\nu}\mu\alpha\tau\alpha$: compare $\theta\epsilon\hat{\omega}\nu$ . . . $\pi\alpha\tau\rho\acute{\omega}\omega\nu$ $\H{\epsilon}\delta\eta$ of Aeschylus' *Persians*, 404), as a significant aspect in his famous account of the Athenians' definition of $\tau\grave{o}$ $E\lambda\lambda\eta\nu\iota\kappa\acute{o}\nu$ in the winter of 480–479 BC, where shared religious traditions appear alongside and on a par with shared descent and a shared language as determinants of 'Greekness' (8. 144. 2). I have extended this account of Herodotus' reports of sacrificial ritual to make two points: the first, obviously, is to underline the sheer quantity of information that Herodotus gives us; the second, and more important, to bring out not merely how central to his perception of religion such ritual detail is, but also how 'technical', in their language and their choice of significant data, his reports commonly are. Often they read as almost indistinguishable from entries in the religious 'calendars', for example, of the demes of Attica.[16]

Religion, like culture itself, is in large part defined for Herodotus by shared ritual procedures, of which sacrifice, though the most frequently discussed in his text, is no more than one example. His observations characteristically take in also the complex of rituals surrounding death and burial (so with the Persians, the Egyptians, the Scythians, the Ethiopians, the Thracians, the Massagetae, and the Issedones: again we may compare $\theta\acute{\eta}\kappa\alpha s$ $\tau\epsilon$ $\pi\rho o\gamma\acute{o}\nu\omega\nu$ in the messenger speech of *Persians*,

---

[16] See, for example, the Thorikos and Erchia calendars, with the comments of Robert Parker, 'The Festivals of the Attic Demes', in *Boreas* 15 (1987; Symposium called *Gifts for the Gods*), 137–47.

405), and the traditional techniques of divination and oracular consultation peculiar to each culture, as well as their rituals of oath-taking and
those which serve to create social solidarity. He observes and records the
binding importance of rituals connected with ξενία and supplication,
with purification, and with the cult of ancestors, both generalized and
specifically of the heroized dead.

Our first, interim conclusion, then, should be this: in many respects
the most important source of greater understanding of ancient religion
that we can find in Herodotus' treatment of it in his text is to acknowledge how strikingly it underlines for us the extent to which he and, one
might guess, the majority of Greeks, defined their own religion to themselves and understood its significance largely in ritual terms.[17] This is a
point of absolutely central importance to any attempt on our part to
grasp the sense of ancient religion and to take the measure of it; and there
are few sources as informative and as perceptive as the text of Herodotus.

Why, then, the sense of disappointment that I suggested earlier
attaches in the minds of his readers to Herodotus' treatment of religion?
What are we missing? And are we right to miss it? What we are missing, I
think, whether rightly or wrongly, is any convincing indication, firstly
that he understood or took adequate account of the significance of
iconography in understanding religion or, secondly and yet more
seriously, that he had any grasp of what we might call the differing 'ideo-
logies' or world-views of different religious traditions.

For anyone brought up in the ambience of Protestant northern Europe,
with its emphasis on 'the spirit' and on 'spiritual states', a religion or a
religious understanding which seemingly offers no more than the
'outward show', the 'vain pomps and glory', of ritual is almost by definition not to be given serious consideration. So Herodotus' apparent failing in this direction is enough in itself to disqualify him at once from
acceptance as a serious interpreter of religion.

On the first point, Herodotus is in fact by no means unaware of
religious iconography nor, in some sense at least, of its significance: he
refers to images of divinity in religious contexts more than sixty times; he
comments explicitly on the absence of (anthropomorphic) imagery in
Persian religion (1. 131. 1), and he notes the theriomorphic characteristics of some, at least, Egyptian religious iconography (2. 41. 2 (Isis); 2.
42. 4 and 6, 4. 181. 2 ('Zeus' at Egyptian Thebes); 2. 46. 2 ('Pan')). But

---

[17] So, too, Burkert ('Herodot als Historiker . . .', 19 ff.) points out that Herodotus'
accounts of non-Greek sacrifice serve to define Greek sacrificial procedures for us by
contrast and inversion.

his comments and interpretations almost never involve him in measuring the distance from his own religious perceptions that such phenomena may imply. He comments, by way of explanation for Persian avoidance of anthropomorphic iconography, that he supposes that the Persians do not imagine their divinities as 'human in nature (ἀνθρωποφυέες), as the Greeks do', but he says no more and it is not at all clear quite what he intends his own phrase to imply. He makes comparisons between Egyptian theriomorphic iconography and 'parallels' in Greek religious imagery which seem confused and reductionist in tendency: Isis is portrayed as a woman with cow's horns, 'as the Greeks depict Io'; 'Pan' is painted and carved with the head and legs of a goat, 'as he is by the Greeks . . . although they [the Mendesians] do not think he is like that but rather like the other gods'; and he explains the ram-headed 'Zeus' of Thebes by telling an equally reductionist story of 'Zeus' masking himself with a ram's head in order not to reveal his true appearance to the importunate Herakles.

None of this conveys any suggestion that the Egyptians or the Persians perceived the world, in any serious sense, differently from Herodotus and his fellow Greeks. A prime reason, clearly, for this is that Herodotus finds no difficulty, in Egypt or elsewhere, in making straightforward equations between the powers worshipped there and those familiar in Greek cult. Thus the Persians worship and make sacrifices to 'Zeus', even though Herodotus acknowledges that that is a name they give to the circuit of the sky (1. 131. 2), and a notorious confusion enables him to equate the Persian male divinity Mithras with Aphrodite, via other equations involving the Assyrian Mylitta and the Arabic Alilat (1. 131. 3). The Egyptians worship 'Dionysus' (calling him Osiris), 'Demeter' (calling her Isis), 'Pan' (calling him Mendes), 'Zeus', 'Athena', 'Apollo' (called Horus), 'Artemis' (called Boubastis), 'Leto', 'Aphrodite', 'Ares', 'Hephaistos', 'Hermes', 'Herakles', and 'Epaphos' (called Apis). The Scythians worship 'Hestia', 'Zeus', 'Ge', 'Apollo', 'Aphrodite', and 'Poseidon', called respectively Tabiti, Papaios, Api, Goitosyros, Argimpasa, and Thazimasades (4. 59. 1–2); they also worship 'Ares', whose image is an ancient iron sword planted on top of a great mound of brushwood (4. 62. 1–2).[18]

Herodotus does observe that the cult of heroes has no place in

[18] In this, Herodotus is no innovator: see Burkert, 'Herodot als Historiker . . .', 5–8, for the long-standing tradition of equating one's own with foreign gods in Near Eastern cultures and in early Greek texts; and for the exception that 'proves the rule', Homer, *Iliad* 3. 276 ff.

Egyptian religion (2. 50. 3), but in general almost his only significant concession to the alienness of what he is describing is his acknowledgement that other cultures use names for their divinities that differ systematically from their Greek names. That is itself a characteristic point: Herodotus is everywhere interested in the names of gods and discusses at length the process by which the Greeks themselves came to call their gods by their familiar names (2. 50. 1–3, 52. 1–3). They were inherited, he believes, from the Pelasgians, who in turn took the bulk of them from Egypt, having previously long sacrificed and prayed to them anonymously, using only the collective name θεοί. The Egyptian names were ultimately adopted on the authority of Dodona, with some exceptions: Dionysus was only known and named later than the other gods, and Poseidon, unknown to the Egyptians, was learnt of from the Libyans (elsewhere he notes that the names of Poseidon and the Dioskouroi were unknown to the Egyptians: 2. 43. 2). Herodotus' authority for some at least of this remarkable passage is again specifically given as the priestesses of Dodona (2. 52. 1, 53. 1).

This passage leads immediately into another, equally, if not more, remarkable (2. 53. 1–3), which needs to be quoted in full:

Whence each of the gods derived their existence or whether they had always existed; what they were like in appearance—these are things the Greeks came to know [ἠπιστέατο: contrast ἐπύθοντο in 2. 50. 2, 52. 2] only yesterday or the day before, to exaggerate a little. For it is my belief that Hesiod and Homer lived 400 years before my time, and no more; and it is they who by their poetry gave the Greeks a theogony and gave the gods their titles; they who assigned to them their statues and skills and gave an indication of their appearance. The poets who are said to be earlier than these men are, in my view, later. . . . All this last, about Hesiod and Homer, I say on my own authority.

This passage in turn raises important questions, the most important that we have yet encountered in this enquiry; questions not only about what precisely Herodotus means but also about the kind of evidence on which he is basing his assertions. I will leave aside for now the question of Herodotus' dating of Hesiod and Homer, but as soon as we raise the question of evidence and ask: how is he able to be so sure, so emphatic, about what he believes Hesiod and Homer to have contributed to Greek religion and religious understanding?—the answer throws a sudden and brilliant light on the issues with which this essay is concerned. For the answer is surely that there was no other or earlier source that Herodotus could think of for the shared religious perceptions and imagery of the

Greeks, wherever they might live. There was nothing in the primary source of Greek religious awareness, namely ritual practice, that could possibly have suggested the complex imagery of descent and kinship that created a sense of coherence and community out of the huge plurality of powers worshipped in ritual; nothing that could have created either the shared perception of what we can best call the 'character' and 'history' of the various powers. What Herodotus is saying, in other words, is that what I have loosely called the 'ideology' of Greek religion is entirely due to the 'commentary' on ritual provided by the tradition of epic narrative poetry.

And he is surely right. In a religion without any tradition of defining divinity by the codification of belief into creeds and without the authority of divinity for its narratives of divine action in human affairs, without (that is to say) a revealed text dealing with the actions of the gods, the fictional narrative tradition of poetry alone creates a 'world' and a 'history' which makes identifiable and describable beings out of the recipients of sacrifice and prayer. Beyond that history, religion is defined entirely by ritual action and by names, the non-human names of the powers with whom ritual action creates and maintains a relationship of reciprocity.

If this line of argument is right, then we have an explanation both for Herodotus' remarks about Hesiod and Homer and for his almost obsessive concern with the names of divinity, but more importantly also for the apparent limitation of his understanding of other religions. Since, after all, his evidence for these, the only source of understanding that he possessed, is precisely the observation of ritual action and the recording of divine names, and as he apparently had no access within the alien culture to any kind of commentary on such evidence, he lacks what Hesiod and Homer gave him for the religion of his own culture. He knows that what was chanted over Persian sacrificial victims was, as his Persian informants told him, a 'theogony' (1. 132. 3), but he shows no knowledge of its contents. He knows, he tells us, 'sacred stories' told by the Egyptians in the context of ritual action (2. 62. 2, 81. 2) and in the course of the Samothracian mysteries (2. 51. 4), but religious inhibition keeps him from communicating them, though it is likely that such stories were an important factor in some at least of his equations between Egyptian and Greek divinities (most obviously in the case of Osiris = Dionysus). We have to imagine him closely observing ritual behaviour, in the light and through the lens, I have argued, provided by his close familiarity with Greek ritual practice, and interpreting what he observed by analogy with the only framework of understanding that he possessed, which was once again that given him by Greek religion. 'Analogy' may include the idea of inversion (as with the absence of anthropomorphic imagery in Persian religion and with much in Herodotus' account of the religion of Egypt) but it could not help him to an understanding, for example, of Persian religious dualism or the significance of the divinity of the Egyptian pharaoh.[19]

I asked earlier whether we were right to feel that Herodotus somehow lacks understanding of religion and I think that the answer must clearly be that we are not. I come back to end with the role of Protestant religious tradition in hindering us from a just appreciation of Herodotus' dealings with religion. The ease and absence of discomfort, for example, with which Herodotus seems almost to identify religion with ritual process and the great concern he shows for exactitude over the names and titles of divinity are things much more easily accommodated to Catholic religious tradition, above all that of southern, Mediterranean countries, or to Orthodox tradition than to those of the Protestant north. I hope, by way of coda, that I do not disqualify myself, as an observer of Herodotus' observations on religion, by confessing to having been brought up in northern Europe but in the traditions of Catholicism.

---

[19] Burkert ('Herodot als Historiker . . .', 25) also points to Herodotus' inability to appreciate the importance of the myths of Osiris and Seth, of which he shows an awareness, for the Egyptian notion of kingship.

4

Herodotus on Alexander I of Macedon:
A Study in Some Subtle Silences

E. BADIAN

To Charl Naudé

Der Aufgabenkreis eines Historikers sollte sich nicht nur auf eine kritisch geläuterte Rezeption des in den Quellen wortwörtlich Berichteten beschränken, sondern müßte zugleich das Bemühen umfassen, sich von den einstigen Vorgängen . . . auch eine lebendige Vorstellung zu machen.   Fritz Schachermeyr

There is not much Macedonian history in Herodotus, and what there is is difficult to interpret, for it touches upon Persian and Athenian history, and it must be viewed in the light of Herodotus' habits and prejudices. It is in any case marginal to his interest in the great war between Greeks and barbarians and therefore comes up only incidentally. There is, on the other hand, a good deal of myth, but I shall not be concerned with it here, since it has been ably sorted out by N. G. L. Hammond. Herodotus seems to repeat it as he heard it, as a backdrop to his treatment of Alexander I, who is the only Macedonian relevant to his narrative in the books about the war. Alexander does indeed appear at some crucial points in the war, and his character clearly aroused the historian's interest: he fashioned an elaborate prelude in order to introduce him, and the themes stated in that prelude are developed on the occasions of Alexander's appearances in the war. I am not here concerned with the character of Alexander, but rather with the way in which Herodotus depicts that character. We shall find that he employs a great deal of subtle art even on this incidental actor in his drama, and the very fact that the author's treatment of him on each of his appearances after the prelude is brief makes it easier to study that art. The focus of this study is historiographic and not historical; but as is usually the case, there will be some incidental accretion of historical knowledge—or at least conjecture.[1]

---

[1] I propose, on the whole, to refer only to the sources (to Herodotus normally only by

We start with the extended prelude to Alexander's part in the war, the story of his killing the seven Persians (5. 17–21). Darius, after his return from Scythia, leaves Megabazus to conquer Thrace while he himself returns to Sardis (where he stays for some time, we are not told why). Megabazus sends seven of the most eminent Persians on an embassy to Amyntas, king of Macedon and Alexander's father, to demand 'earth and water' (i.e. formal submission to the King's suzerainty). Amyntas agrees to this and gives a banquet to entertain the envoys. At the banquet the envoys misbehave towards some noble Macedonian women, and Alexander seizes this opportunity to have them all killed by a trick. He then has all their property removed, so that no trace of them remains. When a Persian army comes to investigate their disappearance, Alexander buys off their commander Bubares by giving him his sister Gygaea in marriage (the version picked up by Justin 7. 8. 9 improves on this by reporting that Bubares had fallen in love with her before there was any fighting), and a huge sum of money as well. So far Herodotus.

Fortunately, no one has believed the tale. Among other obvious marks of fiction, it was long ago noticed that the seven envoys, the noblest in Megabazus' army, parallel the seven great Persian families whom Herodotus knows about, and that he does not name the men, even though he normally delights in naming prominent men on lists: thus the companions of Darius in his *coup d'état* (3. 70) or the suitors of Agariste (6. 127). What we must ask is: what are the facts behind the story? It has

number of book, chapter, and, where appropriate, section according to the Oxford text), and to a few recent works dealing precisely with the topic here discussed, where references to earlier treatments will be found. But let me at once express my appreciation of the article by Ross Scaife, 'Alexander I in the Histories of Herodotos', *Hermes* 117 (1989), 129–37 (with useful bibliography in note 1). He sees, as few have done, the ambivalence in Herodotus' presentation (summed up at p. 135: 'a marginal man') and appreciates Herodotus' critical ability (e.g. pp. 129 f.: '. . . there is no evidence to suggest that Alexander charmed away Herodotus' critical faculties' (as is maintained by N. G. L. Hammond, in N. G. L. Hammond and G. T. Griffith, *A History of Macedonia*, ii (Oxford, 1979), 99). It will be seen that I agree with him in much of his interpretation of the facts, but I am more concerned with the conclusions that Herodotus' presentation of these facts allows us to draw regarding more general aspects of his method and with some of the historical conclusions that recognition of that method allows us to draw in this instance. On the 'semiautonomous' status of Macedonia under Amyntas and the status of Thrace within the Persian kingdom (never a true satrapy, but governed by officers sent by the satrap of Sardis), see J. M. Balcer, *Historia* 37 (1988), 1–21, especially pp. 5 f. He compares Amyntas' status to the status of Cilicia and the Phoenician cities. It will be seen that I would agree with much of this, except that I would accept Herodotus' description of Amyntas as in fact a 'native satrap', as later known in the fourth century, and I would regard the 'large sum of money' given to Bubares as a skilful reinterpretation of tribute.

usually been thought[2] that Amyntas gave earth and water to the King, and his daughter to an eminent Persian noble, and that Alexander later had to put the best face on it, especially as far as his own involvement in the incident was concerned. Let me say at once that I think this view essentially correct. (I shall come back to it later.) But the matter now needs renewed discussion since it has been argued by R. M. Errington[3] that Amyntas never submitted to Darius at all (for if he ceremonially gave earth and water to men who then disappeared without trace, it follows that no one would ever get to know about it); moreover, since it was Alexander and not Amyntas who (in the story) gave Gygaea to Bubares, he must have been king of Macedon at the time—so the submission took place much later. In fact, so Errington thinks, Alexander invented the whole thing, and the marriage only dates from 492, when, according to another passage in Herodotus (6. 44. 1), difficult to reconcile with the Amyntas story, Macedonia was added to the 'slaves' of the King by Mardonius. The disproof of the act of homage by Amyntas was essentially accepted by E. N. Borza,[4] although he prefers (for reasons I cannot accept) to put the marriage around the end of the sixth century, either still in Amyntas' reign or at most straight after Alexander's accession in 498.

[T]he marriage signifies alliance . . ., but not necessarily vassalage . . . the practice of Argead men and women marrying foreigners for political/diplomatic goals would become common enough. But the existence of such a union is not . . . alone sufficient evidence of determining which party, Macedonian or foreign, was subservient to the other.

It is always useful to have conventional views challenged: too many have been passed down through generations of scholars without critical scrutiny. It will never do any harm to be made to think afresh about what has always been believed. But in this instance, once we do think again, we must conclude that the revisionist view is wrong and that the accepted interpretation should be reinstated.

Borza's implication that we cannot be sure whether Amyntas became a vassal of the King or the King of Amyntas is no doubt unintended. But it brings out the basic weakness of his version of Errington's case. We

---

[2] Let a reference to Hammond *HM* ii. 58f. suffice for what is the accepted interpretation.

[3] 'Alexander the Philhellene and Persia', in *Ancient Macedonian Studies in Honor of Charles F. Edson* (Thessaloniki, 1981), 139–43.

[4] *In the Shadow of Olympus* (Princeton, NJ, 1990). The passage quoted is on p. 103. Cf. n. 10 below.

know perfectly well, if only from the story of the Athenian envoys at Sardis (which will soon occupy us), that, at this time, the only form of alliance the King would consider was subjection. The man who called himself King of the lands (DPa: dahyūnām) or, in fuller form (DNa), King of the lands containing all men (vispazanānām), King on this great earth far and wide (DNa), King on all the earth (DSb),[5] could hardly have allies on equal terms. This (I think) suffices to make Borza's reinterpretation unacceptable. We therefore return to Errington's original challenge.

But now we must ask: if none of this in fact happened—if the plain truth is that in 492 Alexander gave his sister to a Persian noble and only then submitted to the King, when neither he nor his father ever had before—how could he make that marriage the consequence of his father's submission and his own resistance to it, half a generation earlier—and hope to get away with the story when many were alive who would know the lie for what it was? Nor is the actual argument at all strong. As to no one's finding out about the offer of earth and water to the seven: if they had had positive instructions to receive this formal submission from Amyntas and (no matter what happened to the envoys) this had not been forthcoming—would the King or his representative leave it at that? Would not the army sent to look for them have insisted that the ceremony be performed, and been strong enough to enforce it? We may want to reject the whole story a priori, as Errington essentially does, but we cannot show that it is inconsistent on its own terms. If the story as such is accepted, the act of homage cannot be peeled out of it; and in fact, as we have seen, it is the act of homage that is most unlikely to be an invention: it is the irremovable fact around which the story was built. As to its being Alexander and not his father who gave Gygaea to her bridegroom, we shall see that this is not a problem at all.

Before we go on to actual discussion of this and related facts, it will be helpful to engage in a little prosopography: to find out more precisely who these Persians that we are dealing with were. Who was Bubares, husband of Gygaea? In 7. 22. 2 Herodotus informs us, in an entirely different context, that he was a son of Megabazus: he was one of the two men put in charge of digging the Athos canal. Now, it is true, as D. M. Lewis has pointed out,[6] that there are other men called Bakabaduš (= Megabazus) in the Persepolis fortification tablets, at lower social levels; this is not relevant here: there are many men called Cecil, but

[5] I quote some of Darius' titles in his own inscriptions from R. G. Kent, Old Persian[2] (New Haven, Conn., 1953), adapting the translation.
[6] Sparta and Persia (Leiden, 1977), 50 n. 3.

when we are dealing with the highest aristocracy, only one family can be considered. And here we are indeed moving in the highest circles. Bubares' colleague at Athos was Artachaees son of Artaeus—the tallest Persian, the man with the loudest voice on earth, and, more important to us, an Achaemenid (7. 117), as we are casually told at the time of his death. So this Megabazus must surely be identical with the commander in Thrace, whom Darius esteemed more than any other man (4. 143).

A Megabazus son of Megabates is one of Xerxes' four admirals in the invasion of Greece (7. 97), again in the best possible company: two sons of Darius and a son of one of Darius' associates in the *coup d'état*; a Megabates, presumably his father and undoubtedly a member of the family and an Achaemenid, was a cousin of Artaphernes and of Darius; he had been in charge of the expedition against Naxos (5. 32). The Megabazus so highly esteemed by Darius and commanding his forces in Thrace cannot but be a member of that family: perhaps a brother of Megabates the commander. We are throughout moving in a circle of the highest aristocracy, close to Darius both by birth and by association— indeed a distinguished connection for a Macedonian.

It should now be clear that the most likely time for this extraordinary marriage is surely the time when Megabazus, Bubares' father, was commander in Thrace—not twenty years later, when Mardonius son of Gobryas, also closely related to Darius (7. 2. 2, 5. 1), but not recorded in any connection with the family of Megabazus, was sent down to regain Darius' European territories. And we must take it that Darius, who was at Sardis at that very time, was in touch with Megabazus throughout: he must have given his personal approval. The actual date can only be approximately determined. Darius' Scythian campaign should be put *c.*513[7] and it was at the end of it, probably with no time for any campaigning that year, that Megabazus was left in Thrace. His activities there demand at least two campaigning seasons, one in the Hellespont and one for the campaign against the Paeonians, just before he left for Sardis: we might conjecture 512 and 511. He was succeeded in Thrace by Otanes and met Darius at Sardis. All we can know for certain about Darius' movements is that he must have left Sardis by 507, after appointing his brother Artaphernes as satrap; for when the Athenian envoys got there, they saw Artaphernes (5. 73) and there is no mention of Darius. But in view of the urgency of looking after the other parts of his vast kingdom, it is unlikely that he stayed nearly as long as that. He

[7] See A. Sh. Shahbazi's exhaustive investigation, 'Darius in Scythia and Scythians at Persepolis', *AMI* 15 (1982), 189–235. He finds nothing to contradict that ancient date.

probably left soon after his meeting with Megabazus. At any rate, we must after all return to dating the marriage of Bubares and Gygaea to around 511, and with it the formal submission of Macedon to the King's suzerainty.[8]

But Alexander's part in this still needs further investigation. Is there any merit in the late chronographer Syncellus' version, according to which it was Alexander (not, as is implied in Herodotus, Amyntas) who gave earth and water to the King? And how did Alexander come to give away Gygaea when his father was still king? These two points have been stressed by Errington and Borza. However, there is a very simple solution. After all, it was not the archon or the *boule* of Athens that gave earth and water to the King in 507: it was Athenian envoys at Sardis. (For that matter, it was not the King himself who received the tokens of submission, but the satrap at Sardis.) All that we need assume is that Alexander met Megabazus as his father's envoy. When Alexander actually succeeded Amyntas we cannot tell, within very wide limits. All we know is that Amyntas was still king in (probably) 506/5, when he offered Anthemus to Hippias as a retreat after Hippias' expulsion (5. 94. 1), and that Alexander was king when Xerxes came. Justin's statement, that Amyntas died and Alexander succeeded 'after Bubares left Macedonia' (Just. 7. 4. 1), tells us nothing useful. The whole Bubares story in Justin is ultimately based on Herodotus, with nothing but romantic elaboration added (some of which we have noted). The passage at the beginning of 7. 4, which we have quoted, is plainly a linking passage with no specific chronological implication. It was clearly inserted by an author (whether Justin or Trogus before him) who had no precise chronological information.[9]

---

[8] The appearance of Skudra (whatever its precise significance) and the shield-bearing (or, as the Babylonians understood it, 'shield-wearing') Greeks, whoever they are, in the Persian lists unfortunately cannot be dated, except within limits too wide to be helpful to us here. I shall therefore not discuss the various problems it poses. Let me briefly note that, in Walser's identification (following Herzfeld's final opinion), three of the throne-bearers of Darius I (nos. 23, 25, 26) wear the *petasos* and none carry shields (*Völkerschaften* 54 and Falttafel 1: that could hardly be expected of throne-bearers who have both their arms raised). In the procession of 'tribute-bearers' no one wears the *petasos* and only one delegation (no. 19: Skudra, according to Walser, op. cit. 96f.) has two round shields (wickerwork *peltae*, according to Walser).

[9] Hammond (*HM* ii. 59), misinterpreting Justin, makes Bubares govern Macedonia for a decade! How well informed the late chronographers were on the regnal dates of the Argead kings is, in general, difficult to say. (Their dates are usefully set out in Beloch's table, *GG* iii(2), 50f.). Their unreliability on the reign of Perdiccas is demonstrated by the large variation in early sources regarding his dates, which they do not mention and which cannot be either explained or ignored. (The *ad hoc* explanation that he fought against pretenders is

We can now begin to put together the story of what actually happened. When Megabazus was subduing Thrace, and perhaps even before he reached the Macedonian border (wherever it then was), Amyntas, impressed by Persia as clearly the coming power in Europe, as it was in Asia, and hoping to profit if he acted in time, sent his son to negotiate terms of friendship. Alexander, inevitably, was told that he would have to offer earth and water to the King, as the Athenian envoys later were at Sardis: the King could not contemplate association on any other terms. It was easy for him to consult with his father, if he did not dare to take this step on his own. We cannot tell whether he did, and it does not matter. He agreed to do so, on his father's behalf, and, in return, must have got reassurances that his father would remain in undisturbed possession of Macedonia. Megabazus too could easily consult his King, who was not far away, and he got permission to agree. The terms worked out make it clear that the King had not previously had any thoughts of annexing Macedonia, and Megabazus had no instructions to do so. He and the King were clearly delighted to seize a chance that offered, and were prepared to be generous provided the King's suzerainty was recognized in full form. But payment of tribute seems to have been part of that recognition: Herodotus, as he himself got the story, disarmingly tells us that Alexander paid the Persians χρήματα πολλά, of course to hush up the total disappearance of the seven distinguished Persians.

The tale of the seven Persians now begins to make sense. As we have seen, the terms given to Amyntas in return for his homage (and we shall have more to say about them) make it clear that the initiative in the whole business must have come from him, and that he was not merely obeying an order from the King which he dared not resist. Alexander, when the time came to defend his philhellenic purity, had more to explain than the marriage of his sister to an Achaemenid—more, even, than his father's offering (and Alexander's actually giving) earth and water to the Barbarian. He had to concoct a story that would account for, and disguise, the fact that the act of homage had not been extorted, but spontaneous and

disproved by the fact that none of them seem to have coined money.) It is therefore in principle impossible to arrive at a date for Alexander's accession by using their figures either for his reign or for Perdiccas', even though some may in fact be correct. Borza 103 n. 15 cites G. L. Huxley for the reliability of Syncellus, who reports that it was Alexander (and not Amyntas) who did homage to the King, 'although the date of Alexander's submission is uncertain'. Unfortunately, Syncellus in fact dates it to the time of Xerxes' crossing (p. 316 Mosshammer), which neither Huxley nor Borza would presumably defend. However, the fact that it was Alexander may well be genuine tradition, with the presumed occasion added by a late author as the only contact between Alexander and the Achaemenids that he knew of. As we have seen, that part of the report makes good sense.

calculated, in the hope of benefits to come which duly came; and the fact—presumably widely known—that Macedonia had actually paid tribute to the King. As there was no record or memory of any demand for earth and water, the messengers transmitting that demand had to be invented and then had to disappear without trace; and the payment of tribute, as well as the distinguished marriage connection (by then an embarrassment), became a sacrifice necessitated by the disappearance of the messengers—who, in view of the enormity of that sacrifice (a large sum of money and the marriage of a royal lady of Temenid descent to a relative of the Barbarian), had to become the most distinguished Persians one could imagine for the occasion. Needless to say, there was now a chance to depict young Alexander as an impetuous hero, whose truly heroic deed ultimately had to be paid for in a suitably extravagant manner. In Alexander, the seven characters had indeed found a splendid author.

It is fortunate that Herodotus chose to confer immortality on that drama. We have already learnt a great deal from it—history that would otherwise have disappeared even from the realms of conjecture. And there is more we can learn by further scrutiny. It seems that, in return for his timely offer of allegiance and as a reward for the considerable enlargement of the King's dominions in an area of rich future promise, Amyntas remained as what we might call a native satrap. This is in fact the term that, in very technical language, Alexander (in Herodotus' account, undoubtedly received from Alexander himself or one very close to him) applies to his father (5. 20. 4): 'a Greek man, satrap of the Macedonians' ($\dot{a}\nu\dot{\eta}\rho$ "$E\lambda\lambda\eta\nu$ $M\alpha\kappa\epsilon\delta\acute{o}\nu\omega\nu$ $\ddot{v}\pi\alpha\rho\chi os$—the word $\ddot{v}\pi\alpha\rho\chi os$ is Herodotus' standard term for a satrap, a word he does not actually use). I think that the phrase embodies part of the kernel of truth which the rest of the story was developed to encase. And with this settlement went the outstanding distinction we have noted: a connection with the royal blood of the Achaemenids for the royal satrap's daughter.

That this was a common way of ensuring a powerful satrap's loyalty is well known: there is no need for illustration. Of course, our examples tend to be later. We have very little evidence on Persian provincial administration before the end of the fifth century; above all, none whatsoever (as far as I am aware) on the administration of the eastern satrapies, indeed of any beyond the sphere of interest of Greek writers. We do, of course, have Herodotus' prosopographically invaluable list of contingents with their commanders in Xerxes' invasion force (7. 61–82). But we do not know how many of those commanders were the actual

satraps of the provinces concerned and how many had been specially
appointed as commanders (Herodotus' term is always some variant of
ἄρχων or ἡγεμών) for the purpose of this expedition. That at least some
were cannot be doubted. Thus the Persians have a commander, one of
Xerxes' fathers-in-law (7. 61. 2), yet we know that Persia did not have a
satrap. In fact, the units under separate commanders do not correspond
either to the tribute districts in Herodotus' list or to the lists of subject
nations on the Persian monuments: to take an obvious example, the
'island tribes from the Red Sea and the islands of the deportees' (7. 80)
cannot ever have formed a satrapy, yet they appear with their own
Persian commander. At least some of the commands (and we cannot tell
which) were evidently created for the great expedition. Nor can we tell
the administrative status of the miscellaneous 'rabble' briefly dismissed
by Herodotus in 7. 81.

The first native ruler appointed satrap of whom we actually know
appears to be Hecatomnus, a century later. It does not follow that he was
necessarily the first man anywhere to acquire that status. We must again
stress our all but total ignorance of how the eastern parts of the kingdom
were administered before Alexander's invasion of Asia; and what we find
then, in a much weakened kingdom, can obviously not be used as
detailed evidence for what had been there in the time of Darius I. But we
do at that time, as also in the case of Hecatomnus, see a flexibility in
administrative arrangements that may well have been traditional. But
whether or not it was, there is no reason to disbelieve the evidence
offered by Herodotus in this case: we have seen that Amyntas, owing to
his readiness to worship the rising sun, got the rewards he expected. We
might compare the Syennesis dynasty (see *RE*, s.v. 1) in Cilicia.

We must next note that the son of Bubares and Gygaea, linking the
Achaemenids and the 'Temenids' of Macedonia, received the name of his
maternal grandfather, Amyntas. Herodotus tells us (8. 136. 1) that at
some time (which Herodotus does not specify) the King gave him the
'great city of Alabanda in Phrygia' as his appanage. Whatever and
wherever that city may be (whether Herodotus means Carian Alabanda
or whether the name ought to be changed), it is not a particularly distinguished property for a grandson of Megabazus (not to mention of a
Macedonian king). But it is important for various reasons[10] to insist that

---

[10] Borza 103 n. 19 interprets that aorist as implying that the gift preceded the date (480)
at which it is mentioned. This leads him to reject Errington's date of 492 for the marriage
(but not the rest of Errington's arguments)—correctly, as we saw, but for an invalid reason.
(He is right, n. 16, to reject Hammond's date for Amyntas' death.)

Herodotus does not specify the time of that gift: the aorist is used merely to make a statement, and the phrase defines that Amyntas as the one 'to whom the King gave' Alabanda.[11] It is very unlikely that this was all that was planned for the offspring of that union. We must regard the gift as a consolation prize, at a time when the union no longer had any political purpose and Macedonia had permanently slipped from the King's grasp. Indeed, we cannot help speculating about the Macedonian name given to the son of an Achaemenid. It is an obvious suggestion that, after Xerxes' final conquest of Greece and with Macedonia totally in his power, that child was intended to succeed Alexander as satrap-king—acceptable to the Macedonians and reliably loyal to the empire of his Achaemenid relatives. He would certainly have been much more in place there than in 'Phrygian Alabanda'.

But things did not work out as planned. The first setback, of course, was the Ionian Revolt. It has been claimed that it made no difference to the Persian holdings in Europe, but the evidence suggests the opposite. In Chersonese, Miltiades re-established an independent dynastic territory, and when Mardonius appeared after the defeat of the Revolt, in 493, even Thrace was apparently not safe: all we really know is that the fortress of Doriscus continued to be held (7. 59. 1); yet Persian control was either weak or interrupted even there, for when the Paeonians were conveyed to Doriscus in order to escape Persian forces pursuing them, that was apparently regarded as a safe staging-post: they had no difficulty in making their way home from there, as the Lesbians who carried them there had clearly foreseen (5. 98). In view of this, it is by no means surprising that Macedonia was lost: indeed, it would be surprising if it had been retained. With Thrace slipping from the Persians' grasp, there was no way of enforcing the tribute: it is likely that Amyntas or Alexander stopped it as soon as he received news of the allies' successes in their first campaign, but at the very latest when he saw that Miltiades, much closer to Asia, had succeeded in establishing his principality. (Much would depend on the date of Alexander's accession: we may assume that he was on the throne when Mardonius arrived.) Needless to say, at Mardonius' approach, Alexander, who had no intention of fighting the King's forces, at once submitted and again agreed to pay tribute:

[11] The Loeb translator gets it right: 'to whom the king gave Alabanda.' The Budé's translator seems to avoid committing himself. It should be noted that Amyntas installed one of his sons as husband of the heiress to the throne of Elimeia. That son, no doubt to conciliate the natives, called his son after his wife's father, who must have been king of Elimeia when the marriage took place. For detailed argument on this complex issue see the Appendix (pp. 127–30 below).

later, he could point to the fact that Macedonia had been free when Mardonius came; that he had reversed his father's submission to the King and had renewed it only under dire compulsion. It is interesting that Herodotus, who can hardly have forgotten the elaborate story of Amyntas' submission that he had told in the previous book, does not reproduce any apologetic version: he merely makes the statement that the Macedonians were now added to the King's slaves (6. 44. 1).

Mardonius soon had to withdraw, defeated in part by Thracians who had presumably been the King's slaves at an earlier time. Had he succeeded in advancing as he had planned, it is quite likely that the faithless *hyparchos* would not have escaped unpunished. As it was, no Persian army came back for a decade, and Alexander had ample time to show how loyal and how useful he would be to a new King. In fact, Alexander appears on four further occasions, each time pretending to do the Greeks a service, each time (except possibly the last) in fact behaving as a loyal subject of the King.

First, at Tempe (7. 173. 3), he sends a message to the Greeks pointing out the size of the King's forces (he must be one of the ultimate sources for the figures we find in Herodotus): it would be lunacy to resist. The clear implication is that resistance *anywhere* would be foolish: the Greeks could never hope to match the numbers the King was bringing. He was to make this perfectly clear in Athens later: the King's power was more than human (8. 140β. 2). Herodotus was told (presumably by his Macedonian sources) that the Greeks thought Alexander's advice good and regarded Alexander as their friend; and so he tells the story. But he at once undercuts it by giving it as his own opinion that Alexander's advice in fact had nothing to do with the Greeks' withdrawal: he ascribes it to their finding out, from other sources, that there was another invasion route into Thessaly which they could not cover. Alexander's messengers had not mentioned that, for it had clearly been their aim, as we saw, not merely to persuade the Greeks to withdraw from Tempe, but to persuade them to give up all thought of resistance to the King. The reader is left to judge for himself.

We next hear of Alexander when we are told (8. 34) that he had garrisoned the cities of Boeotia and saved them by making clear to Xerxes that the Boeotians were friends to the Medes just as he himself was. Again, there is no explicit comment. But Herodotus uses an odd and rather striking stylistic device to attract our attention to the oddity of the actual fact: 'Their cities were saved by Macedonian men distributed among them, whom Alexander had sent, *and the way they saved them was as follows* (ἔσωζον δὲ τῇδε): by wanting to make it clear to Xerxes that it was the Medes' side that the Boeotians favoured.' Again, explicit comment has been avoided, but the reader has been alerted.[12]

We shall return to the episode in Athens (8. 136ff.). But let us now look at Alexander's last appearance (9. 44–5), when he reveals Mardonius' plans to the Athenians and asks them to remember him and help him attain his freedom if the Greeks win. It is a dramatic scene, with Alexander, in disguise, riding up to the Athenian lines and calling the sentries by their names without revealing his identity until he speaks to their generals, who hurry to meet the anonymous horseman outside their lines. Borza dismisses the whole story as a myth—not unreasonably, we must agree. It is certainly dressed up for effect, as we have it. Nonetheless, I think it must basically be true. Let us consider Alexander's position at that point. He had nothing to lose by this bold manoeuvre. He had invested heavily in a Persian victory, demonstrating his love for the Greeks solely by advising them against useless resistance. Meanwhile, he had watched them inflict a decisive defeat on the Persian fleet and march out to meet the Persian army with a larger contingent of hoplites (and especially of Lacedaemonian hoplites) than he would have thought possible. He knew that, if the Greek army remained united, Mardonius had nothing of anywhere near equal quality to oppose to it. It had suddenly become quite possible that the Greeks would win the impending battle, and with it the war—and Alexander's role would come under close scrutiny.

If Mardonius won, no one would ever hear of the nocturnal incident: Alexander could make sure of that. Moreover, Macedonian loyalty in the battle itself would be striking evidence against any Greek informer. For it *is* striking that Herodotus nowhere mentions any desertion by the Macedonians, let alone their turning against the masters their king was eager to be rid of. But if the Greeks won, nothing but a spectacular gesture at this point would give him a chance to avoid retaliation after the war: at the worst, perhaps even an attempt to enforce the 'tithing' decree against Greeks who medized—since he had conspicuously and repeat-

---

[12] The word τῇδε is omitted in the 'Roman' group of manuscripts, which are characterized by 'a series of seemingly accidental gaps' in books 2–9 (R. A. McNeal, *AC* 52 (1983), 120). That it must be retained can in any case hardly be doubted: why should anyone have inserted it in this particular position? It should also be made clear that the word does indeed point forward (to what follows) and not backward (to what precedes). Of the citations of ὅδε in Powell's lexicon, 14 are said to have a backward reference, as against 485 that point forward. For the actual adverbial use of τῇδε he lists 11 instances, all of them assigned a forward reference.

edly insisted on his Greek descent. As a matter of fact, even after the decisive battle and the apparent disintegration of the Persian forces Alexander took care not to break openly with his old master, the King: the Macedonians did nothing to harass the forces that Artabazus had gathered and was leading back to Asia; he found Thessaly friendly (as yet unaware of the outcome of the battle, so the Thessalians later said), and was not attacked until he got to Thrace (9. 89). Yet it could not be claimed that the king of Macedon was so out of touch with his country that he had not bothered to send a courier to Aegae with the news. Herodotus, while mentioning the friendly reception in Thessaly and the attacks by the Thracians, has nothing to say about Macedonia except that Artabazus passed through Macedonia by the shortest route, and (as we have noted) that he was not attacked until he got to Thrace. It is unlikely that the comment that *we* feel bound to make on this did not force itself upon Herodotus. I would suggest that, far from having forgotten Alexander's spectacular appearance before the battle (any more than, at 6. 44. 1, he had forgotten his long account of Amyntas' submission to the King and its immediate consequences), Herodotus is once more leaving the reader to draw the obvious conclusion, as at 8. 34, without giving positive offence to a powerful ruler.

Let us now look in more detail at the context of Alexander's appearance at Athens in his master's service (8. 136ff.). He was chosen, so Herodotus says, because his sister had married a Persian (Herodotus does not seem to know that Alexander's brother-in-law was an Achaemenid, hence distantly related to Mardonius himself) and because he was *proxenos* and *euergetes* of Athens. (Herodotus, having briefly explained the Persian connection, does not explain the Athenian: we shall come back to this.) It is precisely here (8. 137ff.) that Herodotus inserts the tale of Alexander's Temenid descent and his ancestors' conquest of Macedonia, which is nowhere claimed to be Greek.[13]

---

[13] We do not know of any attempt, by the kings of Macedon or anyone else at this time, to claim that the Macedonians were Greek. In view of some reactions to an earlier work of mine touching on this topic, I must repeat that I express no opinion as to whether they were. (See *Studies in the History of Art* 10 (1982), 33–51. This note in part retracts some misinterpretations regarding the 'Hellenic' status of the Macedonian people of which I was guilty on pp. 36–7.) It would not have suited the kings to have their subjects recognized as their equals in Hellenic descent. The consequences were easily foreseeable: e.g. some local baron, not too reliably attached to the Argead monarchy, might have won an Olympic victory. To satisfy Macedonian ambitions, Archelaus later instituted what I have called the Macedonian counter-Olympics (p. 35). Even the Greek status of the kings was questioned by their Greek enemies: see Thrasym. F 2 (D–K) and Plato, *Gorg.* 524 on Archelaus (cf. op. cit. 46 nn. 17–19) and, of course, Demosthenes on Philip II (best known 3. 24). Monarchic

It is clearly not by chance that the story appears precisely before Alexander's most conspicuous act of medism, for which his Athenian friends
angrily threw him out of their country (8. 143. 3). If there were any doubt
about this (for one still comes across talk of Herodotus' naivete), it can be
quickly removed if we remember that we have been told of Alexander's
pure Greek descent before: in fact, that account immediately follows the
story of Alexander's giving his sister in marriage to a noble Persian,
which itself follows the story of Amyntas' submission to the King. The
story of Alexander's dealings with the Persians ends, at 5. 21, with the
marriage and a gift of a large amount of money. And 5. 22 starts with an
emphatic assertion of the Hellenic descent of the Macedonian royal
house: 'But that these descendants of Perdiccas are Greeks, as they themselves say, . . .'[14] And we know what Herodotus thought of Greeks who
medized. So he here tells us the story of Alexander's official acceptance as
a Greek at the Olympic Games. Needless to say, he must have known,
even at this point, that the Argive descent, as claimed and now recognized, was in fact Temenid. But that part of the full story could be saved
up for another occasion, and a better one: Alexander's conspicuous
loyalty to the Persian masters to whom he had earlier allied himself (and
whom he later proclaimed a wish to shake off). There was surely no
better place for stressing the fact that Alexander claimed to be, and was
in fact, a Temenid descended from Heracles himself. There was nothing
in the story itself (and Herodotus, of course, fully accepted the claim)
that compelled him to divide it into two parts: his usual practice is to
follow his excursuses through to the end. It is clear that he chose to do so:
the reader would twice be able to draw his own conclusion, with Alexander's appearance as an actual tool of the Persians followed by the
enhanced account of his descent. Yet Herodotus had nowhere given
positive offence to the Temenid kings of Macedon.

Herodotus took great care not to give offence to the powerful, or those
who might be: the gods, of course, Egyptian no less than Greek; Sparta,
especially in one famous passage (7. 139. 1–5, carefully modified by
insisting that Sparta's allies would never have willingly deserted her and

rule over Greeks (as distinct from the position of the Spartan kings—itself known to be
anomalous) was not considered acceptable in Classical times. See (decisively) Isocrates, in
his praise of Philip's ancestors for having left Greek lands in order to rule as kings over non-
Greeks who needed royal rule (quoted in full, op. cit., p. 50 n. 69). We have an amusing
sidelight in Solinus' largely fictitious praise of Archelaus (*Coll. rer. mem.* p. 65 Mommsen),
where the king is described (*i.a.*) as acting 'Graeco potius animo quam regali'.

[14] The strikingly emphatic word order (Ἕλληνας δὲ εἶναι . . .) cannot be rendered in
English. And Herodotus insists that this is not a mere myth: he knows it to be true.

that Sparta would have nobly fought to an inevitable end); even Argos (7. 152—explicitly rejecting the story that the Argives invited Xerxes to invade Greece and expressing sympathy with them); and, of course, the Alcmaeonids (6. 121–4, easily misinterpreted by those who fail to appreciate the author's practice); not to mention Miltiades, whose fine is mitigated by an account of his achievements and at once followed by a striking excursus leading up to Miltiades' capture of Lemnos (6. 136–40; note also the insistence on the Delphic response, by implication exculpating Miltiades, that immediately precedes the trial (6. 135. 3)).[15] It is in this context that we must see his abstention from any overt blame for Alexander, whose son was at that time king of Macedon (and, incidentally, probably an ally of Athens when Herodotus wrote about Alexander). But he has used his literary art to guide the reader to what seemed to him a just verdict on Alexander's duplicity by stylistic and compositional devices that provide as good examples as can be found of his skill and sophistication. The technique deserves detailed investigation in other contexts.

But let us now return to Alexander's visit to Athens. It may have more to teach us, both about Herodotus and about history. We noted that Herodotus explains the choice of Alexander for the mission to Athens by two considerations: his close relationship with the Persian aristocracy and his close relations with Athens; and that he reminds the reader in detail of the former but does nothing to clarify the latter. How did Alexander come to be *proxenos* and *euergetes* of Athens? Since we are not told, we can only try to guess; so our answer must be speculative. But if it is to have any chance of being the right answer, it will have to explain both the origins of Alexander's relationship with Athens and Herodotus' unwillingness to make it as clear as he had made his relationship with the Persians.

The *proxenia* is most easily interpreted as inherited from Amyntas' relationship of *xenia* with the tyrants. This appears in Amyntas' offer of Anthemus to Hippias after his expulsion (5. 94. 1). The new aristocratic republic needed all the international support it could get and would be

---

[15] Failure to recognize this pervasive feature in Herodotus has led to assertions making him a mouthpiece of Alcmaeonid propaganda, or even of Alexander's propaganda. What is rather striking, however, is his openly expressed contempt for the Athenian *demos* (as distinct from the democracy, which he admires). See e.g. the story of Phye, which he goes out of his way to describe as deception of the naive Athenian masses (1. 60: in fact it is more likely to have been deliberate symbolism), and the account of Aristagoras' appeal at Athens (5. 97, with Herodotus' comment s. 2). Since there is no reason to doubt that these passages were recited at Athens, it must be assumed that the prejudice they express was safely shared by his (largely upper-class) audience.

eager to continue friendly relations established, on a personal level, by the tyrants, in the case of Amyntas no less than (e.g.) in the case of Corinth. (It would be all the easier because Hippias had not in fact settled at Anthemus as a vassal of Amyntas.) Thus *xenia* easily became *proxenia*. Herodotus might not want to make it clear because Hippias obviously did not provide a wholesome precedent for the republic that was at this point saving Greece from the Barbarian.

Euergesia is not so simple. What particular benefit had Alexander conferred on Athens to earn the appellation? (For it is clearly understood to be a formal title formally conferred.) I must again stress that the absence of explanation is astonishing. One would certainly expect Herodotus to explain it just as he explains the first of Mardonius' motives. And even if Herodotus chose not to explain directly, one would all the more expect to hear of it in Alexander's speech to the Athenians. Although I have not actually checked, I think it is probably unparalleled for a speaker whom a historian depicts as trying to persuade a person or a body of persons not to scrape together all conceivable instances of benefits he may claim to have conferred. Alexander, on the other hand, carefully *refrains* from specifying (8. 140β. 1): 'As regards the good will for you that I bear, *I shall say nothing*, for this would not be the first time that you would discover it.' The standard appeal to past benefactions is deliberately withheld. Yet it is surely inconceivable that Herodotus was told that Alexander was *euergetes* of Athens but was not told why. Surely the inventor of *historie* was not so devoid of curiosity that he would not have asked, even if the information had not been spontaneously offered. We are forced to conclude that Herodotus deliberately chose to withhold the information.

This important historiographic point has usually been ignored when an explanation for the actual award of the title has been sought. The standard explanation is in itself obvious and has often been advanced:[16] his benefaction would be that he supplied Athens with timber and pitch (as we know some of his successors did) when she built her great fleet just before the Persian War. One might wonder how one who behaved at all times as a loyal vassal of the King, and who had in fact been forgiven for past disloyalty when he renewed his submission a decade earlier, would come to help prepare a fleet for use against his master. That question can be answered, either by his duplicity or by noting that the fleet was intended for the war against Aegina (7. 144. 1) and just happened to be

---

[16] Last by Borza 109f., regarding this (probably rightly) as at least partly explaining Alexander's Athenian title ('his timber grants . . . may have contributed to his receipt of honors at Athens', with references to others less cautious).

available when the Persians came. Whether or not this is true (and I see no reason to doubt that it was the motive put to the Assembly, for whom the long-lasting and, on the whole, unsuccessful war against their neighbour was a much better reason for financial sacrifice than the mere possibility of war against the Persians), it was certainly the official purpose of that fleet. Alexander, even if he knew about Xerxes' preparations, certainly already going on at the time, could claim to be acting in good faith.

This difficulty can therefore be by-passed. If we go on to ask the real question, why Herodotus deliberately omitted any reference to this—it would have been an obvious point for Alexander to make in the circumstances—it might even be possible to suggest an answer. We have seen that Herodotus did not approve of Alexander's medism. When he appears as a Persian spokesman at Athens, Herodotus may have been unwilling to give him credit for contributing as much as anyone to Xerxes' defeat by enabling Athens to build her fleet.

That reconstruction should perhaps not be excluded. But I do not find it fully satisfactory. I feel uneasy about believing that Herodotus went to such lengths to make his Alexander conspicuously suppress such an obvious and relevant benefit conferred on Athens. If he did not want to give him credit for aiding in the defeat of the barbarian, he could have explained in his introduction that Alexander thought he was assisting Athens against Aegina, and that this merely happened (by divine intervention?) to lead to the salvation of Greece. There were ways of undercutting Alexander's claim rather than omitting it, even if Alexander were allowed to claim the credit for the building of Athens' fleet. I cannot see that the conspicuous omission was necessary, on this hypothesis.

But we cannot help being struck by a parallel occasion when we find Herodotus mystifying the reader by suppressing information that he must have had. When an Argive embassy was at Susa, asking the new King whether he still regarded the Argives as his friends, an Athenian embassy headed by Callias happened to be there 'on other business' (7. 151). What he was concealing on that occasion (though it is inconceivable that he did not know) was that the Athenians were there to negotiate about peace with the King.[17] The parallel between the Argives and the Athenians, on this occasion, would have been embarrassing. Was he

---

[17] See my article 'The Peace of Callias', *JHS* 107 (1987), 1–39, reprinted (with corrections and an appendix) in *From Plataea to Potidaea* (Baltimore, 1993). The point here relevant does not involve adherence to the case made about the Peace in that article, nor indeed even acceptance of the authenticity of the Peace.

suppressing information of the same kind in Alexander's speech? The question is worth pursuing.

In 508/7, after the expulsion of Cleomenes and the archon Isagoras, the Athenians, now without allies and threatened with attack on all sides, turned to the satrap of Sardis for help. Their ambassadors offered earth and water to the King's representative in return for a promise of protection (5. 73). How did they come to do so? How obvious was an appeal to the King? Under the tyranny, relations between Athens and the King had been remote and not noticeably friendly until just before the end. Miltiades, clearly supported by the tyrant in Athens, had been an ally of Croesus; although he had not actually assisted him, that would not endear Athens to Cyrus. In Asia itself, which the King no doubt then, as later, regarded as his property, the Athenians had seized Sigeum and were holding on to it right to the end of the tyranny (5. 94). When Darius crossed to Europe, another Miltiades apparently did homage to him (Herodotus, of course, does not actually record this) and was swept up into the Scythian expedition. But in the famous scene at the Danube bridge he saw a chance of ridding himself of his new suzerain. Unlike the commanders of the Greek contingents from Asia, Miltiades (and no doubt the Greek and Thracian warriors he commanded) felt the King's suzerainty to be merely burdensome, without recompense. But Miltiades failed in his attempt and must have fled on the King's return from across the Danube.[18] There is no mention of him when Darius gets to Sestos (4. 143. 1). Nor may we assume that he did homage and was forgiven: the reward given to Histiaeus for opposing Miltiades shows how seriously Darius took the incident. Had Miltiades been caught, he would have had to be impaled, *pour encourager les autres*. He can only have shown himself again when the Ionian Revolt (as we have seen) forced the Persians to give up (at least *de facto*) control of most or all of their European territories. It was only then that, sailing south before the Etesian wind, now openly as a representative of Athens, he captured Lemnos (6. 140), which had apparently shaken off Persian rule—but which the Persians would still claim. Miltiades' attitude was consistently anti-Persian, even when, against his will, he was for a short time a slave of the King.

It was about the very time of Darius' Scythian expedition, three years before his fall, that Hippias established a marriage alliance with Hippoclus, the tyrant of Lampsacus (Thuc. 6. 59. 3), who accompanied the

---

[18] The authenticity of the story of Miltiades at the Danube bridge has often been rejected, but as I attempt to show in the text, its basic veracity (not, of course, the details in the speech) is needed to explain several unimpeachable later developments.

King on the expedition (Hdt. 4. 138. 1)—probably just after the expedition, when Miltiades, whose *dynasteia* had been in perpetual conflict with Lampsacus (6. 37 ff.) and protected by the Pisistratids, had departed and become irrelevant (see above). The aim of the connection was to acquire Darius' support (thus, very credibly, Thucydides, since after the death of Hipparchus and with the Alcmaeonids agitating against him, Hippias must have felt in need of powerful allies). The immediate result, however, was not actual Persian support, but the fact that Hippias' possession of Sigeum now remained uncontested: he decided to take refuge there, in preference to other offers, when he had to leave Athens (5. 94. 1). There can be no doubt that he now became a loyal slave of the King. The Athenians who had expelled him and wanted to be protected against having his return forced upon them would not have found it easy to impress Darius' satrap and brother at Sardis.

There was nothing, therefore, that would have made the satrap of Sardis look with favour on the Athenians in 507. Since they must have been well aware of this, one has to wonder what gave them the idea of applying to him for aid. And this leads us back to Alexander.

It was only about three years earlier, as we saw, that Amyntas and Alexander had established close relations with the King, marrying a lady of their house to an Achaemenid related to the King and to the satrap of Sardis. It is an attractive hypothesis that Alexander, on a realistic estimate of Hippias' chances of returning, decided to transfer his family connection with the Athenian tyrants to the new Athenian *demos*: at one stroke, he could earn the gratitude of the Athenians desperately looking for help and of his suzerain, who would thus acquire a foothold in the very heart of Greece. I suggest that Alexander will have given the Athenians the idea of turning to the King and, for obvious purposes of his own, will have supported their improbable plea at Sardis—subject, of course, to their following his own example and accepting the King's suzerainty. Whether the *demos*, when it despatched the ambassadors, knew what was involved, is a question we cannot answer. Herodotus, of course, implies that they did not: that the act was the result of the ambassadors' spontaneous decision when they found that they could not carry out their instructions (to gain the King's alliance) in any other way (5. 73). We are told that the envoys 'received great blame' when, on their return, they reported what they had done. But if that was so, it is surprising that Herodotus does not (presumably because he could not) add that the Athenians disavowed the action. There can be no doubt that he would have preferred to absolve the saviours of Hellas from the charge that they had offered earth and water to the King long before any other European Greeks considered doing so. The fact that he does not say so, but uses a vague phrase about their blaming the envoys, once more reveals his technique when he has something to hide. We must take it that the Athenians approved of the action, even though there will certainly have been some who expressed strong disapproval: we note that Herodotus nowhere says that 'the Athenians' disapproved, only that (in the passive) the envoys were blamed.[19]

It was in this way that Athens became, in the now famous phrase, 'a city of the Great King'.[20] If this was the benefit that Alexander had conferred on the city in the hour of her need, we do have a very good reason why it is not mentioned in the speech Herodotus assigns him. It would have been very inappropriate, just at this point, to remind the Athenians of their submission to the King through his mediation—an act perhaps never technically revoked, though long since overtaken by events. And it would be no less inappropriate for Herodotus to explain the nature of Alexander's service, precisely at the point where the Athenians were about to make the heroic decision that ultimately saved Hellas. The conspicuous vagueness of Alexander's reference is at last fully explained.

According to Errington,[21] 'the Philhellene Alexander was . . . the first Greek ruler of importance to betray the Greek cause.' Whether or not we link Alexander with his father in this act of submission, it seems too early to speak of a Greek cause either in the last decade of the sixth century or in the first decade of the fifth. That concept did not begin to acquire a definition until Xerxes' invasion; and even then it was by no means all European Greeks who 'had the better disposition about Hellas' (as,

[19] It is surprising how the less than careful reader of Herodotus can misinterpret this quite typical Herodotean instance of *suppressio veri*—perhaps through failure to look at Herodotus' technique in similar cases, to which this study has tried to draw attention. M. Ostwald, in the new edition of the *Cambridge Ancient History* (iv². 308), opines that Herodotus' wording [that the envoys 'were greatly blamed'—we are not even told by whom, as I point out in the text!] 'probably means that their acceptance was disavowed'. (By p. 338 this has become established fact.)

[20] See Fritz Schachermeyr's article 'Athen als Stadt des Großkönigs', *Grazer Beiträge* 1 (1973), 211–20. (My epigraph forms the beginning of that article.) His comment on Herodotus' phrase about the envoys is worth contrasting with Ostwald's (see n. 19): 'Für Herodot ist die Angelegenheit mit Hilfe der erwähnten αἰτίαι allerdings schon erledigt. *Athen ist für ihn reingewaschen*. . . .' (My emphasis.) He stresses that there is no mention of any actual punishment of the envoys, let alone disavowal of their action. In his view, Athens' victories over the Lacedaemonian invasion and, later, over the Boeotians and Euboeans, are due to knowledge of the King's acceptance of Athenian submission; and the submission is only renounced when the satrap orders the Athenians to receive Hippias back as their ruler. (The article, clearly unknown to Ostwald, is not listed in *CAH* iv².)

[21] Op. cit. (n. 3 above), 143.

much later, Herodotus could describe it (7. 145. 1)), but essentially only the alliance led by Sparta, now joined by Athens and one or two others. It would, correspondingly, be anachronistic to charge Athens with being the first European Greek city to betray the Greek cause. But it is plain fact that it was Athens that first invited the King into the heart of European Greece, about three years after Macedonia's submission to him. I have merely suggested that another example of Herodotus' reticence about unpalatable facts permits the conclusion that the two acts were not unrelated. It was only much later, when a 'Greek cause' did emerge with the European Greeks' victory over the Barbarian, that both Alexander and Athens, and not least the historian of these events, thought it best to combine reinterpretation with vagueness over inconvenient details regarding those early actions.[22]

APPENDIX

Amyntas and Elimeia

It was N. G. L. Hammond (*HM* ii. 99 ff.) who chiefly drew attention to the profit that Macedon derived from timely homage to the Persian King. But he was chiefly concerned with Alexander I and overlooked (indeed, misinterpreted) a crucial item of evidence concerning Amyntas.

When Philip, son of Alexander, and Derdas 'join in opposition' to Perdiccas, the Athenians, although at the time allies of Perdiccas, who had committed no act of disloyalty, make a full alliance (ξυμμαχία) with them (Thuc. 1. 57: for scrutiny of the way Thucydides tells the story see my study 'Thucydides and the *Archē* of Philip', in *From Plataea to Potidaea* (1993), 172 ff.). On this occasion Thucydides informs us that Philip was the brother of Perdiccas, but he tells us nothing about Derdas. A scholiast on the Thucydides passage (p. 48 Hude) managed to elicit the information (which could hardly be fictitious) that Derdas was a son of Arrhidaeus and a cousin of Perdiccas and of Philip. A reference in Xenophon and a fragmentary inscription (see Beloch, *GG* iii(2), 72 f., followed by Hammond p. 18) enable us to establish that Derdas called himself king of Elimeia.

Before Beloch, it was generally thought (although little was made of it) that the

---

[22] I should like to thank Dr Hornblower, whose invitation to address his seminar on Greek historiography made me pull together some ideas long propounded in my own seminar; also the Institute for Advanced Study, where the sketchy seminar paper was written up in this form, as a *parergon* of a stay in principle devoted to other things.

father of Derdas was a son of Amyntas (i.e. a brother of Alexander) who married
a daughter of the king of Elimeia. Let us call this Hypothesis I. Beloch (62f.)
pointed out that the scholiast's statement would equally support the conjecture
that an Arrhidaeus of Elimeia married a daughter of Amyntas, thus becoming a
brother-in-law of Alexander. Let us call this Hypothesis II. He thinks this hypothesis more probable, for two reasons: (1) the royal Macedonian names do not
reappear in the royal family of Elimeia (let us call this Argument 1); (2) it is difficult to believe that Elimeia was subject to the Argead kings as early as the late
sixth century (let us call this Argument 2). Hammond (l.c.) added the further
possibility (let us call it Hypothesis III) that Alexander married a sister of a
(presumably Elimiot) Arrhidaeus. (He accuses Beloch of not mentioning this possibility, while himself overlooking Hypothesis I, which was the usual explanation. But he quite unjustly accuses Beloch, who in fact argues at length, of
'assuming' the hypothesis he favours.) Borza (op. cit. (n. 4) p. 124) follows
Hammond without actually setting out the possibilities. As a result, Hypothesis I
is now in danger of being forgotten. I shall proceed to argue that it remains much
the most probable of the three, and that it allows us to retrieve some lost history.

We know very little about the royal family of Elimeia and their names through
the generations. But as far as our knowledge goes, Beloch's Argument 1 must in
fact be extended so as to refute his conclusion. For the name Arrhidaeus itself
nowhere reappears in our record of that family, and one would certainly expect it
to, if the Arrhidaeus who was Derdas' father had been an Elimiot king who
married an Argead princess and passed his title on to his son. On this hypothesis,
the total disappearance of the name in that family seems inexplicable. Once this
is seen, Hammond's Hypothesis III becomes as improbable as Beloch's Hypothesis II.

On the other hand, Hypothesis I can be positively supported. We must note
that the name Arrhidaeus does indeed appear in the Argead royal family, first in a
grandson of Alexander I: as shown in every stemma (see e.g. Beloch p. 73 and
Hammond, facing p. 176), this man is both the son and the father of an Amyntas,
and in fact the grandfather of Philip II. (A half-brother of Philip also bears the
name.)

It is in theory possible that, if Alexander had married a sister of an Elimiot
Arrhidaeus and his son Amyntas was the offspring of that particular marriage,
that Amyntas then decided to name his son (the only son we hear of) after his
Elimiot uncle, instead of using one of the many Argead names. But that would
imply some claim to the Elimiot succession, and we are told by Syncellus (p. 316
Mosshammer) that he lived as a private man (ἰδιωτικῶς) all his life, i.e. did not
even take part in the scramble for the Argead throne and Argead principalities
that followed Alexander's death. Moreover, such a conjecture would still not
account for the absence of the name from our Elimiot records. It is much simpler
to stay with the obvious possibility that Amyntas, in naming his son, used a traditional Argead name.

Nor need we be puzzled by the fact that we do not come across the name before Alexander. For down to Alexander I, we know of only one royal name in each generation: the one of the man (always his precedessor's son, we are told) who became king; and for this purpose we need not distinguish between a mythical and a historical period. (Hdt. 8. 139 gives the traditional king list as known in the fifth century.) As a result of this (no doubt), many royal names appear rather late in our record; thus Philip and Menelaus do not appear until the generation of Alexander's sons. And among his eight grandsons, i.e. in the generation that gives us the first indubitable Arrhidaeus, we meet two and probably three more names not met with before: Agerrhus, Agelaus, and surely even Archelaus (if we exclude Euripides' court fiction: see Hammond p. 5). Out of the eight names known, exactly half, therefore, cannot be documented before. Yet it is very likely that most, if not all, of them were by then traditional in the family.

There is thus no difficulty in positing a brother of Alexander I, not elsewhere recorded, who bore the name Arrhidaeus. Full investigation of Beloch's Argument 1 seems to lead to this result, strongly supporting Hypothesis I, rather than Hypothesis II, which he thought it supported.

Beloch's Argument 2 does not need such long discussion; consideration of it does, however, lead to a conclusion of considerable historical interest—though, like everything in this shadowy area of history, it can only be propounded as a reasonable conjecture. Beloch was right to note that one would not normally expect an Argead to be discovered as the father of an Elimiot king (which implies a claim to Argead suzerainty) as early as this. But he did not recognize that something far from normal had intervened: Amyntas had become a vassal of Darius I. Among his rewards, we may now say, was indirect control of Elimeia, which was surely unable to resist the King's mandate, through the marriage of one of his sons to the daughter of the Elimiot king, apparently the heiress to the throne. (We must presume that she had no eligible brothers.) This, incidentally, helps to confirm that it was indeed Amyntas and not Alexander who offered submission to the King: had it been Alexander (acting as ruler of Macedon), it would surely have been one of *his* numerous sons, and not a brother, who would have received the royal bride.

Let us pursue the investigation. The son of that marriage, named Derdas (no doubt after his maternal grandfather), was intended to embody the link between two dynasties and to rule his mother's country as the vassal of his father's family. As we have seen on independent grounds, it seems very likely that the son of another marriage linking two dynasties, Amyntas, son of Bubares and Gygaea and also named after his maternal grandfather, was intended to embody the link between Achaemenids and Argeads and ultimately to rule his mother's country as the vassal-satrap of his father's relative, the King.

Had Xerxes not lost a war that, on all rational calculation, he ought to have won, a Greater Macedonia (enlarged by control of Elimeia) would have settled down to the status of a satrapy of the Achaemenid empire, ruled by satraps linked

E. Badian

by blood to the King's family. We have, I think, discovered two related instances of what we may call classic *Heiratspolitik* worthy of the Habsburgs at their zenith, which has not so far been known as a political instrument of Achaemenid expansion. It was due to the unforeseeable incompetence, and the bad luck, of Darius' successor that his well-designed scheme fell apart.

5

Narratology and Narrative Techniques in Thucydides*

SIMON HORNBLOWER

In his classic book *The Rhetoric of Fiction*, Wayne Booth speaks of 'the rhetoric that makes me believe in Thucydides' *History* as a report of actual events'.[1] The remark is an aside, specifically a disclaimer: Booth is actually saying that he has not discussed, and does not propose to discuss, the rhetoric which is found in narrative history. But he implies that such an analysis is possible. By contrast, Gérard Genette, in his *Narrative Discourse*, scarcely glanced at historiography at all. Genette returns in his more recent *Fiction et diction* to the difference between fictional and factual writing and concludes that fictional writing is parasitic on factual; and that narratology should be willing to cross the boundary between fiction and fact.[2]

By narratology I mean 'the theory that deals with the general principles underlying narrative texts'. That is the definition offered by Irene de Jong in what is surely a strong candidate for any prize for the most successful application of narratology to an ancient text, her excellent if algebraically written book *Narrators and Focalizers: The Presentation of the Story in the Iliad*.[3] I want in the course of this paper to address the basic

* I gladly acknowledge the help I received from members of the Oxford Philological Society after I read this paper to them on 14 February 1992, especially the following: Ewen Bowie, Lesley Brown, Miriam Griffin, Leofranc Holford-Strevens, David Lewis, Simon Price, Nicholas Richardson, and Richard Rutherford. David Lewis helped me in correspondence afterwards, as did Robert Parker. A revised version of the paper was subsequently read by Jasper Griffin, from whose comments I benefited greatly and gratefully, as also from the written comments of Christopher Pelling and Oliver Taplin. References in the form e.g. '7. 42' are to Thucydides (Th.) unless otherwise stated. Hdt. = Herodotus.

[1] W. Booth, *The Rhetoric of Fiction*[2] (Harmondsworth, 1983), 408.

[2] G. Genette, *Narrative Discourse* (Oxford, 1980)=Eng. tr. by J. E. Lewin of *Figures III* (Paris, 1972), 67–286, section entitled 'Discours du récit: Essai de méthode' (I cite by the English version); G. Genette, *Fiction et diction* (Paris, 1991), 65–93, section entitled 'Récit fictionnel, récit factuel'. See also R. Barthes, 'The Discourse of History', in M. Lang (ed.), *Structuralism: A Reader* (London, 1970), 145–55, also reprinted in Barthes, *The Rustle of Language*, tr. R. Howard (Oxford, 1986), 127–41. I cite by the latter.

[See p. 132 for n. 3]

question whether narratology can be simply transferred from poetic or fictional texts to historical ones, as Booth asserts and Genette now seems to agree. There is an even more fundamental issue at stake: are history and fiction separate genres? Modern historians, not to mention biographers like Peter Ackroyd in his *Dickens*, are readier than ancient historians to run history and fiction together. Simon Schama's amusing *Dead Certainties* is a good example; Schama incidentally writes in his Afterword that since ancient Greek times, 'historians have . . . differed on the implications of the term [*historia*], sometimes imagining themselves lined up behind opposing platoons commanded by Herodotus or Thucydides', Thucydides representing here objectivity, Herodotus representing gossip, hearsay, and the fantastic.[4]

But these questions of genre are very broad and deep. My theme is a (slightly) narrower one, the sense in which there is an identifiable rhetoric of history; but above all I am interested in the *differences* between the way historians and fictional writers use narrative devices. The relevant chapter (3) of Genette's *Fiction et Diction* is less illuminating on this second topic than might have been hoped. It is largely concerned to examine schematically the relation between author, narrator, and character or *personnage*, in a way which has to be qualified immediately to cater for writers like Thucydides, Xenophon, and Caesar who feature as historical agents in their own narratives. In fairness, Genette does make this qualification. But the schemata, qualified or not, do not seem to get us very far.

The shape of my paper is as follows. First, I will discuss some of the general narrative techniques relevant to Thucydides' rhetoric of history. Second, I will look at narratological devices in particular, trying where appropriate to ask how Thucydides' use differs from a poet's or a fiction

---

[3] I. J. F. de Jong, *Narrators and Focalizers: The Presentation of the Story in the* Iliad (Amsterdam, 1987). Some of de Jong's narratological insights are exploited by M. W. Edwards, in G. S. Kirk (general ed.), *The Iliad: A Commentary*, v: *Books 17–20* (Cambridge, 1991), 1–10; see also R. B. Rutherford (ed.), *Homer: Odyssey Books XIX and XX* (Cambridge, 1992), 67–8, and Scott Richardson, *The Homeric Narrator* (Nashville, 1990). The issue of Homer's 'objectivity' is approached by J. Griffin, *Homer on Life and Death* (Oxford, 1980), without de Jong's arsenal of technical narratological terms; but the upshot is similar: see Griffin p. 139 for emotional effects not spelt out but produced in the course of the narrative or in speech by one of the poet's characters.

[4] P. Ackroyd, *Dickens* (London, 1990); S. Schama, *Dead Certainties (Unwarranted Speculations)* (London and New York, 1991), 325. But history written like a novel is no novelty; see already Thomas Carlyle's *History of the French Revolution* (London, 1837). Not to mention the procedures of Shakespeare in the history plays.

writer's, and why. Third, finally and briefly, I will sum up by recapitulating those differences.

Let me begin as if Booth is in the right to say there is such a thing as a rhetoric of history and see where that gets us. Remarks like Booth's are capable of arousing strong emotions among professional students of ancient history. Books or articles which treat ancient historiography as *nothing more* than a branch of rhetoric are unsettling,[5] clearly because they may tend to suggest, or may be taken to suggest, that the events narrated by the ancient historical writer *did not happen at all* and that would put ancient historians out of a job. We may agree that there is a problem: can we apply, to ancient historical writers, techniques of analysis successfully applied to poetry and fiction, without thereby committing ourselves to the view that the history *is* fiction? Put like that, the fallacy becomes obvious. By examining the *techniques* of historical presentation we do not necessarily imply that the subject-matter of that presentation is true or false. True facts can be presented rhetorically or non-rhetorically. Or rather, true facts may be presented with a rhetoric which is more or less *obtrusive*. (I put it that way so as to make a gesture in the direction of those fundamentalists who refuse to believe in the possibility of totally objective descriptions, or as Genette calls them, 'zero-focalized statements'.[6]) The historian who recounts facts which are on any common-sense view demonstrably true may still have a problem. We may call it Cassandra's problem. How to get people to believe the true things you are saying? *Magna est veritas, et praevalebit* is a noble doctrine, but in a highly agonistic culture like the fifth century BC *veritas*

[5] Strong emotions: see e.g. the report of T. P. Wiseman, *CR* 38 (1988), 263, in the course of a review of A. J. Woodman, *Rhetoric in Classical Historiography* (London, 1988). See also Momigliano's famous reply to Hayden White, reprinted as A. Momigliano, 'The Rhetoric of History and the History of Rhetoric: On Hayden White's Tropes', *Settimo Contributo alla storia degli studi classici e del mondo antico* (Rome, 1984), 49–59, discussing H. White, *Metahistory* (Baltimore, 1973) and *Tropics of Discourse: Essays in Cultural Criticism* (Baltimore, 1978). Note esp. Momigliano p. 49: 'I fear the consequences of [White's] approach to historiography because he has eliminated the research for truth as the main task of the historian. He treats historians, like any other narrators, as rhetoricians, to be characterized by their modes of speech.' See also P. Green, *TLS* for 19 July 1991, 4, reviewing Momigliano's *Classical Foundations of Modern Historiography* (Berkeley, 1990). Momigliano is there quoted as saying to Hayden White at a seminar 'after all, we do have these inscriptions and these artefacts. They are there. We cannot disregard them. What then are we to do about them?' With this cp. *Settimo Contributo*, 51. See also O. Murray, 'Arnaldo Momigliano in England', in Michael P. Steinberg (ed.), *The Presence of the Historian: Essays in Memory of Arnaldo Momigliano* (*History and Theory*, Beiheft 30; Chicago, 1991), 49–64 at 63: Momigliano 'visibly distressed at the view of his American colleague Hayden White that history was a form of rhetoric'.

[6] 'Zero-focalized statements': Genette, *Narrative Discourse*, 189.

surely needed all the rhetorical help she could get. So in what follows, I
am not particularly concerned with the truth or falsity, with what
philosophers call the truth-function, of the Thucydidean texts I shall be
discussing. Instead, I shall try to apply some of the insights of narratology.

The chief, or one chief, contribution of narratology is the rigorous and
scientific study of what has been called focalization,[7] that is, the different
perspectives or points of view from which events are viewed and interpreted. The narrator is the person narrating. The focalizer is the person
who orders and interprets the events and experiences which are being
narrated. Secondary or embedded focalization is when the first focalizer
or interpreter quotes or refers to a focalization or interpretation by a
person other than him- or herself. This may be explicit or implicit
depending on whether the word for thinking is included. A simple
explicit example is in Thucydides book 5: the Boiotians and Megarians
thought Argive democracy would be less congenial than Spartan
oligarchy. A simple implicit example is in book 6, where, as often, an
embedded focalization is introduced by γάρ, 'for . . .': 'the generals did
so and so, for otherwise the Syracusan cavalry would do damage to their
own light-armed troops'. Here the embedding is implicit, i.e. there is no
word like 'the generals saw that'. But sometimes γάρ introduces material
whose focalizer is really Thucydides himself, the obvious example being
that at 2. 13. 3, from the account of Athenian finances ostensibly taking
the form of encouragement by Pericles. A more controversial and much-
discussed case is 7. 42: Demosthenes on his arrival in Sicily didn't want
to suffer what Nikias had suffered, followed by a long explanatory
bracket. The problems arise because scholars disagree about whether the
explanatory bracket represents Thucydides' reasoning or that of Demosthenes.[8] Dover has however shown from the nominative and finite

---

[7] For focalization generally, see Mieke Bal, *Narratology: Introduction to the Theory of
Narrative* (Toronto–Buffalo–London, 1985), 100–15.

[8] Focalization in Th.: 5. 31 (Boiotians and Megarians); 6. 64 (Syracusan cavalry); 2. 13
(finances); 7. 42 (Demosthenes' arrival), on which see K. J. Dover, 'Thucydides' Historical
Judgment: Athens and Sicily', *The Greeks and their Legacy* ii (Oxford, 1988), 74–82,
reprinted from *Proceedings of the Royal Irish Academy* 81c (1981). On the relevant point,
this supersedes Dover's own discussion of 7. 42 at *HCT* iv. 419ff. Th. 2. 20 (which
purports to give Archidamus' thinking) is an intermediate case, in that *oratio obliqua* and
indicative construction are mixed. See the interesting discussion of C. B. R. Pelling, 'Thucydides' Archidamus and Herodotus' Artabanus', in M. Flower and M. Toher (eds.),
Georgica: Greek Studies in Honour of George Cawkwell (*BICS* suppl. 58; London, 1991),
120–42 at 127 n. 27. In general, my statements in the text are dogmatic; Pelling is no doubt
right to protest to me that things are greyer than that, and that 7. 42 does *also* give Demosthenes' view, just as 2. 13 does *also* give the eventual view of the Athenians on whom

tenses that the reasoning is Thucydides' not Demosthenes'; contrast the accusatives and infinitives in the book 6 passage about the cavalry. That is, the focalization in book 7 is *not* embedded—Thucydides is the only and primary focalizer. It would be nice if all focalization problems could be so neatly solved.

So much for the technical terminology, which as a matter of fact I shall, apart from plain focalization itself, make little use of. But this is not out of obscurantism, or because results have not been achieved by the application of narratology to ancient texts. On the contrary, the brilliant work of de Jong on Homer and now on the messenger-speeches of Euripides[9] has revealed previously unnoticed subtleties. Nevertheless I think we have to admit (for instance) that sometimes we cannot determine, and perhaps should not ask, whether a certain statement or expression of Thucydides represents his own feeling or that of his agents. So for instance (a simple example) at 3. 49, the first lot of Athenian sailors on their way to execute the Mytileneans were not hurrying on their horrible mission, ἐπὶ πρᾶγμα ἀλλόκοτον.[10] Who thought the mission horrible, Thucydides or just the sailors? This problem resembles what has been called deviant focalization,[11] which is when the narrator is made to say things which really belong, so to speak, to the focalizer. But we can call the Thucydidean example deviant only if we think (as I do not) that Thucydides—or rather what Wayne Booth would call the implied Thucydides as opposed to the man Thucydides—was too detached to be capable of saying on his own account that the mission was horrible.

To be sure, Thucydides' narrative technique did not lack students, even before the arrival of narratology. Momigliano once wrote witheringly of the present 'ridiculous adoration of so-called prosopography (which as we all know claims to have irrefutably established the

Pericles' persuasive power was exercised—and of Pericles himself (I mean, Pericles the literary and rhetorical construct. Note the way that Pericles omits to mention the financial trierarchies, surely because this would be depressing to individual rich Athenians.) And what are we to make of simple statements like 'a bit of the wall was weak' at 7. 4. 2? (with which compare the weak wall at *Iliad* 6. 434: Andromache? or Homer?).

[9] I. J. F. de Jong, *Narrative in Drama: The Art of the Euripidean Messenger-Speech* (*Mnemosyne* suppl. 116; Leiden, 1991).

[10] πρᾶγμα ἀλλόκοτον: 3. 49. 4. Close analysis of the exact words chosen does not resolve the ambiguity: the adjective ἀλλόκοτον carries the idea of 'unusual' (an objective, factual judgement, cp. Plato, *Theaetetus* 182a, though at Sophocles, *Philoctetes* 1191 the idea of 'unwelcome' is present); but the statement that the sailors were not hurrying (οὐ σπουδῇ) implies lack of enthusiasm (subjective).

[11] D. Fowler, 'Deviant Focalisation in Virgil's *Aeneid*', *PCPhS* 36 (1990), 43–63.

previously unknown phenomenon of family ties)'.[12] In the same way the sceptic, aware, as we shall see, that there is for instance some good narratology in Longinus, might want to say that narratology is new words for old insights, and that the new words will not last whereas the insights will. Actually Genette engagingly faces this possibility in the Afterword to *Narrative Discourse*. But he was too modest. Narratology, like prosopography, is based on detail and minutiae, and for such work, precise instruments are needed. But we can admit that narratology does not exhaust narrative technique, hence both terms feature in my title. Mention of Longinus prompts a word about Dionysius of Halicarnassus' *On Thucydides*, which I shall not refer to much because Dionysius was more interested in the speeches and the great set pieces than in the routine narrative which in chapters 13 and 14 of his treatise he objects to as cursory.

Returning to modern work, particularly notable, as straightforward jargon-free analysis, are two books by Stahl and de Romilly.[13] And some of the findings of Fehling on Herodotus[14] have a bearing on Thucydides; though direct *Quellenangaben*, or citation of sources with a view to inviting the reader's or listener's belief or complicity, are much rarer in Thucydides. The usual example is in book 2: 'the Thebans say this, but the Plataians don't agree.'[15]

Moreover there are two monographs on Thucydides, both getting on for twenty years old now, which address some issues which narratologists have subsequently treated, though under newer and more technical names. I refer to the books by Schneider and Hunter, to which

---

[12] Prosopography: A. Momigliano, *Studies in Historiography* (London, 1966), 103. Momigliano's reply to Hayden White (above, n. 5) similarly takes the line 'there is nothing new under the sun', see esp. p. 58 on the '*rediscovery* of rhetoric' (my italics). This is relevant to the advocacy of a return to rhetoric by Terry Eagleton, *Literary Theory: An Introduction* (Oxford, 1983), 205–6; Eagleton's discussion of narratology is at 104–5, but he does not make it quite clear enough that narratology, which is one of the new approaches implicitly rejected in the conclusion to the book, is itself, in effect, the study of a branch of rhetoric.

[13] H.-P. Stahl, *Thukydides: Die Stellung des Menschen im geschlichtlichen Prozess* (Munich, 1966); J. de Romilly, *Histoire et raison chez Thucydide* (Paris, 1956). Note also H. D. F. Kitto, *Poiesis: Structure and Thought* (Berkeley–Los Angeles–London, 1966), ch. 6, and A. W. Gomme, *The Greek Attitude to Poetry and History* (Berkeley and Los Angeles, 1954), chs. 6–7. Of older studies see F. M. Cornford, *Thucydides Mythistoricus* (London, 1907). The late Colin Macleod's work on Thucydides was of the first importance, but he concentrated mainly on the *speeches*, except in ch. 13 of his *Collected Essays* (Oxford, 1983): 'Thucydides and Tragedy').

[14] D. Fehling, *Herodotus and his 'Sources'* (Liverpool, 1989).

[15] 'The Thebans say this, but the Plataians . . .': 2. 5. 6. The discrepancy between the two versions is itself an important fact, and that is no doubt why Th. includes it.

should now be added an excellent new chapter by Westlake (1989). They all deal with the problem of inferred motivation. How, to take an extreme example, did Thucydides know what was going on in the heads of hypothetical Minoan pirates?[16]

To expand on inferred motivation: it is clear that Thucydides made inferences about motive all the time. In this respect he is no different from any modern historian. John Ehrman's life of Pitt the Younger, of which two volumes so far have been published, is full of inferred motivation, and so is Woodward and Bernstein's *Final Days*, about the end of the Nixon presidency. It has on the cover the words 'history as gripping as a novel'.[17]

This issue, on which I will say more later, is what narratologists call the problem of *restricted access* to information or knowledge. It has been discussed by de Jong in her narratological study of the Euripidean messenger-speech. Indeed, she glances in passing at the relevant recent work on Thucydides, notably Schneider, to make her point that Euripides' messengers are like Thucydides in that they often have to guess what was in the mind of the agent or speaker: compare the messenger in Euripides' *Phoenissae*.[18] That is because these messengers are posing as omniscient narrators, to put the matter narratologically.

But narrative technique in Thucydides is too broad a topic and I propose to concentrate on those techniques which I suggest have been too little studied or (as far as I can find) not studied at all, but which *either* make us believe that what Thucydides says is true, *or* make us accept his account of them as convincing. Sometimes, obviously, these two things, truth and convincingness, coincide, for instance over the

[16] C. Schneider, *Information und Absicht bei Thukydides* (Göttingen, 1974); V. Hunter, *Thucydides the Artful Reporter* (Toronto, 1973); H. D. Westlake, *Studies in Thucydides and Greek History* (Bristol, 1989), ch. 14, 'Personal Motives, Aims and Feelings in Thucydides', with E. Badian, *From Plataea to Potidaea* (Baltimore, 1993), 231 n. 45, adding the important case of Sthenelaidas at 1. 86.

[17] J. Ehrman, *The Younger Pitt*, 2 vols. (London, 1969–83); see for instance ii. 19–20 for Pitt's thinking at the time of the Ochakov affair; B. Woodward and C. Bernstein, *Final Days* (London, 1976).

[18] De Jong, *Narrative in Drama*, 25, citing Eur. *Phoen.* 1187ff. For Homeric characters knowing things they have not heard see O. Taplin, *Homeric Soundings: The Shaping of the Iliad* (Oxford, 1992), 70 n. 36, 150, 223, and B. Hainsworth, *The Iliad: A Commentary*, iii: *Books 9–12* (Cambridge, 1993), 102, cp. 114; R. Janko, *The Iliad: A Commentary*, iv: *Books 13–16* (Cambridge, 1992), 404. Modern students of ancient historiography are sometimes uncomfortable when their author claims implausible access, see e.g. G. M. Paul, *A Historical Commentary on Sallust's Bellum Jugurthinum* (Liverpool, 1984), 5: Sall.'s claims to privileged access amount to a 'novelistic freedom' which is a 'disservice to sober history'. This is very strong, given that as we have seen (nn. 16 and 17) so many ancient and modern historians do the same.

question (to which I shall return) whether the number of hoplites who died from the plague was indeed 4,400.[19] On other occasions we shall be concerned with, for instance the causes of the Peloponnesian War or the reasons for the Sicilian Expedition, we would not I think (despite Thucydides' use both times of the word ἀληθεστάτη) want to say, at least this side of paradise, that Thucydides' accounts of the causes of the war or the expedition were true or false. We would say that they were convincing or unconvincing. In this second kind of case Thucydides' narrative techniques may add to the convincingness of his case.

To repeat, I shall not try to deal with techniques which have already been well studied. For instance Connor[20] in an excellent short article in 1985 called 'Narrative Discourse in Thucydides', looked at the alternation between abstraction and vividness. He also as we shall see later, discussed multiple perspective as part of the secret of Thucydides' often-remarked authority, though he did this without using the language of focalization. But he did not really explain *why* multiple perspective confers authority. Again, it was noted by Kitto that Thucydides can achieve results by pacing his narrative differently at different points—e.g. 1. 106, the very detailed death of the Corinthians by stoning, gains much of its effect by being placed as it is inside the most telegraphically brief of all Thucydides' digressions, viz. the *Pentekontaetia*.[21] This pacing is what narratologists like Mieke Bal call rhythm.[22] Or there are techniques of closure, e.g. 'so perished Plataia, in the ninety-third year of its alliance with Athens'. Here the date is more than a date, it has pathetic effect. Compare Odysseus' dog Argos dying 'in the twentieth year'. Or Thucydides may end an episode by trailing a coat, compare the end of the *Kerkuraika*, casually mentioning the return of the prisoners who will introduce the main stasis section two books later.[23] Openings in Thucydides are equally worth study: the opening of the whole work can be seen as an act of simultaneous linguistic homage to and rebellion against

---

[19] 3. 87 (hoplite losses in plague).

[20] W. R. Connor, 'Narrative Discourse in Thucydides', in M. H. Jameson (ed.), *The Greek Historians, Literature and History: Papers Presented to A. E. Raubitschek* (Stanford, Calif., 1985), 1–17.

[21] 1. 106 with Kitto, *Poiesis*, 271.

[22] Bal, *Narratology*, 68–76.

[23] Closure: 3. 68 (Plataia), cp. *Odyssey* 17. 327 (Odysseus' dog Argos), with Russo, in J. Russo, M. Fernandez-Galiano, and A. Heubeck, *A Commentary on Homer's Odyssey*, iii: *Books XVII–XXIV* (Oxford, 1992); Th. 1. 55 (Corcyra). For 'endings' in Hdt. see D. Lateiner, *The Historical Method of Herodotus* (*Phoenix* suppl. 23; Toronto–Buffalo–London, 1989), 44–50. See generally D. Levene, 'Sallust's *Jugurtha*: An "historical fragment"', *JRS* 82 (1992), 53–70 at 53 and n. 3; D. Roberts, *OCD*[3] s.v. 'Closure'.

Herodotus, a relationship comparable to that between Virgil's *Aeneid* and Homer's *Iliad*, as expressed in the first lines of the two epics. Thucydides' second main opening, at 1. 24, is itself Homeric in manner.[24]

All these techniques are there and are brilliantly used by Thucydides. But some of them have *been* studied, and anyway they are emotional effects; and even Connor does not quite satisfy my purpose which is to explore the way in which—to return to Wayne Booth—Thucydides makes us believe that his actually subjective reports are objectively accurate and (I would add) his interpretations convincing-looking.

So the first topic I wish to discuss is what I shall call *narrative displacement*. This resembles what Bal and Genette call anachrony (or even achrony), i.e. chronological deviation.[25] It is the technique by which an item in Thucydides loses or occasionally gains[26] (but much more often loses) its impact by being placed at a point other than we'd expect it. I suggest that a difference between historical writing and other, i.e. fictional, kinds is that the fiction writer or poet is usually concerned to *gain* impact by such displacement; it is the historian who may need to *lose* uncomfortable facts by putting them in the wrong file or box. The closest fictional equivalent to this deliberate losing and playing down of an item is the detective writer who leaves a clue in an unexpected place so as to reduce its impact. (What the film-maker Alfred Hitchcock called the McGuffin.) But that is different because the essence of, say, a Sherlock Holmes story, to use a favourite de Jong example, is that the importance of the stray item *does* eventually get revealed.

It is reasonable to speak, where Thucydides is concerned, of a thing being put 'other than where we'd expect it' because Thucydides is by and large a linear or serial sort of writer. It would not be nearly so reasonable

[24] For 'openings' generally see A. D. Nuttall, *Openings: Narrative Beginnings from the Epic to the Novel* (Oxford, 1992); also *YCS* 29 (1992), a special number on 'Beginnings in Classical Literature'; Lateiner (above, n. 23), 35–43, for 'beginnings' in Hdt. On the opening of the whole of Th.'s history (1. 1. 1) and on 1. 24, see the nn. in my *Commentary on Thucydides*, i: *Books I–III* (Oxford, 1991); on 1. 1. 1 and on the similarities to and differences from Hdt. see my particular nn. on Θουκυδίδης Ἀθηναῖος, on ξυνέγραψε, on τὸν πόλεμον . . ., and on ὡς ἐπολέμησαν. . . . For the relation of Virgil's opening to Homer's see Nuttall, p. 3.

[25] For achrony and anachrony see Bal, *Narratology*, 53, 66–8; Genette, *Narrative Discourse, passim*.

[26] For simple Thucydidean examples of *gain* in emphasis, achieved by displacement, see the two passages which anticipate the eventual fall of Athens, 2. 65. 12 and, even more emphatic, 5. 26. 1. Note also 6. 15. 4. There is also an interesting forward reference at Th. 4. 81. 2 to the main Sicilian expedition. A question from Lesley Brown, after the original delivery of this paper, forced me to think of examples of emphasis by displacement. My paper had concentrated on 'de-emphasis', in an ugly modern word. Cp. also 2. 31. 2 (plague).

to complain about such dislocation in Herodotus, because *his* narrative is already structured in a much more complicated and richer way; see John Gould's brilliant recent book[27] for the complicated reciprocity network. That is, Herodotus is more like Proust, whose anachronies were studied by Genette.[28] Another way of making the point about linear and non-linear writers is to say that in this respect as in others, the *Odyssey* is like Herodotus and the more serial *Iliad* is like Thucydides. But Bal actually takes the first 20 lines or so of the *Iliad* to illustrate anachrony, of a rather stylized sort: the order of themes is DCBA compared to the real-life ABCD.[29]

Let us return to narrative dislocation in Thucydides. I shall take a simple example from book 1 (1. 50).[30] The Athenians have decided, half

[27] J. Gould, *Herodotus* (London, 1989).

[28] This was the feature of Proust's novel to which Evelyn Waugh took such exception: see E. Waugh, *The Letters of Evelyn Waugh*, ed. M. Amory (London, 1980), 270 (letter of February? 1948 to John Betjeman): 'I am reading Proust for the first time. Very poor stuff. I think he was mentally defective. I remember how small I used to feel when people talked about him & I didn't dare admit I couldnt get through him. Well I can get through him now—in English of course—because I can read anything that isn't about politics. Well the chap was plain barmy. He never tells you the age of the hero and on one page he is being taken to the WC in the Champs Elysées by his nurse and the next page he is going to a brothel. Such a lot of nonsense . . .'. And again at pp. 273–4 (letter of a few days later, 16 March [1948] to Nancy Mitford, also about Proust): 'I . . . am surprised to find him a mental defective. No one warned me of that. He has absolutely no sense of time. He cant remember anyone's age . . .' etc. One can imagine Waugh's reaction if he could have known that precisely this feature of Proust would one day be hailed as an ultra-sophisticated narrative device. Compare Gary Taylor, *Reinventing Shakespeare: A Cultural History from the Restoration to the Present* (London, 1990), 38–9, discussing the 'embarrassing incoherence of temporal sequence in *Othello*', and quoting the late 17th-cent. critic Thomas Rymer's complaint that Shakespeare cannot decide whether Act Three 'contains the compass of one day, of seven days, or of seven years, or of all together'. Taylor adds: 'modern criticism dignifies and neutralizes this impossibility by calling it a "double time-scheme" . . .'.

[29] For the *Iliad* as Thucydides and the *Odyssey* as Herodotus see J. Griffin, *The Odyssey* (Cambridge, 1987), 99, and my *Commentary* (above, n. 24), 524. Barthes (above, n. 2), 129 calls Herodotus 'zigzag history'. It has to be said, however, that book 1 of Thucydides also zigzags a good deal; this helps to soften or conceal its anachronies, cp. n. 38 below. The other 7 books are more linear, though there are sudden jumps forward like 2. 100 on Archelaus of Macedon (reigned 413–399), a remarkable prolepsis given that the narrative has reached only the year 429/8. On the opening of the *Iliad* see Bal, *Narratology*, 54–5. But even Bal's analysis (DCBA) is too simple because the narrative returns to D. For the principles which may govern such complexities of arrangement see E. Fraenkel, *Aeschylus: Agamemnon* (Oxford, 1950), 119f. on line 205.

[30] In view of the importance of this example for my argument, I give the relevant Thucydidean passages in full, with translations: Ἀθηναῖοι δὲ ἀκούσαντες ἀμφοτέρων, γενομένης καὶ δὶς ἐκκλησίας, τῇ μὲν προτέρᾳ οὐχ ἧσσον τῶν Κορινθίων ἀπεδέξαντο τοὺς λόγους, ἐν δὲ τῇ ὑστεραίᾳ μετέγνωσαν Κερκυραίοις ξυμμαχίαν μὲν μὴ ποιήσασθαι ὥστε τοὺς αὐτοὺς ἐχθροὺς καὶ φίλους νομίζειν (εἰ γὰρ ἐπὶ Κόρινθον ἐκέλευον σφίσιν οἱ Κερκυραῖοι

a dozen chapters earlier, to send just ten ships to help the Corcyraeans (1. 44, cp. 45). That decision was taken at Assembly meeting no. 2. At meeting no. 1 they had actually inclined to favour the Corinthians. But the pro-Corcyraean decision at no. 2 was the last decision we were told about, and we were also told that this modest commitment of forces was the result of a very punctilious desire not to break the Thirty Years Peace. So far so good. But in the ensuing battle narrative, we are suddenly confronted with a fresh Athenian squadron of twenty ships looming up over the horizon, which (Thucydides says) the Athenians had sent out in addition to the first ten, fearing that the ten would not be sufficient. Let us call this decision no. 3. This is very interesting, though as a piece of narrative construction it has not attracted attention from commentators: there is nothing whatsoever in Gomme. One of the interesting things is the implication that there had been another debate in the Assembly, a debate totally unrecorded by Thucydides, at which decision no. 3 was taken. It is inconceivable that some executive authority like the *boule* or the *strategoi* (or Pericles alone, as Plutarch apparently thought) audaciously took the decision without the Assembly's authorization.[31] Historically,

---

ξυμπλεῖν, ἐλύοντ᾽ ἂν αὐτοῖς αἱ πρὸς Πελοποννησίους σπονδαί), ἐπιμαχίαν δ᾽ ἐποιήσαντο τῇ ἀλλήλων βοηθεῖν, ἐάν τις ἐπὶ Κέρκυραν ἴῃ ἢ Ἀθήνας ἢ τοὺς τούτων ξυμμάχους (1. 44. 1). 'The Athenians heard both sides, and they held two assemblies; in the first of them they were no less influenced by the arguments of the Corinthians, but in the second they changed their minds and inclined towards the Corcyraeans. They would not go so far as to make a full offensive and defensive alliance with them; for then, if the Corcyraeans had required them to join in an expedition against Corinth, the treaty with the Peloponnesians would have been broken. But they concluded a purely defensive alliance by which the two states promised to help each other if an attack were made on the territory or the allies of either.'

προεῖπον δὲ αὐτοῖς μὴ ναυμαχεῖν Κορινθίοις, ἢν μὴ ἐπὶ Κέρκυραν πλέωσι καὶ μέλλωσιν ἀποβαίνειν ἢ ἐς τῶν ἐκείνων τι χωρίων· οὕτω δὲ κωλύειν κατὰ δύναμιν. προεῖπον δὲ ταῦτα τοῦ μὴ λύειν ἕνεκα τὰς σπονδάς (1. 45. 3). 'The commanders received orders not to fight the Corinthians unless they sailed against Corcyra or to any place belonging to the Corcyraeans, and tried to land there, in which case they were to resist them to the best of their ability. These orders were intended to prevent a breach of the treaty.'

ἤδη δὲ ἦν ὀψὲ καὶ ἐπεπαιάνιστο αὐτοῖς ὡς ἐς ἐπίπλουν, καὶ οἱ Κορίνθιοι ἐξαπίνης πρύμναν ἐκρούοντο κατιδόντες εἴκοσι ναῦς Ἀθηναίων προσπλεούσας, ἃς ὕστερον τῶν δέκα βοηθοὺς ἐξέπεμψαν οἱ Ἀθηναῖοι, δείσαντες, ὅπερ ἐγένετο, μὴ νικηθῶσιν οἱ Κερκυραῖοι καὶ αἱ σφέτεραι δέκα νῆες ὀλίγαι ἀμύνειν ὦσιν (1. 50. 5). 'It was now late in the day and the paian had already been sounded for the attack, when the Corinthians suddenly began to back water. They had seen sailing towards them twenty ships which the Athenians had sent to reinforce the previous ten, fearing what actually happened, that the Corcyraeans would be defeated, and that the original squadron would be insufficient to protect them.' (Translations by Jowett, adapted.)

[31] ML 61, 19ff. (official accounts of the second squadron) rules out surreptitious explanations. I add a word about the slightly different accounts of (*a*) Plutarch (*Per.* 29. 1 and 3), who personalizes things (after a general admission that 'the people voted help') by saying 'Pericles sent' too small a force at first, out of malice towards Lacedaemonius, and

I find this alternative totally unacceptable. Either way there are things we are not told.[32]

To some extent, it must first be said, this is an attempt to get round the difficulty of what Don Fowler in his recent article on Ekphrasis[33] has called linearization, where an author has to decide what order to present details in. Film-makers talk of the need to discard everything but the *spine* of a story when filming it. Some apparent anachronies in Thucydides are perhaps attempts to solve the linearization problem, like the important scene-switch signalled by 'interea Manlius' in Sallust's *Catiline*; though perhaps the better analogy, for the understanding of the Thucydides passage, is with the archaic use of *delay* as an effective narrative device.[34] At 1. 50 Thucydides was faced with a problem of presentation. Having got the Athenians to Corcyra he was reluctant to go back to Athens to describe the assembly meeting at which decision no. 3

---

then that he (third person singular again) stiffened this force later after criticism; and (*b*) Diodorus (12. 33. 2), who smooths over the awkwardness in his basically Thucydides-derived account by saying that the Athenians 'sent ten triremes *and promised to send more later if necessary*', thus facilitating his mention of the subsequent twenty ships a few lines later at para. 4. Even if these accounts preserve true and/or independent traditions (rather than merely being reworkings of Thucydides) they are irrelevant to the question of the presentation adopted by Thucydides, the earliest extant account.

[32] In a helpful letter he sent me after the original delivery of this paper, David Lewis observed, 'as far as 1. 50. 5 is concerned, there is some preparation in 1. 44. 2, which asserts that attitudes to the Corcyra situation have moved to a new phase and doesn't encourage the belief that they are now trying to avoid war'. There is some force in this; though 'asserts . . . that attitudes have moved to a new phase' is a little too strong for the simple account—nothing about new thinking—in 44. 2 of Athenian worries and motives, viz. (i) they thought war was inevitable, (ii) they did not want Corcyra's navy to fall into Corinth's hands, (iii) they wanted to embroil them more and more with each other, and (iv) they reasoned that Corcyra was conveniently situated. In any case, note that (*a*) 44. 2 *precedes* 45. 1 with its emphatic statement (see n. 30) about Athens' wish to avoid a breach of the Thirty Years Peace, and (*b*) Lewis's point does not remove the main oddities I am concerned with, viz. the suppression of the third debate, which surely occurred, and the postponement, until a very different sort of narrative context, of any statement of the result of that debate. Harriet Flower, 'Thucydides and the Pylos Debate (4. 27–29)', *Historia* 41 (1992), 40–57, now argues that Th. suppressed completely another Assembly debate, in 425. Cp. also, for 415, ML 78, comm.

[33] D. Fowler, 'Narrate and Describe: The Problem of Ekphrasis', *JRS* 81 (1991), 25–35.

[34] On Sall. *Cat.* 28. 4 see R. Syme, *Sallust* (Cambridge, 1964), 79 f. For archaic narrative use of delay see above all Fraenkel, *Agamemnon*, app. A at 805, discussing Hdt. 1. 110–12: a significant detail is deliberately held back until the point in the story when it is most necessary and important. (This is not just an 'archaic' poetic device. In Tennyson's *Morte d'Arthur*, the narrative of Sir Bedivere's third visit to the lake, when he actually does throw the sword Excalibur in, suppresses the fact that he turned his eyes away when throwing. The fact is disclosed only later in Sir Bedivere's report to King Arthur. This is more effective, if only because otherwise the poet might have had to repeat the item.) In technical language (see below) delay is a sort of analepsis. For anticipation (prolepsis) see below n. 38.

was taken. I now give two innocuous examples of events narrated, for reasons of linearization, out of natural sequence. (i) From book 3 we learn under the year 427 BC that Itamenes and his barbarians had occupied Notium in 430 BC, the time of the second Peloponnesian invasion of Attica. This event really belonged in book 2. (ii) In book 4 we are told in the middle of the main Megarian narrative that Brasidas happened to be in the region preparing a hitherto unheard-of expedition to Thrace: a 'flashback'.[35]

By contrast, the effect of the dislocation at 1. 50 is profound. Thucydides here reveals only incidentally, and in a non-political context, that the Athenians had in fact trebled their commitment to Corcyra. He thus leaves artfully undisturbed the impression, clearly stated at 1. 44–5, that the behaviour of the Athenians had been scrupulous throughout, i.e. they had been anxious not to break the Thirty Years Peace. I am not concerned now with the question, *did* the Athenians break the peace, but with the belligerency of their psychology: were their actions likely to lead to a breach of the peace? This, we may reasonably say, is manipulation of narrative to suit a political thesis. The thesis is that argued for by E. Badian in 1990:[36] Thucydides systematically understated Athenian aggressiveness in the run-up to the war. Once we start looking for other examples we soon find them. There are items in books 2, 3, and 4[37] which ought, in the narrative sense of ought, to have been mentioned in book 1 (I mean, that they belonged there chronologically), and would

[35] Innocuous dislocation: 3. 34. 1 (Notium); 4. 70 (Brasidas). This second example is surprise technique with no obvious other motive for suppressing a previous decision. The explanation is delayed until ch. 79. With 1. 50 compare 4. 96. 8, arrival of the Lokrians (trivial but comparable).

[36] E. Badian, 'Thucydides and the Outbreak of the Peloponnesian War: A Historian's Brief', in *Plataea to Potidaea*, 125–62, esp. 141: Th. 'had to give up any full treatment of that notable incident [the Spartan decision about Samos] and decided to bring it in obliquely, from the perspective of the Corinthians.' Cp. 141–2: 'One of the most important incidents in the Pentecontaetia, which ought to have been central to the account of it on any reasonable assessment, had to be totally banished from it.'

[37] Malign anachronies: 2. 68 (Akarnania); 3. 2. 1 (Mytilene appeal); 4. 102 (foundation of Amphipolis); 1. 40. 5 (Samos). Another way of making my point would be to say that book 1 argues that the war was inevitable, whereas Sparta's behaviour over Mytilene showed that this was not quite true. So the rebuff to Mytilene is kept out of book 1 and Th. conceals the vulnerability of his book 1 thesis. I accept that *one* of Th.'s reasons for displacing his material from one context to another is to give the material greater explanatory power: Th. has an economical mind and uses things where they will do the most work. This does not I think dispose of 1. 44 and 50, but it may go some way to account for 1. 40. 5 (the Samos debate), and perhaps also (as Pelling suggests to me) for the fact, if it is a fact, that Th. gives a fuller exploration of the character of the Athenian empire in the Mytilenean context of book 3 than in book 1. But structural considerations are relevant to the latter example: book 1 was already very heavily freighted.

have had a different impact there. In fact they would have tended to put the pre-war behaviour of the Athenians in a more aggressive light, or the Spartans' behaviour in a more favourable one. The items are, the Akarnanian alliance left timeless or achronic in book 2; the refusal of the Spartans, at the beginning of book 3, to respond positively to the Mytilenean appeal which we are specifically told had been made well before the war broke out; and the foundation of Amphipolis referred back to in book 4, an event which from other evidence we can date to 437. But perhaps the most spectacular dislocation is the extraordinary treatment of a crucial incident in the *Pentekontaetia*. The incident is the Spartan decision in 440 to try to go to war with Athens over Samos, an action from which they were prevented by the Corinthians. We learn about this not from the *Pentekontaetia* narrative, where we will be thoroughly alert to such flashpoints in the history of the Athenian empire, but from a Corinthian speech much earlier in the book where the topics being discussed are different. The Corinthian reference is, in technical language, an external analepsis. That is, it looks back to an event not inside but outside Thucydides' own narrative. Compare the story of Odysseus' scar in *Odyssey* 19, or, from a speech, the Meleager story told by Phoenix in *Iliad* 9.[38] Almost all other analepses in Thucydides' speeches are either *internal* or they draw on Herodotus. I have discussed this last topic elsewhere.[39]

Badian, with no particular interest in narratology, has recently shown (n. 36) how this distortion actually helps Thucydides' general picture of Spartan aggressiveness. He has argued that (i) there was a general autonomy clause in the Thirty Years Peace, (ii) that it was in virtue of this clause that the Spartans were tempted to act, and (iii) that Thucydides by dislocating the incident was able to report it less than fully and thus to suppress the autonomy aspect, which would have done the Spartans credit.

---

[38] 'External analepsis': Genette, *Narrative Discourse*, 49, and de Jong, *Narrators and Focalizers*,85, cp. *Iliad* 9. 527 (Meleager)=external analepsis in speech, and *Odyssey* 19. 363 (Odysseus' scar)=external analepsis in narrative. Most of the Thucydidean dislocations I discuss in the present section are analepses (references back), but for examples of prolepsis (anticipation) see n. 26 above; note that 1. 40 is in one sense an anticipation (events of 440 'really' belong in the *Pentekontaetia* which at 1. 40 is still to come, see n. 29 above for book 1 as a zigzag) and in another more straightforward sense, the sense I have adopted in my text, a reference back or analepsis: in 431, 440 lay in the past. For prolepsis in another historian see *Ath. Pol.* 13: the post-Pisistratid *diapsephismos* or review of the citizen body is mentioned well in advance of its natural context.

[39] See my paper 'Thucydides' Use of Herodotus', in J. Sanders (ed.), *ΦΙΛΟΛΑΚΩΝ: Lakonian Studies in Honour of Hector Catling* (Athens, 1992), 141–54.

It may be objected that if Thucydides did not want to highlight an incident, he had the option of simply not mentioning it at all, whether in narrative or speech. There is no simple answer to this; it is not enough to insist that Thucydides was not a novelist and so could not take such liberties. That will not do, because he fails to mention other events which we think or know did happen, such as the peace of Kallias. My analysis takes Thucydides' treatment as we have it and seeks to identify and explain its peculiarities. In this regard I would insist once again (see n. 39) on the singularity of the Samos item. It is unique in being the only important past event which we are told about only in a Thucydidean speech.

A more radical objection would urge that Thucydides' displacement of the Spartan decision over Samos in 440 has the effect not of de-emphasizing but of emphasizing or highlighting the decision (cp. n. 26 for other examples of this). I would agree that the Corinthians' point has great rhetorical force where it is, and that to that extent the treatment is emphatic. But I would still maintain that a positioning later in the book, specifically somewhere in or at the end of the *Pentekontaetia*, would have highlighted the *legal* issue much more clearly, and drawn attention to a possible line of Spartan justification (cp. Badian in n. 36). In any case we should not forget the possibility that by putting the episode where he does, Thucydides is able to suppress the autonomy aspect altogether, an aspect which favoured Sparta.

Thucydides achieves an effect similar to that achieved by dislocation, through the use of what Bal calls iterative presentation. For instance, it is not until book 4 that we are told that the Athenian invasions of the Megarid took place *twice* yearly; the invasions themselves were mentioned as an *annual* event in book 2.[40] In book 2 we are in effect being told to bear the invasions in mind in the narrative which follows. Fair enough (there are other examples of this shorthand),[41] but why did Thucydides not use the same economical device for the Peloponnesian

---

[40] Iterative presentation: M. Bal, *Narratology*, 78; and see N. J. Richardson, *The Iliad: A Commentary*, vi: *Books 21–24*, 358 on 24. 768–72. For Megara in Th. see 4. 66, contrast 2. 31. It may *also* be true that Th.'s reason for delaying the information that the invasions were twice yearly is the desire to place it where it will make most impact, cp. n. 34 above. But we should also compare Homer's 'technique of increasing precision', for which see O. Taplin, *Homeric Soundings*, 198, and B. Hainsworth, *The Iliad: A Commentary*, iii (above, n. 18), 144. In Th., 5. 43 is more precise than 6. 89 (Alcibiades' ancestor who renounced his proxeny is a mere ancestor in the second passage but grandfather in the first). There are however special reasons for this, see my paper at n. 39 above: speeches (and 6. 89 is from one) are often less precise than narrative. See further n. 43 below.

[41] See e.g. 2. 24. 1 with my *Commentary* (above, n. 24); 2. 34. 7.

invasions of Attica? In any case, why keep back until book 4 the very material point that the invasions were not just once but twice yearly? From the literary point of view this is perhaps an example of what (with reference to Homer, see n. 40) has been called a 'technique of increasing precision'. Historically, the effect is to reduce the impact of the Athenian invasions of the Megarid, and so perhaps to carry through the distortion already effected by the notoriously low profile he accords to the Megarian decrees. David Lewis has recently suggested that Thucydides 'was not all that interested in Megara and may not be a reliable guide'.[42] We can perhaps particularize a little further and say that Thucydides had an intermittent blind spot not just about the scale but perhaps also (since two invasions a year suggest greater commitment than one) the aggressiveness of Athenian designs on Megara. Or is the truth rather that Thucydides knew perfectly well, but wanted *us* not to know? (Or not to know all at once: it is after all Thucydides himself who eventually tells us that the invasions were twice yearly.) This question does not arise in quite the same way when we deal with narrative organization in a poem or a novel.

Another well-known instance of what we may call the Megara phenomenon is the postponement till book 6 of a candid authorial acknowledgement of the huge scale of the Athenian attack on Epidaurus, adverted to in book 2 but without comment. Again, Thucydides has masked Athenian aggression (or rather Athenian failure to stick to the Periclean defensive strategy) by a narrative device, although again we may wish *at the same time* to speak of 'increasing precision'. (Sometimes Thucydidean authorial judgements are attached, as here, not to the relevant slab of narrative but to some later incident. A striking case is the opinion that it was a mistake for Nikias to winter at Katana. The fact is given baldly and briefly in book 6, the judgement half way through 7!) Similarly, the statement at the beginning of book 3 that the Athenians made cavalry forays 'as usual' is a way of playing down an Athenian tactic which meant that Athenian abandonment of Attica was less complete than Pericles had urged at the end of book 1.[43] Defending Attica can

---

[42] Lewis, *CAH* v². 388. But in the discussion after the original delivery of my paper, Lewis wondered whether Thucydides' iterative handling of e.g. the 'Megara phenomenon' might have been his way of signalling that an initiative or a policy did not come to anything. This does not seem to me satisfactory, because of such striking counter-examples as Sitalkes' invasion (see my *Commentary* (above, n. 24), introductory n. on 2. 95–101), not to mention the inconclusive but exhaustively documented Peloponnesian diplomacy in book 5.

[43] 6. 31, contrast 2. 56 (Epidaurus); 6. 72 and 88 with 7. 42 (see n. 8 above); 3. 1. 2

hardly be called 'aggressive' behaviour; but the general effect of Thucydides' presentation is to make Athens seem more pacific and quietist than was really true.

Here I should like to digress briefly and ask whether Badian's view of Thucydides as systematically malicious and mendacious is plausible. In another paper, Badian represents the fifth-century Athenians as cynical treaty-breakers in their dealings with Macedon; this suggestion developed, in a different theatre of diplomacy, the thesis of his earlier and more general 1990 paper on the origins of the Peloponnesian War. In the discussion which followed Badian's second paper at its delivery in Oxford, David Lewis replied by quoting Nikias' implied complaint[44] that the Athenians were incapable of saying 'No' to anybody who asked for an alliance, and were constantly landing themselves with undesirable commitments as a result; Lewis suggested that the Athenians were often only vaguely aware of what their existing treaty commitments were. We might say that the Badian view is a conspiracy theory of Athenian foreign policy, and the Lewis view is a cock-up theory. I should like to suggest a compromise. We can certainly admit that many voters in the Assembly would have hazy or non-existent notions of the up-to-date diplomatic position when they came to vote on a particular treaty. To this extent David Lewis is surely right. But we have to reckon with the existence of

(cavalry) with I. Spence, 'Perikles and the Defence of Attika during the Peloponnesian War', *JHS* 110 (1990), 91–109. At 1. 139. 1 the 'iterative' verb φοιτῶντες masks Spartan readiness to negotiate, for which see Badian, *Plataea to Potidaea*, 157. (*Contrast* the same word at 4. 41. 4, combined with explicit criticism of Athens' greed. But by then Pericles was dead.) I suppose one could say that 6. 31 and 7. 42 are examples of Thucydides delaying something until it is most relevant. The biggest example of this is the suppression of much Sicilian material until the beginning of book 6 (the beginning of the second 'pentad' of Thucydides' work, on the theory discussed above, p. 16. Compare the postponement of a formal introduction of Sejanus till 4. 1, the beginning of the second half of the first 'hexad' of Tacitus' *Annals*.) Some Sicilian (or rather S. Italian) items are delayed very long indeed, e.g. 7. 33. 4–5, the Athenian alliances with Artas of Messapia and the Metapontines. These may date from as early as the 430s. Perhaps Th. is merely keeping them back until they become most relevant; or perhaps he wished to increase the sense that the whole expedition was a mad shot in the dark. The more he revealed early in books 1 or 6 about antecedent Athenian diplomatic relations with Sicily and S. Italy, the more sensible the 415 expedition would appear. But this would take away from the tragic effect. I postpone to my commentary on 6. 6. 2 the implications for Thucydides' narrative technique—and for his veracity—raised by the recent claim that ML 37, Athens' alliance with Egesta, dates not from 458/7 but 418/17, in which case we need to explain why Th. does not mention it: M. H. Chambers *et al.*, 'Athens' Alliance with Egesta in the Year of Antiphon', *ZPE* 83 (1990), 38–63.

[44] Th. 6. 13. 2. Another possibility is that the Athenians were like Bismarck, and liked to 'scatter promises so that they would not have to keep them': A. J. P. Taylor, *The Struggle for Mastery in Europe 1848–1918* (Oxford, 1954), 278. For Badian's 'Thucydides and the *Arche* of Philip' in his *Plataea to Potidaea*, 171–85, see above, Preface.

regional experts, like Diotimos who was a western expert (Euphemos at Th. 6. 75 may be another, if ML 37 dates to 418). For the north, the Assembly and Council would look to men like Hagnon or Thucydides himself, with his Thracian influence. If Athens *did* enter into inconsistent treaty arrangements, men like Diotimos, Hagnon, or Thucydides the *strategos* can reasonably be held to blame. So to that extent Badian is right. It is another matter whether Thucydides the historian was as manipulative as Badian thinks, but there are certainly some serious oddities as I have tried to suggest. And it is sinister that so many of the narrative tricks have the effect of diminishing Athenian duplicity or aggressiveness. I want to stress this. If the explanation was purely literary (i.e. to do with linearization), we would expect a more even distribution in terms of political implication, some oddities tending to favour Athens, some not. But, with one apparent exception, that does not seem to be so.[45]

Agreed, not all suppression by means of iteration signifies politically; for instance there is Nikias' reference to his own 'other letters' in book 7, and there is the narrative reference to other Spartan embassies to Persia in book 4.[46] These are, we may say, innocuous instances of iteration.

I now pass on to try to discuss the light thrown by narratology on the supposedly godlike objectivity or pseudo-objectivity of Thucydides. For de Jong in her book on Homer, the main target is the school of Homeric thought which holds that Homer is an objective narrator. (De Jong herself actually thinks Homer neither subjective nor objective but multiple.) We could profitably apply to Thucydides some of de Jong's detailed techniques for unmasking the complexities of epic narrative, and in particular for demonstrating that there is a concealed subjective personality behind

---

[45] For Diotimos see 1. 45 with *FGrHist* 566 F 98 and on ML 37 see Chambers (above, n. 43). For Hagnon see 2. 67. See now Badian, *Plataea to Potidaea*, 242 n. 18. The apparent exception to the rule that the dislocations tend to favour Athens is 7. 18. 2. The Spartans, says Th. under 414/13, came to think that they had been at fault in 431 (σφέτερον τὸ παρανόμημα μᾶλλον) because the Thebans had attacked Plataea in time of peace, *and they themselves had refused arbitration when offered*. His point is that the Spartan retrospective attitude to 431 contrasted with their attitude to the current situation (414) when they considered that *Athens* was the peace-breaker. But this is not a dislocation; it is a report of a new fact about the way the psychological situation had changed by 414. As Badian, *Plataea to Potidaea*, 143, puts it, 'we are meant to see the Spartans as developing a conscience only when things begin to go wrong'. There is absolutely no doubt that Th. here has in mind not Spartan attitudes as they had been in 431 (that would indeed make 7. 18 to that extent an anachrony or delayed report) but as they developed many years later than 431. This is proved by his mention, at the end of 7. 18. 2, of Pylos (425) and other disasters. I labour this point but it is sometimes misunderstood. I would only add, by way of comment on or qualification to Badian, that 5. 32. 1 shows that Th. is impartial in that his Athenians have similarly intermittent consciences.

[46] Innocuous iteration: 7. 11; 4. 50. (Not malign: already prepared for at 1. 82. 1.)

apparently objective statements. There is, arguably, nothing new here except the word narratology. Longinus in *On the Sublime* had already noticed, ch. 26, that Homer 'gives a sense of urgency with the line "you would say that they were tireless . . ." '.[47] My problem as a historian is to ask whether it makes any difference that Thucydides is a historian not an epic poet or (to use a favourite example of the narratologists) Nick Carraway in Scott Fitzgerald's *The Great Gatsby*. There is no actual second person singular in Thucydides—I mean outside speeches—but there is some *implied* second person. I had to look up 1. 10, where he says in effect you would make wrong inferences about Athens and Sparta from their physical remains, before I could be quite sure that there is no second person there. Thucydides uses roundabout expressions like 'if somebody were to look', or 'I think there would be much disbelief'. And we can add the two (there are only two) authorial rhetorical questions in Thucydides. Contrast Herodotus' relatively free use of the second person: 'if *you* look into the matter, you will find that all Persian names without exception end in "S" . . .', and 'you will not be able to sleep with a Babylonian temple-prostitute after she has once gone with a man, however much you pay her'. And there are other passages where Herodotus comes very close to addressing his readers or hearers direct, although the second person remains implicit: 'I shall now say something which will come as a great surprise to those of the Greeks who didn't accept what I said [in book 3] about a Persian advocating democracy.' Here 'those of the Greeks' is close to 'those of you Greeks', but the person is third. So too with the final sentence of Xenophon's *Hellenika*: 'that's as far as I go, somebody else can worry about the sequel'.[48]

Let us look at some particular devices. First, the 'self-conscious

---

[47] Longinus, *On the Sublime* 26, citing *Iliad* 15. 697. For other such second-person verbs in Homer (always singular, and always negative in form) see *Iliad* 4. 85 (γνοίης); 4. 429, 6. 697, and 17. 366 (φαίης); 4. 223 (ἴδοις). Cp. also 17. 397f.; 4. 539; 13. 127.

[48] Second person in Hdt.: see Lateiner, *Historical Method*, 30–1 (an excellent discussion), on e.g. 1. 139, 199. 4; and add 2. 30. 1 and φέρε ('come, now, well': LSJ⁹) at 2. 14. 1, 105. 1. Christopher Pelling interestingly suggests to me that such personal Herodotean interventions are more common in explicitly 'ethnographic' contexts. For rhetorical questions (to the reader or hearer) in Hdt. see Lateiner, p. 64, noting that most of them are in the Egyptian book 2; this fits Pelling's observation. See also the valuable remarks of Mabel Lang, *Herodotean Narrative and Discourse* (Cambridge, Mass. and London, 1984), 39–41. Implied second person in Hdt.: 6. 43. Implied second person singular in Th.: 1. 10. 2, 21. 2, 22. 4 (where even the famous κτῆμα ἐς αἰεί, 'possession for ever', perhaps suggests 'you' will possess it for ever). 'Cross-referencing' falls into this category: on 5. 1 see above, Ch. 1 n. 30. Rhetorical questions in Th.: 7. 44. 1; 8. 96. 2. End of Xenophon's *Hellenika*: 7. 5. 27. For Tacitus' (rare) addresses to his audience see D. C. A. Shotter, *Tacitus: Annals IV* (Bristol, 1989), on 4. 11. 5.

narrator'.[49] One way in which a narrator can inspire belief in categoric-
ally uttered proposition *p* is by at the same time expressing diffidence
about proposition *q*. So in the *Odyssey*, Odysseus, who is pretending to
Penelope to be a Cretan, says with artful diffidence, 'I don't know if
Odysseus had the brooch when you knew him (οἴκοθι)'; that is, 'my
clinching piece of evidence may (I pretend to think) cut no ice with you at
all'—a brilliant piece of bluff. Tom Stinton discussed something like this
a number of years ago in his article '*Si credere dignum est*', citing
Herodotus on Rhampsinitus, 'I personally don't believe the king put his
daughter in a brothel', a way of encouraging belief in the incredible
things which have preceded.[50] I have found an entertaining footnote
example in a modern historian: Braudel, *Mediterranean World*, 'I have
mislaid the precise reference'; this remark, which only a Braudel could
get away with, nicely contrasts with and encourages respect for the
massive documentation in the other thousands of footnotes.[51] In a way,
Homer uses this device in a famous first-person-singular pronouncement, when he says in the *Catalogue of Ships*, 'as for the rank and file
that came to Ilium, I could not name or even count them, not if I had ten
tongues' etc. 'But here are the captains of the fleet and here are the ships
from first to last. First the Boiotians, with Peneleos' etc.[52] Herodotus'
statement, that 'to list all the captains of the Persian side is not necessary
for the *logos* of my inquiry', performs something of the same function
when attached to the very circumstantial list which follows.

With all this compare Thucydides 3. 87: the number of cavalry who
died from the plague was 300, the number of hoplites was 4,400, but the
number of the other ranks could not be ascertained. Here Thucydides
does not invoke the Muses, but the effect of saying 'thetic losses could

[49] For the 'self-conscious narrator' see Booth, *Rhetoric of Fiction*, 205 with n. 28, and
ch. 8 generally; de Jong, *Narrators and Focalizers*, 46.

[50] T. C. W. Stinton, '*Si credere dignum est*: Some Expressions of Disbelief in Euripides
and Others', *PCPhS* 22 (1976), 60 ff. (= *Collected Papers on Greek Tragedy* (Oxford,
1990), 236–64), esp. 61 on Hdt. 2. 121. For Odysseus and the brooch see *Odyssey* 19. 237
with Rutherford (above, n. 3). Tacitus' reservations at *Annals* 4. 10–11 are comparable in a
general sort of way. Part of the humour of the penultimate scene (Act 5, sc. 2) of Sheridan's
School for Scandal derives from this special kind of circumstantiality. The (wholly made up)
story of the duel between Sir Peter Teazle and Charles Surface is narrated in absurd detail,
but note the pseudo-caution of e.g. 'Sir Peter forced Charles to take one [of the pair of
pistols] and they fired *it seems, pretty nearly* together' (my italics).

[51] F. Braudel, *The Mediterranean World in the Age of Philip II* (London, 1972), i. 171
n. 4.

[52] Self-conscious avoidance of numerical precision: Homer, *Iliad* 2. 488 ff. (But see G. S.
Kirk, *The Iliad: A Commentary*, i: *Books 1–4* (Cambridge, 1985), 167: the poet is also saying that he *can* recall the detail if the Muses help.)

not be ascertained' is surely to strengthen our disposition to believe that very circumstantial 4,400 for the hoplites. There is a parallel in Herodotus. He is talking about the massacre of the Tarentines by the Messapians. Their allies, the men of Rhegium, lost 3,000 men, but as for the Tarentines the number was too big to count, οὐκ ἐπῆν ἀριθμός. This expression not only reinforces the chillingness of this greatest of Greek massacres, but adds credence to the preceding 3,000. Returning to Thucydides, the word ἀνεξεύρετος is passive in form, it was not able to be found out about. But the effect is to take the reader into Thucydides' confidence, i.e. *you* could not find out and you could not expect *me* to do so either.[53] It is important incidentally to realize that it now seems agreed[54] that there was no hoplite *katalogos* or register, so the historical fact is that Thucydides' evidence for the figure 4,400 may not have been as good as Gomme, for example, supposed. His note on the passage[55] confined itself to the *katalogos* issue and he simply observes, in effect paraphrasing Thucydides, that there was no muster of the *thetes* etc.

A final example: the death of Lamachos comes as a shock, like the death of Petya Rostov in *War and Peace*. Its stark specificity is I suggest highlighted by the indeterminacy which follows, 'five or six others were killed with him'.[56]

In this connection (indeterminacy) I would emphasize just how rare in Thucydides are such statements of doubt or ignorance, or statements of alternative versions like the Theban–Plataian discrepancy I mentioned earlier, and we can add from book 1 the two-pronged statement that Themistocles died of illness or some say he took poison, λέγουσι δέ τινες καὶ ἑκούσιον φαρμάκῳ ἀποθανεῖν. The flavour in this section is anyway Herodotean. As for expressions of uncertainty or conjecture, phrases like δοκεῖ δέ μοι naturally proliferate in the *Archaeology*, and several are found in the unfinished book 8, e.g. a doubt about the whereabouts of the Phoenician fleet. Elsewhere they are few: a famous piece of diffidence in book 5 about the Spartan numbers at Mantinea and one in book 7

[53] Hdt. 7. 96; Th. 3. 87 (cp. 4. 101. 2); Hdt. 7. 170 (Tarentine massacre). Other artful Herodotean combinations of precision and hesitation are the famous 1. 1. 3 ('on the fifth or sixth day'—said about the mythical abduction of Io!), cp. 1. 30. 1; 3. 42. 1. See Lateiner, *Historical Method*, 32–3, partly drawing on unpublished further work by Rubincam (cp. below, n. 58).

[54] M. H. Hansen, *Demography and Democracy: The Number of Athenian Citizens in the Fourth Century BC* (Herning, 1986), 83–9 = app. V, 'The So-called Hoplite *Katalogos*'; A. Andrewes, 'The Hoplite *Katalogos*', in G. Shrimpton and D. McCargar (eds.), *Classical Contributions: Studies in Honor of M. F. McGregor* (Locust Valley, NY, 1981), 1–3.

[55] See A. W. Gomme, *HCT* ii. 388.

[56] Lamachos' death: 6. 101, cp. L. Tolstoy, *War and Peace* (Penguin edn.), 1252.

where he says that the night made the battle at Epipolae hard to be sure about.[57]

Before I leave this point, let me note another much used or abused numeral in Thucydides. In book 7 he says that 'more than 20,000' slaves deserted from Attica after the Spartans fortified Decelea, and they lost all their cattle. Where on earth did Thucydides get this figure (which is regularly cited in books on ancient slavery as one of the few hard quantitative bits of data that we have)? For my present purpose I note merely the rhetorical effect of saying not 'about 20,000 slaves' but the more precise-sounding 'more than'.[58]

There are other, less obvious, rhetorical devices for producing an emotionally and intellectually satisfying interaction between narrator and narratee. Let me turn to presentation through negation, as de Jong calls it. This is a way in which the poet, under the guise of making an objective statement of fact, actually engages in a sort of dialogue with the listener's expectations. De Jong's examples from the *Iliad* are 'Agamemnon did not stop fighting' (understand 'although you'd have expected him to because he was wounded'); or 'Patroclus did not take Achilles' spear'. Admittedly, some Homeric negatives are not much more than ways of singling out the one thing which *will* be talked about, such as the opening of *Iliad* 2: 'all the other gods were asleep, but not Zeus. . . .'

---

[57] Rarity of expressions of doubt or uncertainty in Th.: 1. 138 (death of Themistocles); 8. 87 (whereabouts of Phoenician fleet); 5. 68 (battle of Mantinea); 7. 44 (Epipolae; cp. Introduction, p. 67 above, for the possible Homeric echo here); cp. Woodman (above, n. 5), 16f. Add 6. 60; 7. 87. Note however the suggestion of Z. M. Packman, 'The Incredible and the Incredulous: The Vocabulary of Disbelief in Herodotus, Thucydides and Xenophon', *Hermes* 119 (1991), 399–414 at 410–11: sometimes Thucydides projects his own disbelief on to the reader or on to one of his characters. Cp. C. Pelling, *Plutarch: Life of Antony* (Cambridge, 1988), 40, on Plutarchan 'characterisation by reaction'. The most famous admission of difficulty is at 1. 22, the chapter on method, a rather special case. Th.'s use of οὐ τοσοῦτον, 'not so much', has some bearing on the question of alternative versions: see below. 4. 122. 6 is a rare and emphatic adjudication between claims.

[58] 7. 27. 5, taken very seriously by e.g. A. H. M. Jones, 'Slavery in the Ancient World', in M. I. Finley (ed.), *Slavery in Classical Antiquity* (Cambridge, 1960), 4, cp. W. Westermann, 'Athenaeus and the Slaves of Athens', in the same volume, 86–8; Y. Garlan, *Slavery in Ancient Greece* (Ithaca, NY, 1988), 66. On this passage, and on numerals in Th. generally, especially those subjected to qualification, see the valuable article by Catherine Reid Rubincam, 'Qualification of Numerals in Thucydides', *AJAH* 4 (1979), 77–95, esp. 85 for 7. 27. 5. On 7. 27. 5 in particular see the interesting Rubincam-influenced study by V. D. Hanson, 'Thucydides and the Desertion of Attic Slaves during the Decelean War', *Classical Antiquity* 11 (1992), 210–28, a good stab at reconciling the beliefs that (i) Th. operated with numerals in a rhetorical way, and (ii) that he did make some sort of genuine calculation. Similar problems are posed by 2. 70 and 7. 48: 2,000 talents begins to look like a conventional Thucydidean figure in siege contexts; but ML 55 = Fornara 113 shows that the order of magnitude is about right.

Nevertheless, de Jong has surely identified an important narrative device.[59]

Presentation through negation is certainly to be found in Thucydides, from the second chapter of the whole work (Greece in very early times 'not regularly settled', οὐ πάλαι βεβαίως οἰκουμένη: the implication is perhaps, 'not as you would expect if you merely read back present conditions into the past'). Further on in book 1, Thucydides describes the situation at the beginning of the First Peloponnesian War, when Athens was laying siege to Aigina. The Corinthians, as a diversionary tactic, sent an expeditionary force into the Megarid, thinking that the Athenians would be unable to deal with both situations and would have to withdraw their force from Aigina; note as usual the inferred motivation. But the Athenians, says Thucydides, did not move their army away from Aigina, but sent out their oldest and youngest to Megara under Myronides. Here the force of the words 'they did not move their army' is comparable to 'Agamemnon did not stop fighting'. The expression, however, is (it may be objected) perfectly natural after νομίζοντες etc., that is, after the statement that the Corinthians were hoping in effect that the Athenians would move their army. But as we saw, that statement of hope is itself merely a Thucydidean inference from the fact that the Corinthians had invaded the Megarid.

Another possible reply might go like this, a reply which draws on the difference between the subject-matter of poetry and the facts in which the historian deals. It can be objected that the statement about what the Athenians did *not* do differs from the statement about Agamemnon because Thucydides' report is a telescoped way of saying the following. There was a debate in the Athenian Assembly. Some people said, 'let's bring the army back from Aigina.' Others said, 'No, don't let's move it, let's leave it where it is and send out the oldest and youngest to Megara under Myronides.' This second view, as a matter of historical fact, prevailed. All Thucydides has done is abbreviate drastically (we are after all in the very skeletal *Pentekontaetia* narrative). Similarly, and more explicitly, book 8 opens with an account of Athenian gloom at the Sicilian Disaster, and Thucydides tells us 'they decided they must not give in, but instead . . .'. Here too it is possible that there really was a defeatist element which favoured an accommodation with Sparta, and

---

[59] Presentation through negation: de Jong, *Narrators and Focalizers*, 61–8, citing *Iliad* 11. 255 and 16. 140. Negatives which single out what will be talked about: *Iliad* 2. 1 ff., cp. 11. 1 ff.; note also 9. 29 ff., cp. 9. 693 ff.: 'all were silent except . . .'; 11. 717, 'my father would not let me fight, but . . .'; and even 1. 22–4.

Thucydides is telling us about it in abbreviated style. (Or is he telling us that surrender might be expected of an ordinary *polis*, but Athens was special?) The Homer passages can't plausibly be unpacked in quite the same way; it seems far-fetched to argue that Homer meant that Agamemnon wondered 'shall I stop fighting?' and decided 'No, certainly not'.[60]

In any case, the passage from Thucydides book 1 is a simple example of Thucydides taking his readers aside and establishing his own credentials: you and I know, as sensible strategists, that the expected thing would be to move back the army, and so on; or perhaps this is again a way of saying that Athens was special: if it were any *other* city you would expect the main army to have been recalled in such a crisis. In any case, the denouement has already been cleverly set up in advance. That is, the statement 'they did not move their army' is made to look innocently factual by another narrative technique, that of inferred motivation—I refer to the statement about what the Corinthians were hoping for. Thucydides' access in all this was in fact very restricted, but the effect is that of a dry military report.

Now take a similar but more complicated instance, from book 3.[61] The Athenians realized that the Peloponnesian preparations described in the previous chapter were the result of a calculation ($\kappa\alpha\tau\acute{\alpha}\gamma\nu\omega\sigma\iota\nu$, a word

---

[60] 1. 2. 1; 1. 105; 8. 1. 3. Note also 1. 90. 5, Themistocles did not present himself officially to the Spartan magistrates [as you would expect under normal circumstances] and 3. 50. 2, Athens did not [as you might expect] impose tribute on Mytilene after its revolt. At 1. 18 Sparta 'did not have a tyrant' (lit. 'was untyranted') implies expectations derived from more normal places (at least, normal in Th.'s view, see my *Commentary* on 1. 13. 1). Again, the statement at 2. 65. 8 that Pericles did *not* take bribes ($\chi\rho\eta\mu\acute{\alpha}\tau\omega\nu$ $\tau\epsilon$ $\delta\iota\alpha\phi\alpha\nu\hat{\omega}\varsigma$ $\grave{\alpha}\delta\omega\rho\acute{o}$-$\tau\alpha\tau\sigma\varsigma$) perhaps suggests the cynical contrary expectation ('as you might expect from a politician'); or it may be polemical (Plut. *Per.* 32 shows that financial accusations were current against Pericles); or there may, as my pupil Kate Emmett suggests to me, be an implied contrast with Kleon. Note in any case that the phrase picks up Pericles' own claim at 2. 60. 6, where the language is less obviously negative ($\chi\rho\eta\mu\acute{\alpha}\tau\omega\nu$ $\kappa\rho\epsilon\acute{\iota}\sigma\sigma\omega\nu$). At 2. 39 Pericles' negatives or implied negatives mean 'unlike the Spartans'. Note also the nice piece of negative presentation at 8. 73. 3: Hyperbolus was *not* ostracized because of his power and influence. Does this imply (wrongly) that this was the usual reason for ostracism? Or is it rather just a strong condemnation of H.?

[61] 3. 16. 1. Again and for the same reason (see n. 30 above) I give the Greek in full, with translation: $\alpha\grave{\iota}\sigma\theta\acute{o}\mu\epsilon\nu\sigma\iota$ $\delta\grave{\epsilon}$ $\alpha\grave{\upsilon}\tau\sigma\grave{\upsilon}\varsigma$ $\sigma\acute{\iota}$ $\grave{A}\theta\eta\nu\alpha\hat{\iota}\sigma\iota$ $\delta\iota\grave{\alpha}$ $\kappa\alpha\tau\acute{\alpha}\gamma\nu\omega\sigma\iota\nu$ $\grave{\alpha}\sigma\theta\epsilon\nu\epsilon\acute{\iota}\alpha\varsigma$ $\sigma\phi\hat{\omega}\nu$ $\pi\alpha\rho\alpha$-$\sigma\kappa\epsilon\upsilon\alpha\zeta\sigma\mu\acute{\epsilon}\nu\sigma\upsilon\varsigma$, $\delta\eta\lambda\hat{\omega}\sigma\alpha\iota$ $\beta\sigma\upsilon\lambda\acute{o}\mu\epsilon\nu\sigma\iota$ $\acute{o}\tau\iota$ $\sigma\grave{\upsilon}\kappa$ $\grave{o}\rho\theta\hat{\omega}\varsigma$ $\grave{\epsilon}\gamma\nu\acute{\omega}\kappa\alpha\sigma\iota\nu$ $\grave{\alpha}\lambda\lambda'$ $\sigma\acute{\iota}\sigma\acute{\iota}$ $\tau\epsilon$ $\epsilon\grave{\iota}\sigma\iota$ $\mu\grave{\eta}$ $\kappa\iota\nu\sigma\hat{\upsilon}\nu\tau\epsilon\varsigma$ $\tau\grave{o}$ $\grave{\epsilon}\pi\grave{\iota}$ $\Lambda\acute{\epsilon}\sigma\beta\omega$ $\nu\alpha\upsilon\tau\iota\kappa\grave{o}\nu$ $\kappa\alpha\grave{\iota}$ $\tau\grave{o}$ $\grave{\alpha}\pi\grave{o}$ $\Pi\epsilon\lambda\sigma\pi\sigma\nu\nu\acute{\eta}\sigma\sigma\upsilon$ $\grave{\epsilon}\pi\iota\grave{o}\nu$ $\rho\alpha\delta\acute{\iota}\omega\varsigma$ $\grave{\alpha}\mu\acute{\upsilon}\nu\epsilon\sigma\theta\alpha\iota$, $\grave{\epsilon}\pi\lambda\acute{\eta}\rho\omega\sigma\alpha\nu$ $\nu\alpha\hat{\upsilon}\varsigma$ $\grave{\epsilon}\kappa\alpha\tau\acute{o}\nu$. ... 'The Athenians realized that the activity of the Spartans was due to a conviction of Athenian weakness. So they decided to show them that they were mistaken, and to prove that, without moving their fleet from Lesbos, they were fully able to deal with this new force which threatened them from the Peloponnese. So they manned a hundred ships. ...' (Translation again by Jowett, adapted.)

implying some contempt) of Athenian weakness. The Athenians there-fore wanted to show that the Peloponnesians' decision, or judgement, was wrong, οὐκ ὀρθῶς ἐγνώκασιν, and that they, the Athenians, were able, *without moving the fleet from Lesbos*, to repel the danger from the Peloponnese. This is very neatly done by Thucydides. The Greek for 'without moving their fleet' is μὴ κινοῦντες τὸ ἐπὶ Λέσβῳ ναυτικόν. Again, I suggest, the implied aside to the reader is 'without moving their fleet as you'd expect'. But this time the Athenians are the focalizers, and the negative 'without moving', μὴ κινοῦντες, with its implied 'as you would expect', does double duty. Partly, it repeats the exchange with the Thucydidean reader which was a feature of the book 1 passage (you and I, says Thucydides, would expect this). This is after all narrative, not speech. But more noticeably this time it describes what the Athenians thought the Corinthians expected (we do after all have μή not οὐ, perhaps because the negative is subjective not objective in sense, or more likely because the whole expression is a sort of conditional, 'even if they didn't move . . .', 'without moving . . .'). Or rather, the negative gives what Thucydides thought the Athenians thought the Corinthians expected. The focalization here is very complex. Compare, from *Iliad* 6, the statement that Hector did not find Andromache at home: as Hector expected, and, surely, as the audience would expect. And, we may add, as Helen just now *was*: the special anxiety of Andromache is thus brought out.[62]

These Thucydidean examples are admittedly not as emphatic as the Homeric ones. Here is an example where emphasis *does* seem to be con-ferred by a comparable phrase, one pointed out to me by Christopher Pelling. In book 8 (the account of the oligarchic revolution of 411), the commissioners, ξυγγραφεῖς, proposed nothing else, ἄλλο μὲν οὐδέν, but just this, αὐτὸ δὲ τοῦτο. . . . This is an emphatic way of saying 'you might have expected something else', probably, 'they held back their specific proposals'. (Andrewes compares a fragment of Theopompus.) The alternative explanation, rejected by Andrewes in his note on the passage, is that the negative is polemic against somebody or other.[63]

That brings me to a very intriguing negative usage: 5. 70, the Spartans

---

[62] *Iliad* 6. 371 (Andromache not at home), contrast 6. 323 ff. (Helen at home).

[63] 8. 67. 2 (the ξυγγραφεῖς); Andrewes cites *FGrHist* 115 F 347b, 'where a similar phrase implies "not as you might expect"'. However, David Lewis remarks to me (n. 32 above) that ἄλλο μὲν οὐδέν 'surely only records that they didn't make the proposals they were asked to make in 67. 1'. But his view and mine are not incompatible. Other examples: 1. 139. 3; 2. 51. 1 (which actually includes a reference to what was usual or ordinary, i.e. what you might expect), 2. 78. 4; 4. 14. 3; 6. 41. 1; 7. 75. 5; 7. 77. 5.

march to the sound of flutes 'not for religious reasons but to keep in step'. This is oddly emphatically put. At first sight this *does* sound like explicit polemic, something surely peculiar to historians as opposed to novelists (though I suppose Jane Austen's celebrated exclamation in *Mansfield Park*, 'let other pens dwell on guilt and misery', is a kind of light-hearted polemic against other more lurid novelists. Or there are Fielding's remarks in *Tom Jones* about reptile critics. And poets are capable of being polemical, a point to which I shall return later.) At Thucydides 5. 70 we look at the Spartan sections of Herodotus to see if it is he who is being got at over these mysterious flutes. But he is not. Of course, there were plenty of fifth-century writers other than Herodotus and one of them, lost to us, may be the target. This possibility is perhaps weakened by Aulus Gellius 1. 11, his only quotation of Thucydides, and a paraphrase of this very passage. Gellius *doesn't* show awareness that the issue was part of a debate between the historians of antiquity. But as Holford-Strevens says, Aulus Gellius mentions very few Greek historians anyway. Again, I suggest, we should consider the possibility that the apparently objective and categorical statement 'not for religious reasons' conceals a more subjective and rhetorical apostrophization of the reader. 'Not, *as you might think*, for religious reasons.'[64]

There is incidentally no 'pathetic apostrophe' in Thucydides of the kind frequently found in *Aeneid* 10. Compare, from the *Iliad*, Homer's narrative interjection, 'the blessed gods did not forget you, Menelaus', or the authorial vocatives used towards other favourites of the poet, like Patroclus in the *Iliad* and Eumaeus in the *Odyssey*.[65] Historians gener-

---

[64] The flutes: 5. 70; Aulus Gellius 1. 11; L. Holford-Strevens, *Aulus Gellius* (London, 1988), 181 f. Note, by way of contrast, some other 'negative flutes' (as we may call them) at Hdt. 1. 132. 1: the Persians do not play flutes at their sacrifices, or use altars, or fire, etc. Here the negatives are (cp. the last words of 131. 1) a shorthand way of saying that in these respects Persian sacrifice is unlike Greek, see W. Burkert, 'Herodot als Historiker fremden Religionen', in G. Nenci and O. Reverdin (eds.), *Hérodote et les peuples non-grecs* (Entretiens Hardt 35; Vandœuvres–Geneva, 1990), 1–39 at 14 and 20; also J. Gould in the present volume: above, p. 98. Finally, there is an interesting piece of 'presentation through negation' in Xenophon's *Hellenika*, 5. 4. 64: Timotheos (on Corcyra) *did not* enslave or exile anybody and he *did not* change the existing laws. Here the negatives are, as Cawkwell suggests (Penguin edn., n. on the passage), an oblique way of drawing a contrast with a 'future' event, Chares' famously disastrous expedition of 361/360, Diod. 15. 95. 3 and Aen. Tact. 11. 13 with D. Whitehead, *Aineias the Tactician: How to Survive under Siege* (Oxford, 1990), 133.

[65] Pathetic apostrophe: Homer, *Iliad* 4. 127 (Menelaus) or *Odyssey* 14. 55, 17. 272, etc. (Eumaeus), with A. Parry, *The Language of Achilles and Other Papers*, ed. H. Lloyd-Jones (Oxford, 1989), 300–26, G. S. Kirk (above, n. 52), 343 on 4. 127, and Russo, in J. Russo, M. Fernandez-Galiano, and A. Heubeck, *A Commentary on Homer's Odyssey*, iii: *Books XVII–XXIV* (Oxford, 1992), 33, n. on 17. 272; Taplin, *Homeric Soundings*,

ally do not have it unless we are to count the vocative near the end of Tacitus' *Agricola*, 'tu vero felix Agricola'.[66] Returning to negatives in Thucydides, another puzzle is the notoriously awkward one at 1. 125. After the Spartan war vote 'not a year elapsed but slightly less' before the war broke out. This may be polemical against lost authors who said that a full year did elapse; or Thucydides may be stressing the relative rapidity of Sparta's mobilization and therefore culpability. But it can hardly mean 'the period of diplomatic to-ing and fro-ing which now elapsed was not, as you might think, a year'.[67]

A related mannerism οὐ τοσοῦτον, 'not so much' *x* as *y*, *is* however specifically and characteristically Thucydidean; and may similarly *both* negate an expectation *and* signal a disagreement with a received view. An example of the second is from the *Archaeology*: it was not so much the oaths of Tyndareus as Agamemnon's power which led to the Trojan expedition. (This device has a further function: it enables Thucydides to mention a second view, economically, in addition to his main one. So the 'not so much' formula can sometimes represent an exception to Thucydides' avoidance of alternative versions, for which see n. 57.) For a very singular instance of a negated expectation see, from Pericles' obituary, the authorial judgement that the Sicilian Expedition failed, not so much because of bad judgement—as you might think from reading my books 6 and 7 which you haven't got to yet—as because it was marred in the execution.[68]

A final sort of negation, implied this time, is the use of 'instead of', ἀντί. Thucydides certainly uses this device: he says of the defeated Athenians in Sicily that instead of doing some enslaving as they (and the reader?) had expected, they were themselves at risk of enslavement. With

---

169 n. 28; S. Harrison, *Commentary on Aeneid 10* (Oxford, 1991), 98 on line 139. Homer uses apostrophe towards, but not quite exclusively towards, the three 'favourites' mentioned in my text. (Melanippos son of Hiketaon, a very minor character, is apostrophized at *Iliad* 15. 582, see Parry, p. 311). See also R. P. Martin, *The Language of Heroes: Speech and Performance in the Iliad* (Ithaca, NY, and London, 1989), 235 f.

[66] Tac. *Agric.* 45. 3.

[67] 'Not a full year, but less': 1. 125. I suppose the interpretation I rule out in the text ('not, as you might think, a year') might work if the whole chapter could be taken as a basic decision to delay, in which case Th. is drawing attention to the relative speed with which war broke out. But things are not so simple, because the chapter also contains the explicit words that they decided (to make proper preparations but) *not* to have any delay, μὴ εἶναι μέλλησιν! See my *Commentary* at p. 238.

[68] For 'not so much' see H. D. Westlake, *Essays on the Greek Historians and Greek History* (Manchester, 1969), 161–7; Thucydidean instances: 1. 9; 2. 65. 11. The latter instance might also carry the suggestion 'as you might think from noting Athens' tendency to recklessness, which my History contains so much about'.

this compare, from the *Odyssey*, Telemachus' jeer at Ktesippos—instead of a marriage feast your father would be getting ready a funeral.[69]

I pass now to the use of 'if . . . not', which de Jong calls another example of interaction between narrator and narratee, and a 'congenial feature of story-telling'. She cites, from *Iliad* 3, 'Paris would have choked on his shoulder strap and Menelaus would have pulled him in, *had not* Aphrodite the daughter of Zeus broken the strap'.[70] There are numerous examples of such counterfactuals in Thucydides, such as 1. 101. 2: the Spartans at the time of the Thasos episode promised to invade Attica and they would have done so (or they intended to do so), καὶ ἔμελλον, but they were prevented by the earthquake. But the most celebrated examples are two practically identical phrases from books 3 and 7, about the peril of Mytilene (book 3) and the arrival of Gylippos (book 7): so close did Mytilene/Syracuse come to fatal danger, παρὰ τοσοῦτον ἦλθον κινδύνου. These are in effect highly rhetorical uses of the epic 'if . . . not' formula. That is, Thucydides is saying that the two places would have fallen had it not been for the arrival of the second trireme and Gylippos respectively. The rhetoric is surely enhanced by the repetition, though oddly neither Gomme in book 3 nor Dover in book 7 comments on either passage in any way at all, let alone refers the reader to the parallel.[71] But in justice to Dover he has (not in the commentary) provided an excellent discussion of counterfactuals in Thucydides,[72] singling out 8. 96, the possible consequences of the seizure of Euboia. In his use of *this* device

---

[69] For 'instead of *x*, *y*' in Th., see 7. 75 (and 4. 62. 3 is comparable, though from a speech; note also 7. 11. 4, from Nikias' letter); cp. *Odyssey* 20. 307.

[70] 'If . . . not' expressions: de Jong, *Narrators and Focalizers*, 68–81, esp. 70 on *Iliad* 3. 373 ff. See however H.-G. Nesselrath, *Ungeschehenes Geschehen: 'Beinahe-Episoden' im griechischen und römischen Epos von Homer bis zum Spätantike* (Stuttgart, 1992), ch. 1, esp. pp. 8–9, doubting whether de Jong's account of the purpose of the 'if . . . not' passages in the *Iliad* (to confirm Homer's 'status as a reliable presentator', de Jong, p. 81) does justice to the whole range of instances, of which he says there are 46 rather than de Jong's 33. The poet, Nesselrath suggests (p. 9 n. 10), wants us to be aware of the whole range of fascinating possibilities. The whole *Iliad* can be seen as one giant 'if . . . not', an episode in the Trojan War which could have produced a totally different outcome to that war: Nesselrath, p. 27. This is an extension not a refutation of de Jong. See also M. Lang, 'Unreal Conditions in Homeric Narrative', *GRBS* 30 (1989), 5–26. Tod 116 has a good 'if . . . not'.

[71] Thucydidean examples: 1. 101. 2 (Thasos and earthquake); 'so close did they come to destruction': 3. 49 and 7. 2 (the latter is more important because it is a way of saying that the Athenian expedition nearly *succeeded*). See Pelling, *Plutarch: Antony* (above, n. 57), 237 on ch. 48.

[72] Dover, *Greeks and Their Legacy*, ii 74 on 8. 96: Euboia. Cp. Pelling (above, n. 57) on Plut. *Ant.* 50. 4. Other relevant Thucydidean texts are 2. 18. 4, 1. 9. 5, and perhaps 1. 11 (the Greeks would have taken Troy sooner if . . .).

(I mean 'if . . . not') Thucydides surely differs not at all from the poets and novelists.

Then there are evaluative and affective words. We have looked at one example already, ἀλλόκοτον in the description of the second trireme sent to Mytilene. Another interesting and for the historian very important case is from book 4, the Pylos/Sphakteria episode. Here the Spartans are said to make a formal protest against the injustice of the unreasonable Athenian attitude to the petty Spartan infringements of the truce made a little earlier. Thucydides' phrase for 'protesting against injustice' is ἀδίκημα ἐπικαλέσαντες. But who is the focalizer here, Thucydides or the Spartans? Or to put it traditionally, is Thucydides endorsing the Spartan view that the Athenians were in the wrong? The Greek incidentally picks up another equally ambiguous phrase a couple of lines earlier, the Athenians pleaded various trivial-seeming infractions of the truce, ἄλλα οὐκ ἀξιόλογα δοκοῦντα εἶναι.[73] To whom did they seem trivial, Thucydides or the Spartans? The word here is δοκῶ which *only* takes the infinitive so the infinitive is no help, contrast φαίνομαι (I was taught at school φαίνομαι ὤν quod sum, quod non sum φαίνομαι εἶναι). I think we have to accept that Thucydides has left the issue suspended, but I also think the heavy concentration of evaluative or affective language ('trivial', 'injustice') leaves in our minds a bad impression of Athens. Or rather, of Kleon, because Thucydides' spite has here got the better of his patriotism. Incidentally we should not mind leaving such issues suspended. There are other places where Thucydides has, I think, left his meaning deliberately fluid or unstable. Adam Parry argued for this instability in an article on the translation of the phrase κατεῖχε τὸ πλῆθος ἐλευθέρως at 2. 65—Pericles 'restrained the people freely'. Does this mean he led them in a liberal way or that he led them like free men? Parry says in effect that translation has its limits and that both meanings are simultaneously there.[74]

---

[73] Evaluative and affective words: see J. Griffin, 'Words and Speakers in Homer', *JHS* 106 (1986), 36–57, esp. 47; de Jong, *Narrators and Focalizers*, 136–46. Thucydidean examples: 3. 49, cp. n. 10 above 4. 23 (Pylos/Sphakteria 'injustices' etc.). On the latter instance, Pelling comments to me, 'I'd have thought that δοκοῦντα grammatically must be dependent on the main verb ἀπεδίδοσαν for its tense, i.e. "seemed οὐκ ἀξιόλογα" at the time. That of course does not exclude an atmosphere of Thucydidean dismissiveness . . .'. Th.'s whole expression for 'protesting against injustice' etc. is relevant: ἀδίκημα ἐπικαλέσαντες τὸ τῶν νεῶν. The definite article means 'the injustice, the one (we all know about) concerning the ships'. That is, Thucydides here subscribes to the view that the injustice was real. (This is more plausible than 'bringing as a charge the affair concerning the ships, and calling it an ἀδίκημα').

[74] On ἐλευθέρως at 2. 65. 8 see Parry, *Language of Achilles* (above, n. 65), 143–7.

Evaluative words like ἀλλόκοτον and ἀδίκημα take us into another
and very interesting area, which has been called attributive discourse or
denomination. The classic example[75] is from Flaubert's *Madame
Bovary*. It has been shown that the heroine is variously referred to as
Emma, Madame Bovary, the young woman, etc. depending on the
rhetorical needs of the situation. So too Homer, and the Euripidean
messengers. The Euripidean *tour de force*, according to de Jong, is the
messenger in the *Ion* who comes up with a new description of one
character every time he mentions him. But de Jong shows[76] that it is not
always easy to see why the variations occur; I would suggest metrical
convenience if I did not know how unpopular this explanation is with
literary scholars. Another equally shameful explanation is simple
variatio. A frivolous example comes to mind here. About the time I was
writing this paper, Robert Maxwell died mysteriously. I heard one radio
broadcast which began by saying that the final inquest on Mr Robert
Maxwell had just taken place. It went on a bit later to say, 'It is still not
clear whether *the publisher* had already had a heart attack before his
body hit the water'. The words 'the publisher' here are surely rather
ludicrous, he hardly fell into the water in his capacity as a publisher. One
can think of better ways of effecting the *variatio*—for instance 'it is still
not clear whether the *overweight 67-year-old* had already had a heart
attack' etc.

This kind of attributive discourse or denomination is certainly a
feature of Thucydidean narrative, though it must be said that on the
whole he sticks to proper names or ethnic descriptions without a lot of
variation or circumlocution. One interesting exception however is the
use of 'Spartiates' in the Pausanias excursus in book 1, to refer not to full
Spartan citizens but to the larger decision-making group whom Thucydides normally calls Lacedaemonians. Here Westlake has suggested a
written source, and if that is right we have here an explanation which is
not available for such variations in poets. It is true that Pindar and
Aeschylus echo Homer, a topic recently explored by Gregory Nagy, so
Homer is a kind of source for other poets. But such echoes are surely

---

[75] Attributive discourse/denomination: de Jong, *Narrative in Drama*, 94 ff. See also now
de Jong, 'Studies in Homeric Denomination', *Mnemosyne* 46 (1993), 289–306, discussing
the significant use of nouns like 'wife', 'king', 'husband' to describe the central figures in the
Odyssey.

[76] De Jong, *Narrative in Drama*, 102 on Eur. *Ion* 1122 ff. See also Pelling (above, n. 57)
on Plut. *Ant*. 25. 3 for the kind of context in which Cleopatra is called 'the Egyptian
woman'. The devices I discuss in this whole section of my paper are what O. Taplin,
Homeric Soundings, 52 aptly calls 'colouring' devices.

more self-conscious. Returning to Spartiates (properly so-called) in Thucydides, denomination varies here too, not every full Spartiate always getting called that. There is in fact a rhetorical aspect to *this* change in denomination if Andrewes was right in the final volume of the *Historical Commentary on Thucydides* to say that Thucydides slightly prefers 'Spartiate' to 'Lacedaemonian' in first introductions; perhaps because the word is heavier and more portentous.[77]

The most promising area is Thucydides' use of patronymics.[78] Pericles and Melesippos are given full patronymics not on the occasions of their first appearance but at their most *solemn* first appearance. Melesippos is the Spartan herald who, when he is being escorted out of Attica, says that this day will be the beginning of troubles for Greece. The reader has in fact already met Melesippos in book 1, where he is merely Melesippos. In book 2 however, the solemn moment, he becomes 'Melesippos the son of Diakritos a Spartiate man'. The treatment of Archidamus is comparable. In book 2 when he leads the Spartan invasion he is Archidamus son of Zeuxidamus and king of Sparta. But at his first appearance in book 1 he is just Archidamus the king, no patronymic. Spartan kings are not, however, always treated even *this* respectfully. There is one and only one mention in all Thucydides of Cleomenes I, the famous late sixth-century king of Sparta. It is not in the *Archaeology* where you might expect it but at 1. 126, and is very incidental. Cleomenes drove out the people who were contaminated as a result of the Kylon affair and dug up their graves. Here Cleomenes is just Κλεομένης ὁ Λακεδαιμόνιος, not even specified as king. What is the explanation for this? Perhaps it is precisely because the reference is so incidental. Thucydides does not want to distract the reader by focusing too much attention on Cleomenes himself. Contrast the first mention of Cleomenes in Herodotus, where he is given as Cleomenes son of Anaxandrides. Cleomenes was of course a big personality in Herodotus and maybe Thucydides reacts against this.

[77] Spartiates: H. D. Westlake, *Studies* (above, n. 16), ch. 1 = CQ 27 (1977), 95–110; A. Andrewes, *HCT* v. 50f.; G. Nagy, *Pindar's Homer* (Baltimore, 1991). Thucydides may have preferred 'Spartiates' for another reason also: it had a more technical sound, Λακεδαιμόνιος being the regular word in poetry—though it is true that Σπαρτιητέων (metrically three long syllables, by two separate synizeses) occurs in Tyrtaeus, POxy. 3316 (1980), line 21. Thucydides is perhaps saying 'I pass up the trite associations of the word *Lacedaemon* in favour of the stricter *Spartiate*'. (At 1. 128. 3 Pausanias is a 'Lacedaemonian', but 'the Spartiates' are mentioned in the next breath. Pelling suggests to me that this may be Th.'s way of hinting that Pausanias found himself at odds with Spartan traditions and institutions.)

[78] G. T. Griffith, 'Some Habits of Thucydides when Introducing Persons', *PCPhS* 187 (1961), 21–33.

Incidentally I note as a curiosity that Xenophon in the *Hellenika* calls King Demaratos, a man never mentioned by Thucydides at all, just Demaratos the Spartan. But Xenophon's use of patronymics is famously arbitrary, thus Agis occurs without patronymic or title presumably because he is assumed to be familiar to us already from Thucydides whom Xenophon is in some sense continuing.[79]

A related issue is demotics. This is easy. There *are* none in Thucydides. Contrast Herodotus; and Xenophon who uses demotics, though only to distinguish the two Thrasybuloi. This Thucydidean feature is an example of his universality.[80] Demotics are parochially Athenian.

This is the place to mention Thucydides' intriguing use of 'the so-called',[81] especially with place-names, such as the so-called Paralos or the two Athenian ones at 6. 57: the Leokoreion and the Kerameikos.That chapter is odd because we have already had the phrase 'so-called Leokoreion' once, at 1. 20; and we have already met the Kerameikos in the context of the Funeral Speech, where however it was periphrastically called the most beautiful suburb. There are about thirty relevant instances of καλούμενον or variants in Thucydides and there are disproportionately many, actually fourteen, Athenian and north Greek places like Krousis. I suggest Thucydides feels coy about showing off his own special knowledge of these areas. But there are others like Peloponnesian Kynouria in book 4. A remarkable instance, because it is the only time Thucydides uses the phrase 'the so-called' of something other

---

[79] Patronymics (or not) in Th.: 1. 112. 12 (Melesippos), cp. 1. 139; 2. 19 (Archidamus), cp. 1. 79. 2; 1. 126. 12 (Cleomenes), cp. Hdt. 3.148 and Xen. *Hell.* 3. 1. 6; 1. 1. 33 (Demaratos, Agis). With such delayed introductions compare *Iliad* 18. 249–52 on Pouludamas, a character who has been active much earlier (e.g. 14. 449), but who is formally introduced just before his really decisive intervention in book 18 (cp. 22. 100). This is like Th.'s treatment of Pericles, whose first, brief, and unceremonious appearance is at 1. 111 but who gets not one but two subsequent, formal, introductions: 1. 127. 3; 139. 4; see my *Commentary* (n. 24) on the various passages. For delayed *description* in Homer see C. W. Macleod (ed.), *Iliad XXIV* (Cambridge, 1982), 123, n. on 448–56.

[80] See S. Hornblower, *Thucydides* (London, 1987), 97 n. 98 with refs. Note however that Thucydides does use some Attic forms, such as νεώς ('temple'), not ναός or νηός, and κόρη ('maiden'), not κούρη. Λεωκόρειον (see below) is a specifically Attic form.

[81] 'The so-called': 2. 55 (Paralos); 6. 57 (Kerameikos and Leokoreion, cp. 2. 34 and 1. 20); 2. 79 (Krousis); 4. 56 (Kynouria); 1. 112 (Second Sacred War). Perhaps some of these would be more fairly rendered 'a place (or whatever) called X' (cp. 4. 70. 1 on Tripodiskos, where the phrase is ὄνομα τοῦτο ἔχουσα, 'a place having the name . . .'); cp. Oxyrhynchus Historian 21. 2 Bartoletti for 'so-called Mysian Olympus' (contrast the simple form at Hdt. 1. 36. 1), and, a little further down, the 'mercenaries called Derkylideioi'; also 21. 3: 'the place called Lions' Heads'. But that innocuous sort of rendering is not appropriate for Th. on the Sacred War, surely.

than a place, is 1. 112, 'the so-called Sacred War'. This *is* I think a distancing device. Thucydides has little use for religion.[82]

Then there's the use of the disparaging τις, 'a certain', 'somebody called . . .' (another buttonholing device, 'the sort of person you might not have heard of'). A fine example is in book 8: Hyperbolus—contrast the more respectful treatment of a more serious demagogue, Kleon. Is τις disparaging? The question has been discussed with reference to Herodotus' introduction of Themistocles, which some scholars think is a slighting use of τις. Fornara however[83] argues the opposite in his 1971 book on Herodotus. He notes that in the *Anabasis* Xenophon calls himself 'a certain Xenophon, an Athenian'; and points to Homer's introduction of Dolon with the word τις. But I am not completely persuaded by this because Dolon is surely not intended as an admirable character. It does seem that Thucydides' use of τις for Hyperbolus helps to leave a negative impression. It is not of course the only thing in that unusually judgemental sentence of Thucydides which does so, he calls him μοχθηρὸς ἄνθρωπος, after all. But it should be added to our collection of devices for buttonholing the reader.[84]

Finally in this area there is the way Thucydides refers to himself. This is a question much discussed by Homerists interested in the poet's objectivity. In fact, Thucydides fluctuates: in the first preface in book 1 he uses the third person, but book 5 has both third and first in the course of a single chapter.[85]

---

[82] See S. Hornblower, 'The Religious Dimension to the Peloponnesian War, Or, What Thucydides Does not Tell Us', *HSCP* 94 (1992), 169–97. As the epigraph to that article I cited P. Veyne on the absence of the gods from Th.'s narrative (*Writing History* (Manchester, 1984), 232); but compare already A. B. Drachmann, *Atheism in Pagan Antiquity* (Copenhagen and London, 1922), 28: Th. 'completely omitted any reference to the gods in his narrative. Such a procedure was at this time unprecedented . . .'. With Thucydides' 'the so-called sacred war', or (a more neutral translation) 'what is called the sacred war', compare another distancing formula, the opening of [Hippocrates] *On the Sacred Disease*: περὶ τῆς ἱερῆς καλεομένης νούσου ὡδ᾽ ἔχει, 'I am about to discuss the disease called "sacred" . . .' (Loeb, *Hippocrates*, tr. W. H. S. Jones, ii. 139).

[83] Hyperbolus: 8. 73. On τις see C. W. Fornara, *Herodotus: An Interpretative Essay* (Oxford, 1971), 68–73, discussing Hdt. 7. 143. 1. Thucydides' use of τις for Hyperbolus may carry the further suggestion that nobody outside Athens would have heard of this miserable creature. Cp. also 8. 92. 5 (Hermon). When Th. does *not* want to be dismissive, he can use ἀνήρ, 'a man', as at 3. 29. 2 (Teutiaplus); cp. also 6. 54. 2 (Aristogeiton the tyrannicide, surely a respectful mention). We have the introductory ἀνήρ at 1. 132. 5 and 6. 64. 2, but 'the man' is not named in either of those passages.

[84] For Dolon see *Iliad* 10. 314 (which Oliver Taplin tells me he thinks 'surely disparaging'); Xen. *Anab.* 3. 1. 4.

[85] Self-reference in Th.: 1. 1; 5. 26. Cp. M. Wheeldon, '"True Stories": The Reception of Historiography in Antiquity', in Averil Cameron (ed.), *History as Text: The Writing of Ancient History* (London, 1989), 45 ff. First person *plurals*: 1. 4 (cp. Hdt. 3. 122); 8. 41. 2.

Lastly, but very important, narrative voices in Thucydides. Events are described 'indirectly', i.e. from a certain viewpoint. This phenomenon has recently been discussed by James Davidson in an interesting article on the 'gaze' in Polybius.[86] But I slightly part company with him when he says that this sort of thing is infrequent in Thucydides, and cites only the great final sea-battle at Syracuse in book 7, a *tour de force* which moved Macaulay and even Dionysius of Halicarnassus to admiration. I would say that for instance much of the Sitalkes invasion in book 2 is not so much narrated as *perceived* by the terrified Greeks and Macedonians who were on Sitalkes' route. But the focalization is subtle because as Hammond showed twenty years ago and Badian has recently reminded us, the actual path of the invasion is described from the point of view of the invaders. Connor in his 'Narrative Discourse' (n. 20) adds the daring Peloponnesian raid on the Piraeus in book 2, where the perspective shifts several times in a paragraph. 'There was no fleet on the look-out' (Peloponnesian point of view), 'and there was no expectation that the enemy would make an attack' (Athenian point of view). Connor briefly reanalyses one or two other famous instances of multiple perspective, notably (Davidson's example) the sea-battle at Syracuse, which is experienced rather than described; the arrival of the news of the Sicilian Disaster. (But Thucydides passed up the nice story in the last chapter of Plutarch's *Nikias* about the arrival of the news of the disaster in a barber's shop in the Piraeus, and the poor man responsible getting tortured by the authorities instead of having a haircut.) Finally I would add that multiple perspective or varied focalization helps with the problem, when did Thucydides think the Peloponnesian War began? Notoriously, he seems to wobble between dates. But the reason why he sometimes hankers for the view, inconsistent with his other indicators, that the invasion of Attica by Sparta and her allies was the decisive event, is surely that he was adopting an Athenian perspective or focalization.[87]

However there is a problem about singling out episodes in Thucydides which seem to be in a particular narrative voice. The problem is that which has run through this paper—the difference between the narrating

---

[86] Narrative voices: J. Davidson, 'The Gaze in Polybius' Histories', *JRS* 81 (1991), 10–24. The careful ordering of the similes at *Iliad* 2. 455–83, conditioned by a Trojan viewpoint, well illustrates this phenomenon.

[87] Narrative voices in Th.: 2. 95 (Sitalkes) with Hammond, *HM* i. 200, and Badian, *Plataea to Potidaea*, 181; 2. 93 (Piraeus raid) with Connor (above, n. 20); 7. 71 (Syracuse sea-battle); 8. 1 (news of Sicily), contrast Plut. *Nik.* 30. On the beginning of the war see my *Commentary* at 236. Adam Smith, *Lectures on Rhetoric and Belles Lettres*, ed. J. Lothian (1963) 88 (contrast 90) praised *Tacitus* over Th. for 'indirect' narration!

of fact and other kinds of narration. An excellent example is the *Potidaiatika*. It was long ago noticed[88] by Westlake that the episode seems to be seen through the eyes of Aristeus son of Adeimantus, who comes out of it and other episodes rather well. But is this narrative skill; or is the truth merely that Thucydides interrogated Aristeus and that his oral account lies behind those chapters? Aristeus is perhaps straightforward enough. More serious difficulties arise when we turn to the messier book 8 and ask how far Alcibiades is the focalizer. Alcibiades was surely an informant whom Thucydides would have been mad not to use; but take a judgemental interjection from book 8: Phrynichus thought that Alcibiades didn't sincerely care for either oligarchy or democracy, 'which was true', ὅπερ καὶ ἦν.[89] This could be Thucydides' own judgement; or it could be just a report of Alcibiades' cynical endorsement of what Phrynichus thought. That is, positing Alcibiades as an oral informant does not this time help us with the problems of focalization or voice.

There is another relevant text in book 4: the account of the battle of Solygeia. This battle has been studied in articles by Sieveking in the 1960s and Ron Stroud in the 1970s;[90] Stroud accepts Sieveking's ingenious suggestion that the unusual imperfect 'the place was steep' rather than the expected 'the place is steep' indicates interrogation by an eyewitness, i.e. participant. But here surely the language of narratology helps. The primary narrator-focalizer is Thucydides but the secondary/embedded focalizers are the participating soldiers.

I end the main section of this paper with a warning, one I have probably disregarded myself from time to time. We should never discuss a Thucydidean habit without asking if Herodotus exemplifies it also.

To conclude, I summarize. There is surely no doubt that Thucydides uses, with great rhetorical skill, narrative devices also found in poetry and fiction, and those devices are interesting enough. Even more interesting, to me at least, are the respects in which narrative is *differently* handled in historical and in fictional texts. I shall recapitulate these.

First, narrative displacement or anachrony (and we can add iterative

---

[88] Aristeus: H. D. Westlake, *Essays* (above, n. 66), ch. 4.

[89] 8. 48. 4.

[90] Solygeia: 4. 43 with F. Sieveking, 'Die Funktion geographischer Mitteilungen im Geschichtswerk des Thukydides', *Klio* 42 (1964), 73–179 at 162; R. Stroud, 'Thucydides and the Battle of Solygeia', *CSCA* 4 (1971), 244. The word for 'steep', πρόσαντες, literally means 'steep-facing' (i.e. facing somebody or other, in this case the Athenians). So the focalization is even more complex than just Th. on the one hand and the soldiers on the other: we have a shift of focalization from the Corinthians who have just retreated (ὑποχωρήσαντες) to the Athenians, because the place was steeply facing *them*.

presentation) are generally used by Thucydides to lessen the impact of an irreducible event or a fact; a novelist by contrast generally uses displacement to highlight or emphasize an item. The difference arises because for novelists there is no such thing as an irreducible fact: they simply need not include anything which does not fit the picture. The technique is, I argued, specially effective in Thucydides because he is normally so serial or linear.

Second, presentation through negation is used by Thucydides (remember those flutes) in a way recognizable from Homer or messenger-speeches; but there is a complicating factor, namely that in a history, unlike a novel, such negation may be a polemical way of flagging a controversy. I am not forgetting that, as Tom Stinton and now Denis Feeney have shown, poets like Stesichorus, Bacchylides, Pindar, or Euripides can be polemical in tone when denying a version of a myth.[91] But for instance Aeschylus does not say, 'Agamemnon was king not, as you all thought, of Mycenae but of Argos', he just transposes the early action of the *Oresteia* to Argos.

Third, presentation through negation may, I suggested, be a telescoped way of reporting an actual debate in which somebody really did advocate the course not followed: 'let's bring the army back from Aigina', etc. This way of looking at the device wouldn't be appropriate if we were dealing with a poetic or fictional text.

Fourth, Thucydides may vary denomination for rhetorical reasons, but this may occasionally, as with the loose use of Spartiates, be an indication of a written source.

Fifth, when Thucydides uses a particular narrative voice, that may be just artistry, but it may also or alternatively be a sign that real people (Aristeus, Alcibiades) were his oral informants.

Narratology is in its infancy.[92] The present paper represents the thoughts which a historian, currently working on the second half of a commentary on Thucydides, has come up with so far.

---

[91] Stinton (n. 50 above) and D. C. Feeney, *The Gods in Epic* (Oxford, 1991), 14–19; cp. e.g. Bacchylides 19, Pindar *Ol.* 1, etc. Passages like Soph. *OC* 374–6 (cp. 1294–5) by their repetitive form, i.e. their emphasis of a departure from the usual version, come close to an explicit statement to the audience that the poet is taking a controversial line. From comedy cp. Ar. *Knights* 514–16 ('not out of stupidity, ὑπ' ἀνοίας'—as you might think) or *Ach.* 514ff.

[92] We shall know more about Thucydidean narratology in particular when we have a doctoral thesis on the subject now being written in the classical languages and literature sub-faculty of Oxford University by Timothy Rood of Oriel College, under the supervision of Christopher Pelling.

Advance copy from

Oxford University Press
200 Madison Avenue
New York, NY 10016

Title: Greek Historiography

Author: Edited by Simon Hornblower

Publication date: 10/27/94

Cloth ISBN: 0-19-814931-X

Cloth Price: 55.00

The Publisher would appreciate observance of the
publication date, and two copies of any review
or article which may appear.

6

The World of Theophrastus

P. M. FRASER

Most of Theophrastus' enormous literary output is lost. Two very detailed works on botany, one on the geography and one on the physiology of plants, a number of opuscula, about minerals, about odours, about fire, and other topics, the *Characters*, some philosophical fragments of varying length including some very important doxography, constitute his surviving legacy in direct transmission. We shall be looking at him mainly as a writer on the geographical aspects of botany.[1]

I propose to limit myself to two aspects of this, what he knew, or at least what he tells us, about the Middle East and beyond, and what he knew about the western world, Italy. I must pass over the picture he

---

[1] I make no attempt to provide a systematic guide to discussions of Theophrastus' botanical writings—a topic which would be beyond the scope of this paper, even if I was capable of it. I may call attention, however, to the recent publication of a vast compendium of Theophrastus-texts, Greek, Latin, and Arabic: *Theophrastus of Eresus: Sources for his Life, Writings, Thought and Influence*, ed. and tr. by William W. Fortenbaugh *et al.*, 2 vols. (Brill; Leiden, 1992), vol. 1, 485 pp., vol. ii, 705 pp. The botanical texts are naturally not included in it, but the testimonia to them are given in vol. ii, pp. 188–237. I have not seen M. G. Sollenberger, *Theophrastus of Eresus, his Life and Work* (Rutgers Univ. Studies in Classical Humanities 2; New Brunswick and Oxford, 1985). The edition on which I have based my paper is that by A. F. Hort in the Loeb series: Theophrastus, *Enquiry into Plants*, 2 vols. (London, 1916), which is provided with botanical indexes. A new edition by S. Amigues in the Budé series: Théophraste, *Histoire des Plantes*, i–ii [i.e. books 1–4] (Paris, 1988–9) is not yet complete at the time of writing. O. Regenbogen's very thorough and clear article in *RE* Suppbd. vii (1940), cols. 1353 ff. (for the botanical works see cols. 1435–79) is by far the best general introduction to Theophrastus, though it has naturally been overtaken by time in some details. For the early history of the text see J. J. Keaney, 'The Early Tradition of Theophrastus' Historia Plantarum', *Hermes* 96 (1968), 293 ff., and for the manuscript tradition see B. Einarson, 'The Manuscripts of Theophrastus' *Historia plantarum*', *CP* 71 (1976), 68 ff. I shall have occasion to refer frequently to H. Bretzl's excellent work, *Botanische Forschungen des Alexander-Zuges* (Leipzig, 1903), to which, as will become clear, I am indebted. For those opuscula which have not been separately published, I refer to the standard, but antiquated, vol. iii of Wimmer's Teubner text (1862), though the reader should now consult the new compendium referred to above, and for the fragments of the physical works P. Steinmetz, *Die Physik des Theophrast* (Palingenesia 1; Bad Homburg–Zurich–Berlin, 1964). Note: the quotations given in the Appendix are indicated by '(Ext. 1)' etc. in the body of the text. In the notes that follow *Ptol. Alex.* refers to my *Ptolemaic Alexandria* (Oxford, 1972).

provides of Greece itself and of western Asia Minor, fascinating though that is.

I need say little about the man, a native of Eresos in the south of Lesbos, the successor of Aristotle as head of the Peripatetic school, who stands always in the shadow of his master. The detailed *Life* of him preserved from Hellenistic sources in Diogenes Laertios (5. 36–57), is rich in personal details, with the complete text of his will, and a more than complete list of his works, which breaks down into three separate lists. One or two anecdotes seem to suggest a rather frigid personality, though we learn from a fragment of Hermippus, who was well informed (Ath. 21a; ext. 1) that he was given to mimetic and theatrical lecturing; and Cicero (*Brut.* 46. 172; ext. 2), who greatly admired him ('Theophrastus meus ... elegantissimus omnium philosophorum et eruditissimus') tells us that though he lived for many years in Athens, he still retained an accent that was noticeably un-Attic, and he could be spotted as a ξένος by an *anicula*, a γραιά, in the Agora. A familiar situation. 'Deutsch?'. His attitude of mind has been well described by a critic of his doxographical work: 'he is less sedulous in discovering the merits than in pointing to the defects of the men over whom his cold eye ranges.'[2]

Though we shall not be primarily concerned with botany as a science, it is important to understand the limitations of Theophrastus' botanical knowledge as contained in the Ἱστορία φυτῶν (*HP*) and the Αἰτίαι φυτῶν (*CP*). He knew nothing, of course, of a hierarchical classification of flora or of the true principles of plant physiology. Although he distinguishes the visible characteristics and forms of plants, and more particularly trees,[3] by their leaf-forms and by the operation of external factors on them—climate, location, rainfall, elevation, etc., and different methods of cultivation—in the absence of any understanding of the reproductive system of flora there is no overall scheme of hereditary relationships, or of the hierarchy of genera and species finally estabished on that basis by Linnaeus. His own categories of plant life may be called horizontal, based essentially on *differences* (διαφοραί), a key word in his comprehension of distinctions of growth and form. Similarly in the *De*

---

[2] See G. M. Stratton, *Theophrastus and Greek Physiological Psychology before Aristotle* (London and NY, 1917; repr. Iowa, n.d.), 53. This contains a re-edition of the *De Sensibus* (from Diels, *DG*), with tr., comm. and introd.

[3] In defining the characteristic parts (μέρη) of plants he says that these apply particularly to trees: *HP* 1. 1. 10–11: (11) ἀλλὰ μάλιστα ταῦτα ὑπάρχει, καθάπερ εἴρηται, τοῖς δένδροις κἀκείνων οἰκειότερος ὁ μερισμός· πρὸς ἃ καὶ τὴν ἀναφορὰν τῶν ἄλλων ποιεῖσθαι δίκαιον.

Lapidibus § 48 he describes his theme as αἱ τῶν λίθων διαφοραὶ καὶ δυνάμεις.[4]

Born in *c.*370 BC, and dying in 288 BC, his prime interest for the student of Greek history, in contrast to Greek thought or science, must be as a mirror of the great changes that the world had recently undergone. His life was passed largely in Athens, and it seems likely that he left that city only under necessity, and only briefly.[5] It has been claimed that he visited Egypt and Cyrene, about which he certainly possessed a rich store of information, but, as we shall see, evidence that he did so is not compelling. He was, then, no precursor of Linnaeus, who travelled some 700 miles on a sorry nag among the Eskimos in 1732 in preparation for his great book, the *Iter Lapponicum*. Linnaeus' pupils endured hardships in China, Arabia, and other remote areas in search of plants, and imposed on the flora of those regions the Linnaean system of classification and nomenclature. There is no clear evidence that Theophrastus despatched students to remote places to collect specimens and to study methods of cultivation and so on, but it is natural to suppose that he did. If he did (a possibility to which we shall return) it is surprising that he apparently left no pupils to carry on his botanical work. For all practical purposes, though the pre-Socratics had undertaken some investigations in the field,[6] he is the first and the last scientific Greek botanist, an isolation perhaps unduly emphasized by the loss of Aristotle's phytological works; it is in any case likely that the master left botany mainly to his pupil. Nevertheless, though we do not have the feeling we have when we read Linnaeus, of travelling with him to see the three kingdoms of nature in all weathers and in all seasons, Theophrastus was of course familiar with the daily agricultural life of Greece, and his accounts, in *CP* in particular, of the fundamental processes of planting, pruning, and harvesting, are clearly based on personal knowledge.[7] He had a χωρίον in Stagira,

---

[4] There are two good current editions of the *De Lapidibus*: that by Earle R. Caley and John F. C. Richards, *Theophrastus, On Stones: Introduction, Greek Text, English Translation and Commentary* (Ohio State University Press; Columbus, 1956), and that by D. E. Eichholz, *Theophrastus: De Lapidibus* (Clarendon Press; Oxford, 1965). Eichholz's edition is a remarkable work of scholarship, but both editions should be consulted. In the introductory chapter of the *De Odoribus* (in the Loeb *Enquiry into Plants*, ii. 326 ff.) Theophrastus emphasizes the limitations of such a classification in that field.

[5] He calls himself σχολαστικός in a letter to his fellow countryman, the philosopher Phanias, Phanias F 4 Wehrli (D.L. 5. 37): ἐν ταύτηι τῆι ἐπιστολῆι σχολαστικὸν ὠνόμακε ⟨ἑαυτόν⟩; cf. Regenbogen, *RE* Suppbd. vii, col. 1359.

[6] See below, pp. 171, 182.

[7] For the *CP* see the elaborate three-volume Loeb edition by Einarson and Link (Harvard, 1976–90). For one example of explicit personal knowledge out of many see his remarks on pruning the vine, 3. 13. 3–4, beginning τὴν δ' ἄμπελον ἄγειν δεῖ κύκλωι περὶ

which he bequeathed to Kallinos of Troizen in his will,[8] and he also had a garden of his own after the death of Aristotle in or near the Lykeion, somewhere near the Ἐθνικός Κῆπος of today, and to be distinguished from the Botanical Garden of the School (D.L. 5. 52);[9] and he presumably cultivated it attentively, though he nowhere refers to personal experiments and only once, almost discreetly, admits to having seen a plant.[10] Contrast the opening statement in the On Weather Signs: 'We have here described the signs of rain, wind, storms, and fair weather, in so far as that has proved possible, some from my own prognostications and others on the authority of trustworthy persons.'[11] In botany we may contrast the approach of Dioscorides of Anazarbos, who states at the beginning of his De Materia Medica that he has tried personally to examine all the herbs to which he refers (prol. § 5; ext. 3): 'We ask you and those who come across this work not to bother about the style of my writing, but to estimate my skill and experience in the subject. For I have for the most part investigated the subject under discussion personally with the greatest care, and where that was not possible have taken every care to employ reliable local information, and thus attempt to employ distinctions of type and to record the qualities of the various species.' As a Roman army doctor he made use of his opportunities: 'as you know, I have lived a soldier's life,' he says to his friend Areios.

Theophrastus lived through the reigns of Philip II, of Alexander, and through most of the restless generation of the Diadochi. But he rarely refers to contemporary events, other than very local ones for purposes of dating (e.g. by Athenian archons). He does not generalize about the

τὸν πυθμένα, πανταχόθεν γὰρ ὁμαλοῦς οὔσης καλλίων καὶ εὐκαρποτέρα· τοῦτο δ' οὐ χαλεπόν, ἐάν τις καταλίπηι μὴ τὰ κάλλισθ' ὡρμηκότα τῶν κλημάτων, ἀλλὰ τὰ ἄριστα πεφυκότα πρὸς τὴν ἀγωγήν. The whole passage, which I cannot quote here, indicates practical experience. Capelle's paper in the Festschrift for E. Zucker (1954), pp. 47 ff., 'Der Garten des Theophrast' is a delightful account of Theophrastus' practical knowledge of husbandry, in his own garden, his ἴδιος κῆπος (D.L. 5. 39).

[8] D.L. 5. 52.

[9] For the approximate location of the Lykeion see C. Edward Ritchie, jun., in Φίλια Ἔπη (Studies in Honour of G. Mylonas), iii (Athens, 1989), 250 ff. It probably occupied the area between, and including, Philhellenon St. and the Ethnikos Kepos.

[10] HP 4. 12. 2: of two types of reed, σχοῖνος, growing from one root: καὶ θαυμαστόν γ' ἦν ἰδεῖν ὅλης κομισθείσης της σχοινιᾶς. The verb presumably indicates that someone had brought the specimen into the lecture room.

[11] The De Signis (Περὶ Σημείων) is also edited at the back of vol. ii of the Loeb Enquiry: Σημεῖα ὑδάτων καὶ πνευμάτων καὶ χειμώνων καὶ εὐδιῶν ὧδε ἐγράψαμεν καθ' ὅσον ἦν ἐφικτόν, ἃ μὲν αὐτοὶ προσσκοπήσαντες ἃ δὲ παρ' ἑτέρων οὐκ ἀδοκίμων λαβόντες. This small treatise, whose authorship is disputed, but which contains Theophrastean material, if not his ipsissima verba, is full of vivid observations of weather signs, still easily recognized in Greece today, and is an excellent vade-mecum for the traveller.

significance of Alexander's campaigns for botanical studies, in the way that Strabo, following the lead of Eratosthenes, refers to the great increase in geographical knowledge resulting first from Alexander's campaigns, and in due course from Rome's Parthian wars—perhaps the perspective was still too close for generalizations.[12] Yet the *HP* presupposes Alexander's campaign at many points. Moreover, Theophrastus names his sources very rarely, and when he does they tend to be pre-Socratics, notably Menestor of Sybaris and Demokritos, who had interested themselves in plant-behaviour and plant-nomenclature. A splendid example of his taciturnity is provided by his statement (*HP* 2. 3. 3) when speaking of the capacity of apparently dead trees to renew themselves in certain circumstances: he says, 'for instance, there was an olive that, after having been totally burnt down, sprang up in its entirety, the tree and all its branches.'[13] That is a reference to the sacred olive-tree on the Acropolis, in full sight of the Lykeion, which the Persians burned down in 479 (Hdt. 8. 55). Taciturnity could go no further. We may contrast the doxographical *De Sensibus*, in which he analyses the arguments of most of the pre-Socratics one by one.[14] The contrast may reflect the lack of predecessors available to Theophrastus in botany.

The two works of botany were conceived along with the *De Odoribus* as a single whole. In the *HP* he frequently refers back to an earlier statement, with some such formula as ὥσπερ ἤδη εἴρηται, ὥσπερ ἐλέχθη, etc., and once at least he refers forward, from book 4 to book 9: περὶ ὧν ἐν ἄλλοις εἴρηται διὰ πλειόνων, though here the tense must make us wonder whether he is referring to a different context. Similarly, in the later work, the *CP*, he refers back very specifically to the *HP*, and in the *De Odoribus* to the *CP*.[15] While, however, it is true that Theophrastus never, I think, refers to previous observations of his own as if they were *written* statements, we should not on that account suppose that the botanical works as they survive are actual lectures, which can even now be grouped according to a systematic and developing range of topics. This view has been advanced at length by one of the most significant of recent writers on Theophrastus, Steinmetz, and even if we cannot safely press a corrupt and interpolated text, composed in an elliptical, terse

---

[12] For Strabo 14 and 48 (Berger, Erat. I B10 and 11).

[13] *HP* 2. 3. 3: οἷον ἐλάα ποτ' ἀποκαυθεῖσα τελέως ἀνεβλάστησεν ὅλη, καὶ αὐτὴ καὶ ἡ θαλία.

[14] See Stratton (n. 2 above), 51 ff.

[15] For references in the *CP* to the *HP* see e.g. *HP* 4. 5. 6, a reference to αἱ ἐν ταῖς ἱστορίαις εἰρημέναι μεταβολαί = *HP* 2. 2. 4–2. 4. 4. For a reference back in the *De Odor.* to the *CP* see *De Odor.* 11, fin.: ὥσπερ καὶ πρότερον ἐλέχθη = *CP* 6. 12. 9 (not precise).

style so far, we must admit that both works have an oral quality about them, and closely resemble in this respect the biological and zoological works of his master.[16] We should also notice the frequency with which he refers to the need for future research to clarify a point: this is almost invariably presented in some such formula as τοῦτο οὖν ἐπισκεπτέον or περὶ μὲν τούτων ἐπισκεπτέον, 'this requires looking into more closely', and other oral rather than written formulae, marking the end of a discussion or the existence of an *aporia*, very characteristically Aristotelian.[17] In the earlier work, the *HP*, he refers for dating purposes to the Athenian archons of 314 and 310, and his latest references are to the campaign of Ophellas in north Africa in 308, and the capture of Megara by Demetrius Poliorcetes in 307.[18] A reference to the same Demetrius' construction of a *hendekere* may point to a slightly later date, during the period of Demetrius' considerable activity at sea; his father Antigonus is not mentioned at this point.[19] So the main text of *HP* must have been completed just about at the turn of the century; the *CP*, undoubtedly later, must belong to the last decade of his life.[20] In assessing the sources of information available to Theophrastus we must bear in mind the late date of the botanical works. As he compiled his notes about the botany of the east, he probably had access not only to the earliest historians of Alexander's campaign, but also, we may suppose, to information collected by the first generation of Greek exploitation of the east under its new rulers. I shall come back to this.

---

[16] See P. Steinmetz, *Die Physik* (n. 1 above), 14ff., for a consideration of the various possibilities. The contributions of G. Senn to the study of the text suffer from over-analysis in respect of interpolation etc. See especially his (posthumous) *Die Pflanzenkunde des Theophrast von Eresos* (herausgeg. u. eingel. von O. Gigon; Basle, 1956).

[17] Examples of all these formulae are very frequent at the end of individual discussions both in *HP* and in *CP*. Their brevity is sometimes quite startling: e.g. *CP* 2. 17. 10: καὶ περὶ μὲν τούτων ἅλις, vel sim. For the ἀπορίαι see e.g. *CP* 1. 11. 3, fin.: αἱ μὲν οὖν ἀπορίαι σχεδὸν αὗται καὶ τοιαῦται εἴρηνται περὶ τούτων. Such formulae are much less frequent in Aristotle's *HA*.

[18] See *HP* 5. 2. 4: ὅπερ [the envelopment of an object enclosed in a cavity in a tree by the growth of surrounding wood] καὶ περὶ τὸν κότινον συνέβη τὸν ἐν Μεγάροις τὸν ἐν τῆι ἀγορᾶι· οὗ καὶ ἐκκοπέντος λόγιον ἦν ἁλῶναι καὶ διασπασθῆναι τὴν πόλιν· ὅπερ ἐγένετο . . . Δημήτριος.

[19] Ibid. 5. 8. 1: μῆκος μὲν ἦν τῶν εἰς τὴν ἐνδεκήρη τὴν Δημητρίου τμηθέντων τρισκαιδεκαόργυιον, αὐτὰ δὲ τὰ ξύλα τῶι μήκει θαυμαστὰ καὶ ἄοζα καὶ λεῖα. The significance of the passage for dating the whole work is not certain: see Regenbogen, *RE* Suppbd. vii, col. 1453; for Antigonus in *HP* 4. 8. 4 see ed. Budé p. xix.

[20] There is a reference in *De Lap.* §59, apropos of a method of washing gold, to the archon Praxiboulos of 315 BC: οὐ παλαιὸν δ' ἐστὶν ἀλλὰ περὶ ἔτη μάλιστ' ἐνενήκοντα εἰς ἄρχοντα Πραξίβουλον Ἀθήνησι.; cf. the discussion by Eichholz in his edn., pp. 8ff. The *Characters* have no authentic internal date: see Regenbogen, op. cit. cols. 1510–11.

We must now consider whom he may have used for his information regarding the east. Characteristically, he does not supply a cut-and-dried answer himself; though he refers to authorities for botanical and other information concerning the east, he always does so in a totally neutral way. He normally says simply φασί, λέγουσι, ἔλεγον, and so on, and, as with other authors (for instance, Arrian), we have to decide how these vague expressions are to be interpreted. I take it as axiomatic that when he uses these and similar verbs he is indicating that he has no first-hand knowledge of the matter; the information he has received may be oral or written, but it is at second hand, and its reliability is not guaranteed by Theophrastus. I similarly interpret the very numerous passages in *oratio obliqua* without a governing verb. Of course, the reverse is not true: we cannot assume that when he writes in *oratio recta* he is invariably recording direct experience: passages describing phenomena that he cannot have seen are often couched in direct speech. These forms of expression are very relevant when we consider the references to Alexander's campaigns, and the panorama of Asia, which constitutes the background of one major section of the *HP*, namely books 4 and 5. It is possible that in some instances so expressed, namely those dealing with phenomena observed on the Greek mainland, these conventional phrases recording anonymous source-material refer to reports transmitted by his pupils, but I do not believe that to be true with regard to eastern botanical data. Let us now look at the books 4 and 5 more closely with the question of sources in mind.

We may begin our enquiry at the other end of the botanical tradition. Pliny frequently refers to material provided by members of Alexander's expedition, whom he simply calls 'Macedones'. He complains that the 'Macedones' did not name all the plants they recorded: 'Genera arborum Macedones narravere ex parte sine nominibus.'[21] The reason was that the Macedonians were seeing many plants and trees for the first time, and did not know what to call them; that is one reason why Theophrastus uses classification by similarity and difference so much: *a*, the unknown, is like *b*, the known, and he himself says in comparing the observation of wild and cultivated species, 'most of the wild species are nameless, and only a few people have direct experience of them, but of cultivated species the majority are identified by name and widely observed';[22] and he explains why there was no recognized nomenclature:

---

[21] *NH* 12. 25.

[22] *HP*. 1. 14. 4: τῶν μὲν ἀγρίων ἀνώνυμα τὰ πλεῖστα καὶ ἔμπειροι ὀλίγοι, τῶν δὲ ἡμέρων καὶ ὠνομασμένα τὰ πλείω καὶ ἡ αἴσθησις κοινοτέρα. A good example of the need

'There are also many other trees which are different from those found in Greece, but have received no name. There is nothing remarkable in the singularity of these trees, for, as some maintain, with a few exceptions, there is scarcely a tree, shrub, or plant that is similar to those in Greece.'[23] These passages show how much in the dark the Macedonian or other ἔμπειροι must have been. How did the information they collected reach Theophrastus?

Hugo Bretzl, in his excellent book *Botanische Forschungen des Alexanderzuges* (1903), maintained that Alexander had been accompanied by a corps of scientists, who had either sent back, or brought back, detailed information for the use of the Peripatos, that is, of Aristotle and Theophrastus.[24] This view, presented with great clarity and cogency, was substantially developed by F. Pfister, who elaborated the idea of a fully organized scientific team, continually sending reports back to the Lykeion, after they had been vetted for accuracy by Alexander himself, and he went so far as to describe Alexander's expedition as a sort of field-expedition on behalf of the Lykeion, a prototype of Smithsonian Expeditions.[25] This view was perhaps prompted by the analogy of the corps of scientists who accompanied Napoleon to Egypt, and subsequently produced the folios of the *Déscription de l'Égypte*. The end-product of Pfister's argument, as also more guardedly of that of Bretzl, was that there was a complete archive of scientific data in Babylon, and another in Athens in the Lykeion, which preserved all the scientific material for posterity. This notion derives basically from the statement of Strabo that, according to Patrocles, the man who explored the Caspian for Seleucus I, a certain Xenocles gave him the material relevant to his expedition that was available in Babylon:[26] '[Patrocles said that] those who had accompanied Alexander had investigated everything in a hurry, and Alexander had scrutinized all items, when his experts had drawn up their records of the entire region; and he said that the whole report was given to him subsequently by Xenocles the keeper of the Treasury (sc. in Babylon).'

As Strabo makes clear, there was no doubt in the Hellenistic age as to the reliability of Patrocles,[27] and his statement should not encourage us

to illustrate unknown by known is provided in the description of the fruit of the banyan-tree (see below, p. 175), which (4. 4. 4) is said to be as small as a chick-pea, but to resemble a fig: καρπὸν δὲ σφόδρα μικρὸν ἡλίκον ἐρέβινθον ὅμοιον δὲ σύκωι.

[23] *HP* 4. 4. 5 (ext. 4).   [24] See pp. 3 ff. and *passim*.

[25] See *Historia* 10 (1961), 30–67, 'Das Alexander-Archiv und die hellenistisch-römische Wissenschaft'.   [26] Strab. 69 = *FGrHist* 119 T 3; 712 F 1 (ext. 5).

[27] Id. 68–9, describing the opposing views of Eratosthenes and Hipparchus regarding Patrocles: see the brief account of Patrocles in *Ptol. Alex.* i. 535, with notes 125 ff. in vol. ii, p. 768.

to accept the extreme hypothesis that there was a staff of trained scientists, subject directly to Alexander's supervision, distinct from the βηματισταί, though there are a number of texts which can be invoked to support it. Let us take the clearest example of the point at issue. Theophrastus' description of the banyan-tree, ἡ Ἰνδικὴ συκῆ, the *Ficus bengalensis*, shows a correct understanding of the way that the accessory, vertical roots develop on the previous year's growth or on even older branches, whereas Aristoboulos and Onesikritos who both described the same tree (ap. Strab. 694; *FGrH* 134 F 22; 139 F 36)—it was indeed the most remarkable tree the expedition encountered—wrongly maintained that the growth was on the new wood.[28] Bretzl says that since Theophrastus had certainly never seen a banyan-tree he must, in these circumstances, have owed his superior information to a member of the Scientific General Staff.[29] But Theophrastus may well have derived his information from contemporary witness(es), Nearchos, perhaps, or someone unknown to us, or from a later traveller, like Patrocles or Megasthenes (though these two may be too late to fill the bill), who had time to study the tree over a cycle of growth. The tradition that Aristotle expressed annoyance that his nephew Kallisthenes did not send him scientific information as he had promised to do, could be, but need not be, taken as an indication that there was no elaborate organization at work, establishing a detailed archive in the Lykeion and in Alexander's headquarters in Babylon.[30] In any case, leaving on one side the direct alternative, was there or was there not a Scientific Headquarters Staff, we may consider what sources Theophrastus could have used, and in some cases did use. There were several accounts of the campaign, or parts of the campaign, which were in circulation at the time Theophrastus was writing *HP* 4 and 5, though we do not know very precisely when they individually became available. Theophrastus himself never refers to any of them by name. We think particularly of Kleitarchos, Aristoboulos,

---

[28] See *HP* 4. 4. 4: ἡ δὲ Ἰνδικὴ χώρα τήν τε καλουμένην ἔχει συκῆν, ἣ καθίησιν ἐκ τῶν κλάδων τὰς ῥίζας ἀν' ἕκαστον ἔτος, ὥσπερ εἴρηται πρότερον· ἀφίησι δὲ οὐκ ἐκ τῶν νέων ἀλλ' ἐκ τῶν ἔνων καὶ ἔτι παλαιοτέρων. For the accounts of Aristoboulos and Onesikritos, both in Str. 694, see *FGrHist* 134 F 22 and 139 F 36; cf. Bretzl, *Botanische Forschungen*, 158ff., who points out (p. 164) that Theophrastus' first reference to the banyan, in 1. 7. 3, indicates the same error as that transmitted by Aristoboulos and Onesikritos: ἰδίᾳ δὲ ῥίζης φύσις καὶ δύναμις ἡ τῆς Ἰνδικῆς συκῆς· ἀπὸ γὰρ τῶν βλαστῶν ἀφίησι, μέχρι οὗ ἂν συνάψῃ τῆι γῆι καὶ ῥιζωθῆι, καὶ γίνεται περὶ τὸ δένδρον κύκλωι συνεχὲς τὸ τῶν ῥιζῶν οὐχ ἁπτόμενον τοῦ στελέχους ἀλλ' ἀφεστηκός. It is surprising that the error, however it arose, remained uncorrected; cf. ed. Budé, ad loc.

[29] Ibid.

[30] See *FGrHist* 124 T 3 (ext. 6).

Onesikritos, Nearchos, and Androsthenes and the bematists, Baton, Diognetos, and the rest, the latter the only ἔμπειροι whom we can be certain to have accompanied Alexander, and the scanty fragments of whose ἀναγραφαί indicate that they went some way beyond mensuration and recorded varied natural observations. See, for example, Amyntas,[31] who though he may belong to the Seleucid period, will serve as an example of what such bematists' reports made to Alexander probably contained: a description of the ἀερόμελι, oak-manna, in some unspecified locality, and a long account (*FGrHist* 122 F 3) of a plague of mice round the Caspian and at Teredon, at the mouth of the Euphrates. Theophrastus was much indebted in one particular item to the two explorers, Nearchos and Androsthenes. The account left by the former of his journey up the Persian Gulf from the Indus Delta to the head of the Gulf, is preserved in an incomplete summary by Arrian in his *Indica*. A section not preserved by Arrian or Strabo, dealing with mangrove swamps in the Gulf of Hormuz, survives in *HP* 4. 7. 4,[32] where it is followed almost immediately by an extract from Androsthenes of Thasos' narrative of his voyage down to Tylos, Bahrain, ibid. § 7,[33] the other important description of the swamps. Both these accounts are in anonymous *oratio obliqua*, though for Nearchos he does say οἱ ἐξ' Ἰνδῶν ἀποσταλέντες. He quotes Androsthenes by name in *CP* 2. 5. 5 for the quality of water in Bahrain (prefacing his quotation with the words εἰ δὲ ἀληθὲς ἔλεγεν Ἀνδροσθένης).[34] The descriptions of the mangrove swamps in Theophrastus, based on these two accounts, remained the essential description of them until the preparation of Admiralty charts early in the last century. Both the voyages were directly ordered by Alexander, and thus have been used to support the notion of his personal supervision of scientific activity; his decision in the one case was an expression of his πόθος,[35] in the other the purpose was apparently exploratory. Sometimes Theophrastus preserves information that has not survived in the Alexander-historians. Such are his statements that Alexander forbade his troops in India to eat mangoes because they caused diarrhoea (ext. 7), and that a certain species of corn in the land of the Pissatoi (otherwise unattested, but possibly to be located somewhere in the eastern Iranian provinces) had a disastrous, indeed fatal, effect on those who ate it: they burst asunder, 'and many of the Macedonians

---

[31] *FGrHist* 122, esp. F 3 on ἀερόμελι, the tree-honey.
[32] Ibid. 133 F 34 (in small print, i.e. unauthenticated as *ipsissima verba*).
[33] Ibid. 711 F 5.
[34] Ibid. F 3.
[35] Ibid. 133 F 20.

suffered this fate'.[36] The information is given in *oratio obliqua*. And again we are told of the effect of a certain Indian barley, good for bread and porridge, but fatal for horses till they became used to it.[37] And yet again, of the asafoetida that grew in Aria (ext. 8; the text is seemingly deficient at one point): 'And there is another shrub as large as a cabbage, the leaf of which resembles that of the bay in shape and size. If anybody eats this he dies forthwith. Consequently if there were horses in its proximity(?) they kept them on a tight rein.' These passages all show the same use of recorded information without even a φασί as protection for himself. This raises the possibility that Theophrastus may have acquired first hand, oral information, both botanical and general. Is it likely?

It is clear on the one hand that there was casual information to be picked up from returning warriors, veterans of Alexander's campaigns, just the sort of people to remember a sharp attack of 'front-line diarrhoea' when about to launch an attack; such a one as the ἀλάζων depicted in *Characters* 23 makes himself out to be, οὐδαμοῦ ἐκ τῆς πόλεως ἀποδεδημηκώς (like Theophrastus himself). Less imaginary are the soldiers and their commander, perhaps an Athenian, who made a dedication at about the time of Alexander's expedition, on the island of Ikaros, Phailaka in the Gulf.[38] At the same time Alexander's campaigns had opened the gates of Asia, and the Greeks came flooding in. Signs of this are much more apparent now than they were twenty or thirty years ago. Greeks have always been very quick to develop new trading contacts, and the existence of new Greek settlements must have opened the way to frequent intercourse between them and the motherland. We think of Ai Khanum on the Oxus, and of the long journey thither from Delphi made by Theophrastus' friend and fellow pupil of Aristotle, Klearchos, author of a Περὶ Ἰνδῶν, written just at the time that Theophrastus was writing his account of eastern botanical geography;[39]

[36] *HP* 8. 4. 5: καὶ περὶ τὴν Ἀσίαν οὐ πόρρω Βάκτρων ἐν μέν τινι τόπωι οὕτως ἁδρὸν εἶναί φασι τὸν σῖτον ὥστε πυρῆνος ἐλαίας μέγεθος λαμβάνειν, ἐν δὲ τοῖς Πισσάτοις καλουμένοις οὕτως ἰσχυρὸν ὥστ' εἴ τις πλεῖον προσενέγκοιτο διαρρήγνυσθαι, καὶ τῶν Μακεδόνων καὶ πολλοὺς τοῦτο παθεῖν. The tribal name has caused difficulty, and Einarson, 'The Manuscripts', 76, emended the Πισσάτοις of the main manuscript, U, to Πεσσόγοις, a Phrygian tribal region known from a Pergamene inscription, *OGIS* 315, A, 6. There seems to me to be no likelihood that Theophrastus had an Anatolian tribe in mind; his immediately preceding example, in the same sentence, is drawn from corn in Bactra, οὐ πόρρω Βάκτρων.

[37] *HP* 4. 4. 9: κριθαὶ δὲ καὶ πυροὶ καὶ ἄλλο τι γένος ἀγρίων κριθῶν, ἐξ ὧν καὶ ἄρτοι ἡδεῖς καὶ χόνδρος καλός. ταύτας οἱ ἵπποι ἐσθίοντες τὸ πρῶτον διεφθείροντο, κατὰ μικρὸν δὲ οὖν ἐθισθέντες ἐν ἀχύροις οὐδὲν ἔπασχον.

[38] This notable dedication has been studied and discussed many times. See Sherwin-White and Roueché, *Chiron* 15 (1985), 4ff. (*SEG* xxxv. 1477).

[See next page for n. 39]

we think of the Greek dedication at Kandahar on the road to Baluch-istan, and of Asoka's Buddhist commandments inscribed in Greek on the rock-face there, perhaps a generation later, to catch the eye of passing Greek merchants.[40] Fresh scientific knowledge quickly accrued from this newly opened world which came into being in the brief but significant span of years between the death of Alexander and the death of Theo-phrastus. The link between Klearchos and Theophrastus is worth noting because Klearchos was particularly interested in the gymnosophist Brahmin philosophers, who went about naked; while Theophrastus for his part tells us[41] that there is a tree, 'of great size with a wonderfully sweet and large fruit', in other words the banana-tree, known to botan-ists by Linnaeus' descriptive appellation, *Musa sapientum*, and, he goes on, 'the Indian Wise Men who wear no clothes eat it'.

That the human and cultural traffic was not all in one direction we may infer from an interesting metrical, or almost metrical, epitaph recently published from the area of Didymoteicho in eastern Thrace, seemingly of the third century, in which a man commemorates the death of his brother, a Greek from Babylon.[42] He says: 'Truly, far from famous

[39]   Again a notable text, published originally in a masterly article full of historical ima-gination by L. Robert, in *CRAI* (1968), 416–57, and reproduced many times subsequently.

[40]   Asoka's texts are best studied in L. Robert, *JA* (1958), 7–18. For the dedication at Old Kandahar see P. M. Fraser, 'The Son of Aristonax at Kandahar', *Afghan Studies* 2 (1979), 9–21.

[41]   *HP* 4. 4. 5, the text of which is rearranged by Bretzl, *Botanische Forschungen*, 171–2, 194–5. The text of Theophrastus gives two separate plants: first that quoted in the text, ἔστι δὲ καὶ ἕτερον δένδρον καὶ τῶι μεγέθει μέγα καὶ ἡδύκαρπον θαυμαστῶς καὶ μεγαλό-καρπον· καὶ χρῶνται τροφῆι τῶν Ἰνδῶν οἱ σοφοὶ καὶ μὴ ἀμπεχόμενοι; and in con-tinuation, ἕτερον δὲ οὗ τὸ φύλλον τὴν μὲν μορφὴν πρόμηκες, τοῖς τῶν στρουθῶν πτεροῖς ὅμοιον, ἃ παρατίθενται παρὰ τὰ κράνη, μῆκος δὲ ὡς διπηχυαῖον. The first of these was tentatively identified by Hort, ii, Appendix, p. 484, no. 3 with the Jacktree, the *Artocarpus integrifolia*, the second with the banana, the *Musa sapientum*. However, Bretzl (pp. 191 ff.) showed beyond any reasonable doubt that the text is confused, and that there is a descrip-tion of only one plant here, namely the banana, Linnaeus' *Musa sapientum*, and that the intervening passage does not refer to a separate tree. Since this important correction is not made by Hort, I give the text below as Bretzl re-established it. It seems certain that Bretzl was also right in regarding the description of the leaf of the banyan-tree, wrongly said in 4. 4. 4 to be of the size of a πέλτη (τὸ δέ γε φύλλον οὐκ ἔλαττον ἔχει πέλτης), as belonging to the banana, though it is not clear how this clause is to be associated with the rest of the passage, and I therefore do not reproduce it below. Bretzl (pp. 191 ff.) shows that this error arose from a confusion, which began with the original observations on the ground by the Greeks (as Bretzl imaginatively supposed, when they encountered the Indian Wise Men) and survived until the period of Portuguese expansion, between the banyan and the banana-tree. The corrected text runs (§ 5): ἔστι δὲ καὶ {ἕτερον} δένδρον καὶ τῶι μεγέθει μέγα καὶ ἡδύκαρπον θαυμαστῶς καὶ μεγαλόκαρπον· καὶ χρῶνται τροφῆι τῶν Ἰνδῶν οἱ σοφοὶ καὶ μὴ ἀμπεχόμενοι, {ἕτερον δὲ} οὗ τὸ φύλλον τὴν μὲν μορφὴν πρόμηκες, τοῖς τῶν στρουθῶν πτεροῖς ὅμοιον, ἃ παρατίθενται παρὰ τὰ κράνη, μῆκος δὲ ὡς διπηχυαῖον.

[See opposite page for n. 42]

Babylon thou liest, Apollonios, in this tomb, where oft thy brother
Agathon sits and mourns thee with grief and tears, unassuagable grief in
his heart.'

The occupation of settlements in conquered territory, scattered and
isolated though they were, must have led to a considerable improvement
in agriculture and fruit-growing, as it did later when the fortunate
mandarins of Ptolemy II were able to exploit the Fayyūm, and bring it to
new levels of fertility. We may wonder whether, when Theophrastus
refers to plant-forms and to methods of planting and so on in the east, he
is thinking of native husbandry, or of that of the Greek settlers, who
inherited, and improved the methods of their largely Iranian predecessors; of the Greek or the Iranian inhabitants of the fertile oases of Herāt
and Kandahar. In this connection we may note the early experiments in
acclimatization recorded by Theophrastus, of which the most notable
was the attempt of Harpalus to grow ivy in the Gardens of Babylon (ext.
9): 'And indeed Harpalus struggled hard in the gardens around Babylon,
making frequent attempts at planting, but achieved nothing. For the ivy
was unable to survive, unlike the other plants from Greece, for the region
will not support it on account of the atmosphere. And it only grudgingly
accepts box and hazel; for the people who work in the parks have to
struggle very hard to keep box and lime alive.' Who are 'they'? Native
gardeners, or Greek or Macedonian supervisors or experimentalists?

Even so, when due allowance has been made for the changed conditions created by the disappearance of the Persian empire and the availability of new botanical and scientific data in the generation after
Alexander's death, it remains likely that most of Theophrastus' information on eastern matters in books 4–5 came from the bematists and
explorers. Beyond that, I find it difficult to accept the notion of Alexander's direct, personal, regular involvement in the sifting of such information, even though Pliny in the index to book 6 refers to 'Alexander
Magnus' as a literary source for the geography of the east. There is, it
may be noted, no reference to such a scientific interest of Alexander's in
Plutarch's *Life*, which does not miss much edifying information about
his hero—or indeed in any ancient source.

---

[42] Ἀρχ. Δελτ. 35 (1980 (1988)), 435 and pl. 256γ (=*SEG* xxxviii. 734): Ἡ μάλα
πατρίδο/ς εὐρυβότου τη/λοῦ Βαβυλῶνος τῶ/ιδ', Ἀπολλῶνι(ε), τύμ/βωι, κέκλισαι, οὔ σε
κ/ασίγνητος πόλλ/ακις ἐζόμενος σ/τενάχησε Ἀγάθω/ν, κατὰ θυμὸν πένθ/ος ἑλὼν ἀλία-
στον/δακρυοέντι γόωι. The stele is a small, rather crudely worked pedimental stele with
three acroteria, the lettering neat, square, and unapicated. A date in the middle or second
half of the third century BC seems likely. The simple epitaph gives no indication of how the
two brothers came to find themselves in Thrace.

I must say a word here about the possibility that Theophrastus had himself visited Egypt and Cyrene. This was suggested by Capelle, a very knowledgeable student of Theophrastus, and an enthusiastic botanist, on the ground that some of his descriptions of the flora and of the agricultural cultivation of these two regions could only be explained by autopsy, and indeed, in the case of Egypt, by long residence.[43] A botanical Herodotus, then. I cannot believe that the tradition that Theophrastus refused an invitation by Ptolemy I to visit Alexandria, and sent the fallen tyrant Demetrius instead, would have gained currency if in fact he had lived for a long period in Egypt; and moreover that, if he did, Diogenes' *Life* of him would contain no hint of it. On the contrary, the *Life* says only that Ptolemy 'sent for him', and emphatically implies that his only absence from Athens, once he took up residence there, was during his brief exile under the Law of Sophocles in 306 BC (ext. **10**): 'Being such a man he yet nevertheless absented himself briefly from Athens, along with all the other philosophers, when Sophocles the son of Amphikleides introduced a bill that none of the philosophers should preside over a school other than with the agreement of the Council, and the Assembly.' Here the implication of 'being such a man, yet nevertheless he absented himself briefly along with the other philosophers' is absolutely clear: his absence on this occasion was an exception in a life lived largely in Athens. Moreover, though in the main sections on Egypt and Cyrene he writes without reference to any sources of information (much as he does in the passages relating to India which I have already discussed) there are descriptions of other items in the cultivation of crops in those two regions that are introduced by φασί. This seems virtually to exclude autopsy, for who would say, 'they say' or 'it is said' if he knew a fact from personal experience? Egypt from Elephantine to the Delta was familiar to Greeks of the most varied callings, but especially the profession of arms in the fifth and fourth centuries BC. They have left their names and ethnics inscribed on a score of temples from the archaic period onwards, from Middle Egypt to Nubia and out into the Eastern Desert, and there is no difficulty in supposing that information on the unchanging native system of husbandry was available to Theophrastus in Athens. It was in 308 BC when he was in the middle of writing or revising the *HP* that a marriage-contract, a συγγραφὴ συνοικισίας, was signed in Elephantine between a soldier from the Aeolic city of Temnos, and a Coan girl.[44] The witnesses are a Sicilian from Gela, three fellow-Temnites

---

[43] See W. Capelle, 'Theophrast in Ägypten', *Wien. Stud.* 69 (1956), 173 ff.

[44] *PEleph.* 1. A will from Elephantine of a slightly later date, 285 BC (ibid. no. 8), is made

of the groom, a Cyrenaean, and a Coan. Other documents of the early Ptolemaic period from Elephantine show us the great variety of Greek settlers and soldiers in Egypt even before Ptolemy II established a pattern of cleruchic settlements and δωρεαί in the Fayyum. Such people are likely to have provided much of the practical information that enabled the later settlements to establish themselves so quickly, and they are also the sort of people by whom information would be brought back to Greece. For example, Theophrastus says that 'they say' that around Elephantine the vines and figs are not deciduous.[45] That surprising phenomenon would be noticed by any Greek in his new climatic environment, and could readily be checked. I also believe that Theophrastus did not need to go to Cyrene to study so staple an item of the materia medica as silphion, which had aroused the interest of the Greek world from an early date. It is true that he makes a very specific reference to the long march of Ophellas from Cyrene to Carthage in 308 BC, on which the troops supported themselves almost entirely on silphion.[46] Capelle maintains that such a precise piece of information could only have been acquired by Theophrastus on the spot. But, if it was, why does Theophrastus use his neutral formula, 'And they say that they subsisted on this for a considerable period in the absence of other supplies'? Such information is more likely to have reached Theophrastus by word of mouth from a returning member of that expedition, than by information collected on the ground; in any case, any visit he made there, for its results to be incorporated in the *HP*, must almost certainly have preceded the march of Ophellas, which is chronologically about the latest event referred to in the work. Once again intercourse, commercial and cultural, was frequent between Cyrene and the rest of the Greek world in the fourth century BC, as long before. Both the use of φασί in the passage just quoted, and the extended use of *oratio obliqua* in several other passages referring to both Egypt and Cyrene seem to me to determine the matter.[47] I see no reason or justification to invoke his hypothetical pupils as hypothetical intermediaries.

by another Temnite in favour of his wife, also a Temnite, and witnesses this time are an Arcadian, a Cretan, a Maronaean, and two more Temnites.

[45] *HP* 1. 3. 5 apropos of the distinctions between fruitless and unfruitful, flowering and flowerless, which he says are due to the τόπος and the ἀήρ: περὶ γὰρ Ἐλεφαντίνην οὐδὲ τοὺς ἀμπέλους οὐδὲ τὰς συκᾶς φασι φυλλοβολεῖν.

[46] 4. 3. 2: πολὺ δὲ τὸ δένδρον [sc. τὸ σίλφιον] καὶ πολύκαρπον· τό γ' οὖν Ὀφέλλου στρατόπεδον, ἡνίκα ἐβάδιζεν εἰς Καρχηδόνα, καὶ τούτωι φασὶ τραφῆναι πλείους ἡμέρας ἐπιλιπόντων τῶν ἐπιτηδείων. For Capelle's view see 'Theophrast in Kyrene?', *Rh. Mus.* 97 (1954), 169 ff.

[47] As particularly in 4. 3. 4–5, a long passage in oratio obliqua introduced by ἔνιοι δέ.

We must turn now to the second and shorter part of this paper, Theophrastus' knowledge of the western Greek world, and of the non-Greek parts of Italy. The first point to note is that there are almost no references to the botany of Magna Graecia and Sicily in the *HP* and only a few in the *CP*, even though among the few quoted sources are the pre-Socratics, Menestor of Sybaris, Empedokles of Akragas, and Hippon of Samos or Metapontion or Rhegion. To the first of these ancient figures (οἱ παλαιοὶ τῶν φυσιολόγων, as he calls them in *CP* 6. 3. 5) we may attribute, in addition to one or two specific quotations, the statement about the oak at Sybaris:[48] ἐν δὲ Συβάρει δρῦς ἐστιν [notice the tense: Sybaris itself was not inhabited in the fourth century] εὐσύνοπτος ἐκ τῆς πόλεως ἢ οὐ φυλλοβολεῖ· φασὶ [we may wonder that he did not use φασί at the beginning of the whole sentence] δὲ οὐ βλαστάνειν αὐτὴν ἅμα ταῖς ἄλλαις ἀλλὰ μετὰ Κύνα (i.e. in July–August). εὐσύνοπτος ἐκ τῆς πόλεως indicates autopsy on somebody's part, Menestor's maybe, but not that of Theophrastus. So, though we have only a tiny part of Theophrastus' output, and although there are a number of isolated references to legal and other customs in some South Italian cities in his lesser works, some of them very strange (as for instance that it was a capital offence to drink wine in Epizephyrian Locri except on doctors' orders),[49] it is clear that in the *HP* he was less concerned with the flora of Italy and Sicily than, let us say, that of Babylonia or the Arabian Gulf. That is to be expected. They were overall largely the same as those of Greece, unlike the eastern plants and trees, and therefore called for no comment unless exceptional in some way; moreover there was no novelty about them, as there was about so much of what Alexander's campaign in the east had revealed.

Nevertheless, there are two isolated passages in the *HP* which, though not botanical, may suggest what notion he had of the non-Greek regions of Italy. The first relates to the well-known Greek or Hellenized legend

[48] I. 11. 5.

[49] fr. 117 (Ath. 429a): παρὰ Λόκροις τοῖς Ἐπιζεφυρίοις εἴ τις ἄκρατον ἔπιε μὴ προστάξαντος ἰατροῦ θεραπείας ἕνεκα θάνατος ἦν ἡ ζημία Ζαλεύκου τὸν νόμον θέντος. παρὰ δὲ Μασσαλιώταις ἄλλος νόμος τὰς γυναῖκας ὑδροποτεῖν· ἐν δὲ Μιλήτωι ἔτι καὶ νῦν φησί Θεόφραστος τοῦτ' εἶναι τὸ νόμιμον. It is not clear from the context that Theophrastus is the source of the whole passage, and Kaibel does not so indicate, as Wimmer does. Nevertheless, in view of Theophrastus' interest in S. Italy, and its legal systems, as attested by his well-known account of the property laws and procedures of Thurioi (fr. 97 W; cf. F. Pringsheim, *Greek Law of Sale* (Weimar, 1950), 134ff., *Ptol. Alex.* ii. 202 n. 148), it seems likely. It is amusing that, perhaps half a millenium later a Trallian doctor at Pertosa, near Locri, proudly announced his profession as that of an οἰνοδότης: see *IG* xiv. 666 (Ritschel, pl. 72, C).

concerning the νῆσοι Διομήδους, off the Apulian coast, north-west of
Monte Gargano, the modern Isole di Tremiti. There was a persistent
legend from the fourth century onwards that Diomede and his followers,
who paid so heavy a price for his actions in the Trojan War, eventually
came to rest on this coast, where he founded a number of Greek cities in
Apulia, and on these islands. The tradition, as thus presented, seems to
be part of a group of legends about Diomede in the Adriatic, which
involves cult-practices in his honour right up the Adriatic as far north as
Trieste, and across the inmost gulf of the sea to Spina, and also at
numerous sites in Italy, Greek and non-Greek.[50] Those legends may have
been known to Mimnermus and Ibycus, but they seem to have been
further developed in the fourth century; they occur in the *periplous* of
pseudo-Scylax written in *c*.340 BC and in Theophrastus himself. The
author of the pseudo-Scylax mentions the cult in relation to the non-
Greek people who lived behind Ancona, the Ὀμβρικοί, the Umbrians.[51]
The story about the Islands of Diomede is part of this much wider nexus
of cult for which various explanations have been offered: that Diomede
was worshipped first in Sybaris, which he is said to have founded; that
the cult was a Greek adaptation of an older Apulian cult, perhaps
connected with a native rider-god; that it was a projection of a Rhodian
cult, for the Rhodians and Coans are said to have founded Salapia or
Elpiai, which is given as a foundation of Diomede. The possibilities are
numerous, and none of them can be substantiated. The story as it affects
the Islands of Diomede was evidently written up and elaborated, it may
be, by Lykos or Timaios, and the fullest accounts we have of it are in the
poet Lykophron, who devotes to it one of his most powerful passages,
and the paradoxographical Περὶ θαυμασίων ἀκουσμάτων.[52] One

[50] The adventures and cult of Diomede have been much studied, and I largely confine
myself here to quoting some of the ancient sources (much has been written about Diomede
in recent years by L. Braccesi: see, *inter alia*, his *Grecità adriatica*[2] (Bologna, 1979), 55 ff.,
with considerable bibliography). Diomede seems to dominate the cults of the east coast of
Italy, and of the Illyrians and Veneti of the Adriatic, in much the same way as Achilles does
that of the Pontic regions. Unfortunately, while we have abundant epigraphical evidence for
the cult of Achilles, for Diomede we have only a few disjointed passages of literature.

[51] See GGM, i, §16: Μετὰ δὲ Σαυνίτας ἔθνος ἐστι Ὀμβρικοί, καὶ πόλις ἐν αὐτῆι
Ἀγκὼν ἐστι. τοῦτο δὲ τὸ ἔθνος τιμᾶι Διομήδην εὐεργετηθὲν ὑπ' αὐτοῦ· καὶ ἱερόν ἐστιν
αὐτοῦ.

[52] Lyc. 592–632; De Mirab. Ausc. §§ 109–10 (ed. Giannini). The former is too long to
quote here, and I quote the latter passage only: (109, a) λέγεται περὶ τὸν ὀνομαζόμενον τῆς
Δαυνίας τόπον ἱερὸν εἶναι Ἀθηνᾶς Ἀχαΐας καλούμενον, ἐν ὧι δὴ πελέκεις χαλκοῦς καὶ
ὅπλα τῶν Διομήδους ἑταίρων καὶ αὐτοῦ ἀνακεῖσθαι. (110) ἐν δὲ τοῖς Πευκετίνοις εἶναί
φασιν Ἀρτέμιτος ἱερόν, ἐν ὧι τὴν διωνομασμένην ἐν ἐκείνοις τοῖς τόποις χαλκῆν ἕλικα
ἀνακεῖσθαι λέγουσιν, ἔχουσαν ἐπίγραμμα· Διομήδης Ἀρτέμιτι.

recurrent feature of the legend and cult in Apulia is that *Argive* Diomedes founded the city of *Arg*yrippa, the Latin Arpi, and that his companions were turned into a species of heron there. Argyrippa, I may add, was part and parcel of the Greek world in the Hellenistic age: the Delphians awarded an Ἀργυριππῖνος proxeny honours, in 191 BC, perhaps because he had served as *thearodokos* for Delphian *theoroi* travelling in Apulia and Picenum.[53] It is not clear whether Theophrastus is referring to a shrine of Diomede on the islands or in one of the Apulian cities, but the Θαυμάσια makes it clear that there was a shrine on the islands. He tells us (ext. 11) that 'they say, φασί, that the plane and other water-loving trees (φίλυδρα καὶ παραποτάμια) are not found in the Adriatic except περὶ τὸ Διομήδους ἱερόν.' He goes on to say that the plane is scarce in Italy; καίτοι πολλοὶ καὶ μεγάλοι ποταμοὶ παρ᾽ ἀμφοῖν, that is, in Italy and in the eastern Adriatic, in Illyria. He speaks, then, of a shrine of Diomede, such as, according to tradition, also existed in Thurii, Spina, etc. The whole section is governed by 'they say', as indeed are the corresponding sections in the *Thaumasia*,[54] so he is not speaking from personally ascertained information. Nevertheless, the source that he is quoting, oral or written, must have mentioned a particular group of plane-trees at the shrine, and neither the trees nor the shrine can be dismissed as fictitious. He evidently considered the information of sufficient botanical significance to record it, though otherwise he is so sparing in giving us information about Italy. Since he is far removed from the world of western cult-legends, from Timaios and Lykophron, he is the best testimony to the cult of Diomede.[55] We cannot tell where his information about the *hieron* came from, whether from his own reading or from hearsay; if the former, it may well have been the Syracusan Philistos, the contemporary narrator of the *res gestae* of his kinsman Dionysius I, who knew the Adriatic coast of Italy.

The second passage concerns Roman history. Theophrastus is discussing the qualities of wood best for specific purposes, and describes the activity of Antigonus the One-Eyed and his son Demetrius in ensuring the acquisition of long unknotted timber for their navies, in Syria and Cyprus.[56] He goes on (ext. 12): 'And they say that the silver-firs and firs

---

[53] *Syll.*³ 585, 64: Σάλσιος Ταγύλλιος Ταγίλου υἱὸς Ἀργυριππανός, next door to Γάιος Στατώριος Γαΐου υἱὸς Βρεντεσῖνος.

[54] λέγεται and φασί in § 109, φασί and μυθολογεῖται in § 110.

[55] In the same passage he goes on to speak of a stand of plane-trees planted by Dionysius I in his park at Rhegion, subsequently transferred to the *gymnasion* there. The whole passage is repeated by Plin. *NH* 12. 6–8.

[56] Cf. above, p. 172.

of Corsica are the largest of all, for though the silver-fir and the fir of Latium are indeed finer and taller than the corresponding trees in South Italy, they cannot be compared with those of Corsica.' There follows an account of an attempted Roman colonization of the island, which failed because of the quantity of heavy timber on the coast, which apparently broke up their ships' masts and other tackle. 'So they gave up the idea of founding the city there.' The story has been much discussed, necessarily inconclusively, for though it is probable that Theophrastus has transmitted erroneous information, it is not clear where the error lies. Did he or rather his source—note that the whole episode is once more governed by a 'they say'—mistake the place or the occasion or both? Mommsen maintained that the reference was to a historical event that occurred shortly after the establishment of the *duoviri navales* in 311 BC,[57] but Theophrastus does not appear to be speaking of a very recent event, within the span of time during which the *HP* was actually being composed; he says 'once upon a time' (ποτε). I do not think that we can do otherwise than regard it as one of those misconceptions about Roman history of which there are other instances in the earlier third century.[58] We may ask, Where did he get his information from? Well, on Corsica as on the banks of the Akesines, he is his usual cautious self: he will not be drawn beyond 'they say'. If we choose to seek an authority, there is that clutch of distinguished names that figures so largely in the paradoxographical and allied traditions, Eudoxos, Theopompos, and Lykos, all later used by Callimachus. There was also the information, much of it concerned with early legends, to be found in the Aristotelian *Politeiai*, also heavily drawn upon by Callimachus, and which Theophrastus must have known.[59] The matter, however, does not end there.

Pliny, in a notoriously obscure passage speaking about the continentalization of Monte Circeo (*NH* 3. 57 = *FGrHist* 840 F 24(a)), a supposed phenomenon which was dealt with by Theophrastus,[60] informs

---

[57] See the discussion of this passage in *Ptol. Alex.* ii. 1068 n. 339.

[58] On these see my remarks in my review of Treves, *Euforione*, *Gnomon* 28 (1956), 582; *Ptol. Alex.* i. 767–8. Other instances occur below. It is not easy to see why such a bizarre episode should have been simply invented, and I believe that the incident as described represents erroneous rather than totally false information. The historical episode remains concealed.

[59] It is by a curious chance that the only accounts that we have of activity in Corsica in the 4th cent. is its capture by Dionysius I (Strab. 226). That is surely historical, for Diod. 5. 13. 3 (=Timaios, *FGrHist* 566 F 164, in small print) tells us that it had an excellent harbour called ὁ λιμὴν ὁ Συρακόσιος.

[60] *HP* 5. 8. 3, in continuation of the passage about Corsica: τὸν δὲ τόπον [sc. Monte Circeo] εἶναι [governed by the preceding λέγειν δὲ τοὺς ἐγχωρίους] νέαν πρόσθεσιν [coni.

us that Theophrastus was the first person to write *diligentius* about Rome (ext. 13).[61] This passage, in so far as it concerns Theophrastus, is puzzling: as the date of the Athenian archon shows, Pliny is referring to this passage in the *HP* regarding Monte Circeo, but Theophrastus defines that feature as being in Latium, and says nothing about Rome. Pliny, then, though he introduces Theophrastus' information about Monte Circeo in the same sentence as that in which he says that he, Theophrastus, was the first to write about Rome *diligentius*, seems to have had another passage in mind. If we are to accept Pliny's statement, in so far as it concerns Theophrastus (an author about whom he knew a great deal, though not necessarily at first hand) it might be supposed that Theophrastus wrote something more consequential in his Ἱστορικὰ Ὑπομνήματα (a work not infrequently quoted as Ἀριστοτέλης ἢ Θεόφραστος ἐν τοῖς Ἱστορικοῖς Ὑπομνήμασιν), but if so it has left no trace.[62] It is, however, clear that the Peripatetic philosophers and historians of the first generation, the immediate pupils of Aristotle, recorded fictions regarding non-Greek Italy and Rome, which have a generic similarity to each other if only by reason of their *bizarrerie*. Beside Theophrastus we may set the description, by his contemporary and fellow Peripatetic, Heraclides Ponticus, who, apropos of the Gaulish descent on Rome in 387 BC, calls Rome a πόλις Ἑλληνίς.[63] Plutarch said of Heraclides' description (which came in his Περὶ Ψυχῆς, rather surprisingly) that it indicated that a 'faint knowledge' of this event percolated rapidly to Greece at the time.[64] He was more close to the truth than those who have believed that the phrase was a penetrating description of contemporary Roman culture.[65] The phrase πόλις Ἑλληνίς is regularly used by pseudo-Scylax to describe the Greek cities on or near

---

Schneider], καὶ πρότερον μὲν οὖν νῆσον εἶναι τὸ Κιρκαῖον, νῦν δὲ ὑπὸ ποταμῶν τινων προσκεχῶσθαι καὶ εἶναι ἠϊόνα.

[61] *NH* 3. 57 = *FGrHist* 840 F24(a). W. W. Tarn, *Alexander the Great* (Cambridge, 1948), ii. 21 ff., with Addenda, p. 451, touches on most of the problems of chronology created by this passage, but it is the relevance of the information regarding Theophrastus that concerns us here, not the larger question whether Pliny was right or wrong about the priority of the writers mentioned. More detailed discussions will be found in the works of Hoffmann and Wikén noted below, n. 65.

[62] For this work, of which only a handful of quasi-paradoxographical quotations survive, see Regenbogen, *RE* Suppbd. vii, col. 1540, b)1.

[63] Wehrli, fr. 102 = *FGrHist* 840 F23, 3.

[64] *Ibid.* τοῦ μέντοι πάθους αὐτοῦ καὶ τῆς ἁλώσεως ἔοικεν ἀμυδρά τις εὐθὺς εἰς τὴν Ἑλλάδα φήμη διελθεῖν.

[65] So E. Wikén, *Die Kunde der Hellenen von dem Lande und den Völkern der Apenninenhalbinsel bis 300 v. Chr.* (Lund, 1937), 171–2; cf. W. Hoffmann, *Rom und die Griechische Welt* (*Philol.* Suppbd. 27(1); Leipzig, 1934), 105 n. 236.

the coast in native territory in Illyria and on the Tyrrhene and Adriatic littorals of Italy, and it is not impossible that Heraclides' error derived originally from a *periplous* which, misled by the obvious etymology of Ῥώμη, dubbed the city Ἑλληνίς.[66] Yet again, take the account by still another pupil of Aristotle, Aristoxenos of Taras, ὁ μουσικός, of the commemorative festivals held by the people of Posidonia-Paestum, at which they recalled their Greek origin, being at that time under Lucanian domination: Aristoxenus said that at the festival 'they come together and recall those ancient names and customs and after lamenting to one another and weeping go away home' (ext. 14).[67] These seemingly fantastic statements of Theophrastus, of Heraclides, and of Aristoxenus probably have a common origin in the works relating to foreign customs, Νόμιμα βαρβαρικά, which, along with the strictly Greek Πολιτεῖαι, formed part of the literary and historical production of the Lykeion. Callimachus, a generation later, shows the direct influence of the same tradition, in the well-known reference to the siege of Rome by the Peucetians, and the bravery shown by Γάϊος ὁ Ῥωμαῖος on that problematical occasion.[68] Dionysius of Halicarnassus, speaking of just this phenomenon, says that chance information must also have contributed.[69] But though that is no doubt true in many cases, these particular items seem a group apart. May we characterize them as representing the Aristotelian view of Italy?

To return to our starting-point, it is clear that the *HP* provides an objective picture of the immense change wrought in the world by Alexander. Theophrastus is full of information about the flora of the east, and had he been a little less reticent he could no doubt have quoted from a dozen or more oral or written sources whence he drew his information. The east filled men's horizons in a way that had never been possible before, and Theophrastus is one of the best witnesses to that. And when we compare his soundly based and authenticated knowledge of the plant-geography of the east with his sketchy notions of the western part of the Greek and the Italic world, it is worth remembering that two generations later the opposite would be true, and it would be 'the cloud from the west' that would fill people's minds.

---

[66] See ps.-Scyl. (cf. above, n. 51), § 17: (Τυρρηνοί) πόλις (Σπῖνα) ἐν αὐτῆι Ἑλληνίς; cf. Strab. 214: ἡ Σπῖνα, νῦν μὲν χωρίον, πάλαι δὲ Ἑλληνὶς πόλις ἔνδοξος. The term πόλις Ἑλληνίς is naturally used also of the old-established Greek colonies (see §§ 12 ff.).

[67] Fr. 124 Wehrli.

[68] Fr. 106. For the interpretation of the passage see *Ptol. Alex.* i. 767–8, and ii. 1073 ff. nn. 360 ff.

[69] *AR* 1. 6: ἐκ τῶν ἐπιτυχόντων ἀκουσμάτων συνθεὶς [ἕκαστος] ἀνέγραψεν.

Let me now summarize what I regard as 'Theophrastus' world'.
(1) Whatever we may think of the idea of there having been officially
accredited botanists and other scientists in Alexander's train, who transmitted information back to Athens, and stored copies of it in Babylon,
Theophrastus evidently had access to the earliest first-hand accounts
either recorded in the course of that campaign or written shortly afterwards, and the *HP* gives us an excellent opportunity of seeing how
decisive, within one limited scientific field, the effect of that campaign
was. Within the terms of reference relevant to Theophrastus, a whole
new range of observations, a new plant-geography, which would never
be surpassed in Antiquity, had been effected. The *HP* thus illustrates
more clearly than any other transmitted text outside the purely military
or biographical authors the immediate consequences of Alexander's
campaign. (2) In addition to that, the enterprise of Greek-speaking
settlers, travellers, and merchants in the years immediately following
Alexander's death led to the rapid development of intercourse with the
Iranian and Indian regions which were brought within the scope of
further scientific observation in the following generation by Patrocles,
Megasthenes, and others and then by the Alexandrian scientists. From
this new situation it seems probable that Theophrastus learned much.
(3) When we turn to the area west of the Adriatic, the picture is very
different. The non-Greek cultural world, which in Asia rapidly became so
familiar a part of Greek awareness, remained shadowy and fabulous, and
the observation of natural phenomena (much of it to be found in
Theophrastus' works on stones and minerals) became the starting-point
for an unscientific literary genre, in which 'exceptions' to rules were
isolated as miraculous ($Παράδοξα$, $Θαυμάσια$). So, just as botanical
analysis gave way to rhizotomy, the handmaid of medicine, so the
general identification of 'differences' ($διαφοραί$), which was an essential
foundation of Theophrastus' observation of nature, swiftly begat the
collection of $Παράδοξα$. Both these developments occurred very soon
after Theophrastus' death, if not during his lifetime, and drove scientific
botanical, and also zoological, investigation out of the field. In astro-
nomy and in the physical and mathematical sciences another generation
was to pass before royal patronage created the opportunity for new
advances. But there were to be no new advances in botany, in which
Theophrastus remained, and remains, the alpha and omega of Greek
achievement.

APPENDIX

Theophrastus: Texts

1. Ath. 21a: Ἕρμιππος δέ φησι Θεόφραστον παραγίνεσθαι εἰς τὸν Περίπατον καθ' ὥραν λαμπρὸν καὶ ἐξησκημένον, εἶτα καθίσαντα διατίθεσθαι τὸν λόγον οὐδεμιᾶς ἀπεχόμενον κινήσεως οὐδὲ σχήματος ἑνός. καί ποτε ὀψοφάγον μιμούμενον ἐξείραντα τὴν γλῶσσαν περιλείχειν τὰ χείλη.

2. Cic. *Brut*. 46. 172: ut ego iam non mirer, illud Theophrasti accidisse, quod dicitur, cum percontaretur ex anicula quadam, quanti aliquid venderet, et respondisset illa atque addidisset 'Hospes, non pote minoris': tulisse enim moleste, se non effugere hospitis speciem, cum aetatem ageret Athenis optimeque loqueretur.

3. Diosc. Anaz. prol. §5: παρακαλοῦμεν δὲ σὲ καὶ τοὺς ἐντευξομένους τοῖς ὑπομνήμασι μὴ τὴν ἐν λόγοις δύναμιν ἡμῶν σκοπεῖν, ἀλλὰ τὴν ἐν τοῖς πράγμασι μετ' ἐμπειρίας ἐπιμέλειαν. μετὰ γὰρ πλείστης ἀκριβείας τὰ μὲν πλεῖστα δι' αὐτοψίας γνόντες, τὰ δὲ ἐξ ἱστορίας τῆς πᾶσι συμφώνου καὶ ἀνακρίσεως τῶν παρ' ἑκάστοις ἐπιχωρίων ἀκριβώσαντες πειρασόμεθα καὶ τῆι τάξει διαφόρωι χρήσασθαι καὶ τὰ γένη κατὰ τὰς δυνάμεις ἑκάστου αὐτῶν ἀναγράψασθαι.

4. HP 4. 4. 5: καὶ ἕτερα δὲ πλείω καὶ διαφέροντα τῶν ἐν τοῖς Ἕλλησιν ἀλλ' ἀνώνυμα. θαυμαστὸν δ' οὐδὲν τῆς ἰδιότητος· σχεδὸν γάρ, ὥς γε τινές φασιν, οὐθὲν ὅλως τῶν δένδρων οὐδὲ τῶν ὑλημάτων οὐδὲ τῶν ποιωδῶν ὅμοιόν ἐστι τοῖς ἐν τῆι Ἑλλάδι πλὴν ὀλίγων.

5. Strab. 69: FGrHist 712 F 1: τοὺς Ἀλεξάνδρωι συστρατεύσαντας ἐπιδρομάδην ἱστορῆσαι ἕκαστα, αὐτὸν δὲ Ἀλέξανδρον ἀκριβῶσαι, ἀναγραψάντων τὴν ὅλην χώραν τῶν ἐμπειροτάτων αὐτῶι· τὴν δ' ἀναγραφὴν αὐτῶι δοθῆναί φησιν ὕστερον ὑπὸ Ξενοκλέους τοῦ γαζοφύλακος.

6. FGrHist 124 T 3: διὰ τὸ μηδέπω τὰς ὑπὸ Καλλισθένους ἐκ Βαβυλῶνος ἐκπεμφθείσας τηρήσεις ἥκειν εἰς τὴν Ἑλλάδα, Ἀριστοτέλους τοῦτο ἐπισκήψαντος αὐτῶι, ὡς ἱστορεῖ Πορφύριος, κ.τ.λ.

7. HP 4. 4. 5: ἄλλο τέ ἐστιν οὗ ὁ καρπὸς μακρὸς καὶ οὐκ εὐθὺς ἀλλὰ σκολιός. ἐσθιόμενος δὲ γλυκύς. οὗτος ἐν τῆι κοιλίαι δηγμὸν ἐμποιεῖ καὶ δυσεντερίαν, δι' ὃ Ἀλέξανδρος ἀπεκήρυξε μὴ ἐσθίειν.

8. *HP* 4. 4. 12: ἄλλο δὲ ὕλημα μέγεθος μὲν ἡλικὸν ῥάφανος, τὸ δὲ φύλλον ὅμοιον δάφνηι καὶ τῶι μεγέθει καὶ τῆι μορφῆι. τοῦτο δ' εἴ τι φάγοι ἐναποθνήσκει. δι' ὃ καὶ ὅπου ἵπποι ⟨ὠσφραίνοντο;⟩ τούτους ἐφύλαττον διὰ χειρῶν.

9. *HP* 4. 4. 1: καί τοι γε διεφιλοτιμήθη Ἅρπαλος ἐν τοῖς παραδείσοις τοῖς περὶ Βαβυλῶνα φυτεύων πολλάκις καὶ πραγματευόμενος, ἀλλ' οὐδὲν ἐποίει πλέον· οὐ γὰρ ἐδύνατο ζῆν ὥσπερ τἆλλα τὰ ἐκ τῆς Ἑλλάδος. τοῦτο μὲν οὖν οὐ δέχεται ἡ χώρα διὰ τὴν τοῦ ἀέρος κρᾶσιν· ἀναγκαίως δὲ δέχεται καὶ πύξον καὶ φιλύραν· καὶ γὰρ περὶ ταῦτα πονοῦσιν οἱ ἐν τοῖς παραδείσοις.

10. *D.L.* 5. 38: τοιοῦτος δ' ὤν, ὅμως πρὸς ὀλίγον ἀπεδήμησε καὶ οὗτος καὶ οἱ λοιποὶ φιλόσοφοι, Σοφοκλέους τοῦ Ἀμφικλείδου νόμον εἰσενέγκοντος μηδένα τῶν φιλοσόφων σχολῆς ἀφηγεῖσθαι ἂν μὴ τῆι βουλῆι καὶ τῶι δήμωι δόξηι.

11. *HP* 4. 5. 6: ἀλλὰ καὶ τὰ φίλυδρα καὶ τὰ παραποτάμια ταῦθ' ὁμοίως· ἐν μὲν γὰρ τῶι Ἀδρίαι πλάτανον οὔ φασιν εἶναι πλὴν περὶ τὸ Διομήδους ἱερόν· σπανίαν δὲ καὶ ἐν Ἰταλίαι πάσηι· καίτοι πολλοὶ καὶ μεγάλοι ποταμοὶ παρ' ἀμφοῖν· ἀλλ' οὐκ ἔοικε φέρειν ὁ τόπος· ἐν Ῥηγίωι γοῦν ἃς Διονύσιος πρεσβύτερος ὁ τύραννος ἐφύτευσεν ἐν τῶι παραδείσωι, αἵ εἰσι νῦν ἐν τῶι γυμνασίωι, φιλοτιμηθεῖσαι, οὐ δεδύνηνται λαβεῖν μέγεθος.

12. *HP* 5. 8. 1: τῶν γὰρ ἐν τῆι Λατίνηι καλῶν γινομένων ὑπερβολῆι καὶ τῶν ἐλατίνων καὶ τῶν πευκίνων—μείζω γὰρ ταῦτα καὶ καλλίω τῶν Ἰταλικῶν—οὐδὲν εἶναι πρὸς τὰ ἐν τῆι Κύρνωι. πλεῦσαι γάρ ποτε τοὺς Ῥωμαίους βουλομένους κατασκευάσασθαι πόλιν ἐν τῆι νήσωι πέντε καὶ εἴκοσι ναυσί, καὶ τηλικοῦτον εἶναι τὸ μέγεθος τῶν δένδρων ὥστε εἰσπλέοντας εἰς κόλπους τινὰς καὶ λιμένας διασχισθεῖσα τοῖς ἱστοῖς ἐπικινδυνεῦσαι. καὶ ὅλως δὲ πᾶσαν τὴν νῆσον δασεῖαν καὶ ὥσπερ ἠγριωμένην τῆι ὕληι· δι' ὃ καὶ ἀποστῆναι τὴν πόλιν οἰκίζειν· διαβάντας δέ τινας ἀποτέμεσθαι πάμπολυ πλῆθος ἐκ τόπου βραχέος, ὥστε τηλικαύτην ποιῆσαι σχεδίαν ἣ ἐχρήσατο πεντήκοντα ἱστίοις· οὐ μὴν ἀλλὰ διαπεσεῖν αὐτὴν ἐν τῶι πελάγει. Κύρνος μὲν οὖν εἴτε διὰ τὴν ἄνεσιν εἴτε διὰ τὸ ἔδαφος καὶ τὸν ἀέρα πολὺ διαφέρειν τῶν ἄλλων.

13. *NH* 3. 57; *FGrHist* 840 F 24(a): mirum est quod de hac re tradere hominum notitiae possumus. Theophrastus, qui primus externorum aliqua de Romanis diligentius scripsit (nam Theopompus, ante quem nemo mentionem habuit, urbem dumtaxat a Gallis captam dixit, Clitarchus ab eo proximus legationem tantum ad Alexandrum missam, hic iam plus quam ex fama), Cerceiorum insulae et mensuram posuit stadia octoginta in eo volumine quod scripsit Nicodoro Atheniensium magistratu, qui fuit urbis nostrae ccccxl.

14. Aristoxenos fr. 124 Wehrli: ὅμοιον ποιοῦμεν Ποσειδωνιάταις τοῖς ἐν τῶι Τυρρηνικῶι κόλπωι κατοικοῦσιν, οἷς συνέβη τὰ μὲν ἐξ ἀρχῆς Ἕλλησιν οὖσιν

ἐκβαρβαρῶσθαι [Τυρρηνοῖς ἢ Ῥωμαίοις] Λευκανοῖς (?) γεγονόσι, καὶ τήν τε φωνὴν μεταβεβληκέναι τά τε λοιπὰ τῶν ἐπιτηδευμάτων. ἄγειν δὲ μίαν τινὰ αὐτοὺς τῶν ἑορτῶν τῶν Ἑλληνικῶν ἔτι καὶ νῦν, ἐν ἧι συνιόντες ἀναμιμνήσκονται τῶν ἀρχαίων ἐκείνων ὀνομάτων τε καὶ νομίμων καὶ ἀπολοφυράμενοι πρὸς ἀλλήλους καὶ ἀποδακρύσαντες ἀπέρχονται.

7

The Tradition about the First Sacred War

JOHN DAVIES

Every historian of archaic Greece has to come to terms with the First
Sacred War. Located with welcome precision in time by the source
material in the 590s and 580s, involving many of the major Greek cities
and powers of the period, and significantly affecting the life and
functions of one of the major sanctuaries and oracles of Greece, the war
has to be incorporated into any narrative or analysis. Yet for the modern
historiography of archaic Greece it presents acute dilemmas of method
and credibility. The problems are not those presented by Athenian
history, where it is the historicity or otherwise of a growing list of documents (or of 'documents') which poses the major challenge, com-
pounded by growing awareness of the rolling nature of oral tradition and
its impact on 'history'.[1] Rather, with the First Sacred War we have to deal
with a tradition which suddenly appears in crystallized written form long
after the event, so that modern scholarship has an unwelcome choice
between cautious acceptance *en bloc* and a quasi-archaeological disinter-
ment of the various layers which inevitably deconstructs the whole. The
exposition which follows here will therefore have to oscillate between the
registers of modern and ancient historiography, I hope without loss of
clarity, in an effort to reach a formulation which stands some chance of
conveying the truth.

It will be helpful to begin with the ancient historiography of the war, by
listing, and in part quoting, the source material relevant to the story of
the war. For reasons which will appear, it is essential to do so in the
chronological order of the sources, in so far as that can be established.

1. Allusions which may or may not be to the war have been detected
in late archaic poetry in the *Homeric Hymn to Apollo*, lines 540–4
(Forrest, 1956), and in the last lines (478–80) of the Hesiodic *Aspis*
(Parke and Boardman, 1957). Otherwise, the first allusions are now seen

---

[1] Habicht (1961) for the basics, to which I intend to return; Thomas (1989 and 1992).

in Isokrates' *Plataikos* of 373–371 BC, where he refers to Theban support in 404 for a proposal that Attika should become 'a sheeprun like the Krisaian plain' ($\mu\eta\lambda\delta\beta$o$\tau$o$\nu$ $\H{\omega}\sigma\pi\epsilon\rho$ $\tau\delta$ $K\rho\iota\sigma\alpha\hat{\iota}o\nu$ $\pi\epsilon\delta\acute{\iota}o\nu$, 14. 31), and (on ostensible date) in the *Presbeutikos*, a work preserved in the Hippocratic corpus as a speech delivered at Athens by Hippokrates' son Thessalos, which gives a detailed narrative of the war. Scholarly opinion is divided on the question whether the speech is a genuine composition of the fourth century BC or a much later Hellenistic creation.[2]

2. A cluster of material dates from the 340s and 330s. The first, in winter 343/2, is probably the brief allusion to the war in Speusippos' *Letter to Philip*, §§ 8–9,[3] quoting from the researches of Antipatros of Magnesia (§ 1) (= *FGrHist* 69 F 2) three precedents for the expulsion of members from the Amphiktyony (the Krisaians being one) and the transfer of their votes to others. Androtion's *Atthis*, with its assertion that the Amphiktyony met at Delphi from the start (*FGrHist* 324 F 58), belongs at much the same date. That was followed soon after by narratives in two works of Kallisthenes. The first was his *Table of Victors at the Pythian Games from Gylidas onwards and of those who managed the contest*, compiled jointly with Aristotle and rewarded with a crown by the Delphian authorities shortly before or after 330:[4] though no fragment survives, there is general awareness that its material underlies the information in the scholia to Pindar's *Pythians* (no. 5 below). Kallisthenes' own monograph *On the Sacred War* concerned the so-called Third War of 356–346, but the one surviving fragment (*FGrHist* 124 F 1) indicates that it mentioned the First War as well.

3. In the same period Aischines twice broached the topic. The preserved version of his 343 speech *On the Embassy* recounted how, as part of his speech at Pella in 346, he raised the question of the sanctuaries, narrated the foundation of the shrine at Delphi and the first session of the Amphiktyony, and quoted the Amphiktyonic oath (2. 114–16). His 330 speech went much further, to give one of the three main surviving narratives of the war (3. 107–13) (see Appendix below, § 1). To this fourth-century profile we may add a fragment of Diodoros

----

[2] Doxography in Robertson (1978: 68 n. 1) and Cassola (1980: 420 n. 8). Text in Hp. ix. 404–26 Littré; Hercher, *Epist. Gr.* 312–18; Pomtow (1918: 317–20); and Smith (1990).

[3] Text in Hercher, *Epist. Gr.* 629–32 as *Socraticorum Epistulae* xxx, replaced by Bickermann and Sykutris (1928 [1929]), who vindicate its authenticity and date in winter 343/2 (pp. 30–8). Pohlenz (1929: 55 n.) prefers the first half of 343.

[4] *FdeD* iii(1), 400 = *Syll*[3] 275 = Tod, ii. 187 = Harding no. 104. The precise year remains unclear (Lewis, 1958; Robertson, 1978: 55 n. 3) but does not affect the argument.

(9. 16), which is taken to indicate that Ephoros described the war, though if so the absence of any identifiable reference to the war in the fragments of Ephoros' own book 6 is noticeable (*FGrHist* 70 F 54–6, 173–9).

4. Thereafter comes a long gap till the second century AD, broken only by two very summary references to the war in Strabo (9. 3. 3–4, pp. 418–19 C, and 9. 3. 10, p. 421 C) and by a brief reference of Frontinus to a stratagem in the war (*Strat.* 3. 7. 6). A chapter in Plutarch's *Solon* (11: Appendix, §2) and two sections in Polyainos (3. 5 and 6. 13, both recording stratagems during the siege of Kirrha) are echoed in the second main narrative of the war, at the end of Pausanias' description of Delphi (10. 37. 4–8: Appendix, §3).

5. The third and most detailed narrative is to be found in the scholia to Pindar, specifically in the *hypothesis* to the *Pythian Odes* extant in four versions a, b, c, and d (II 1–5 ed. Drachmann). The date when these notes reached their present form, and the route by which they did so, are not to my knowledge established. Yet it is this material, and only this, which contains the chronological evaluations, especially synchronisms with the Athenian archon list, which are so valuable for integrating the information about the war into the overall picture of archaic Greek history.

It is that integration which characterizes most of the modern historiography of the First Sacred War. For our purposes it may start with the second edition of 1893 of Georg Busolt's *Griechische Geschichte*, the fullest and most sure-footed, though hardly the liveliest, of the great classic full-dress nineteenth-century Histories of Greece. Volume i², pp. 690–700, gives us:

(*a*) a description of Krisa, its position and wealth as described in the source material (690–2);

(*b*) a sketch of the tensions between Delphi and Krisa because of attacks on pilgrims, tolls being extracted from them, and the priests' desire to escape Krisan control, leading to the oracle quoted in Aischines 3. 108 enjoining war (692);

(*c*) a description of the contingents of the Amphiktyonic forces against Krisa (Thessaly, Athens, Sikyon) (693–4);

(*d*) a narrative of the first creation of the new style Pythian Games and the new regime for running the shrine, including the dedication of the Krisaian Plain to the god (694–6);

(*e*) a narrative of the second phase of the war, against the Krisans on

Mt. Kirphis, and the second creation of the Pythian Games
(696–7); and

(*f*) an estimation of the effect of the war on Thessalian influence in
Central Greece (698–700).

These ten pages are as valuable as they are illuminating. Valuable,
because Busolt was scrupulous to show his working, not merely referring
to but quoting *in extenso* all relevant sources in footnotes: illuminating,
because he was clearly concerned first and foremost to create a Ranke-
style (or Thucydides-style) narrative of events in the public domain
(mostly political and military) as they unfolded. Valuable and illuminating, in that they show the nature and limitations of his critical approach.
He was not indeed uncritical, for much of the detail about the stratagems
of the war was omitted, and he cast severe doubt on the claim attributed
to Kallisthenes (*FGrHist* 124 F 1) that the war lasted for ten years. However, he saw no reason to challenge the general thrust of the information,
and no impediment to inserting it within his overall framework of development of Greek history of the seventh and sixth centuries BC, the general
horizon of dates having been for him satisfactorily given by the scholia to
Pindar's *Pythian Odes*, even if he saw room for doubt and extensive footnotes about some of the detailed problems of reconciliation.[5]

This, then, is the First Sacred War as seen in the fullest of the standard
pictures, a classic of the Age of Innocence. The picture remained much
the same in subsequent scholarship for eighty years, whether painted on
a smaller scale (as by Beloch (1924: 337–8) or De Sanctis (1940:
564ff.)) or on a larger scale (as by Parke and Wormell, 1956). Not, of
course, that scholarship was static. The tradition was *dissected*, as by
Sordi, 1953, who tried to unpack the various inconsistencies in the
sources about who should have the major credit for the victory. It was
sharpened, as by Forrest, 1956, who (*a*) brought the final lines (540–4)
of the *Homeric Hymn to Apollo* properly into the debate for the first
time, and (*b*) tried to locate in the early sixth century a shift of horizon or
attitude at Delphi, both politically (in the tone of oracles) and archae-
ologically (in the tapering off of the connection between Delphi and
Krete). It was *extended*, as by Parke and Boardman in 1957 or by Boardman on his own in 1978, by bringing into the debate evidence from
monuments or themes in vase-painting which could be thought to reflect
the war. None the less, a fair consensus remained concerning the picture
as sketched by Busolt, and debate remained within manageable limits,

[5] Notably op. cit. (Busolt, 1893), 697 n. 1, practically an article in itself.

though slightly disconcerted by the archaeological failure to locate the archaic city of Kirrha/Krisa.

That consensus was challenged root and branch by Noel Robertson's article of 1978, which argued that the whole narrative of the war was a fourth-century invention. In essentials, his case was that there is no reference to the war until the 340s, and conspicuously none in Herodotus or Thucydides; that there was no major town in the Pleistos Valley to be destroyed in the archaic period (whether located at Chrysso, at Xeropigado, or Itea); that there was in any case no distinction between Kirrha and Krisa, both of which names denote the harbour town in the classical period; that the detailed narrative of the Sacred War was created by Antipatros and Kallisthenes, calqued in part on the Third Sacred War (taking thence the motif of two stages of hostilities—p. 65) and in part on the Trojan War (whence the duration of ten years, and the notion of the abduction of noble ladies as the *aition*); that its creation suited the political convenience of Philip II of Macedon, while the name of one of the allies' generals, Eurylochos of Thessaly, recalled that of one of Philip's senior generals (64–5); that the participation of Kleisthenes of Sikyon, absent from the Antipatros/Kallisthenes story, was a later insertion inspired by a monument of Kleisthenes at Delphi (66); that the participation of the Athenian Alkmaion likewise derived from knowledge that the Alkmaionidai had built the post-548 temple (66–8); and that the core of the story derived from a Delphian local legend of a struggle against a set of bandits using as their lair what may be identified as the Mycenaean ruins of Ayios Georghios (68–72).

Robertson's article was profoundly subversive. Historically, if the First Sacred War did not exist, how far would the macro-history of archaic Greece have to be rewritten? Historiographically, if its narrative was a construct, how can we be sure that it was an isolated example, rather than the visible tip of a cancer running through the entire Greek historiographical tradition? Yet it is fair to say that the reaction to Robertson has been muted, as if his objections would go away if ignored. There have indeed been repeated, but I suspect unavailing, attempts to wrest consistent chronological sense out of the Pindaric scholia,[6] but otherwise the response has largely been to fasten on to the weak points of what is essentially an *e silentio* case. Thus, for example, Lehmann, 1980 pointed out that Aischines 2. 114ff. already prefigured in 343 some of the detail which recurs in his 330 speech, and noted that since in his

---

[6] Bennett (1957); Miller (1978); Mosshammer (1982); Brodersen (1990).

Plataikos of the late 370s (14. 31) Isokrates referred to the sterilization of the Krisaian Plain, there was at least some version of the war which pre-dated the 340s. In complement, and as part of a sophisticated rear-guard action, Cassola, 1980 argued that the silence of Herodotus and Thucydides should not be taken too seriously, and that it was absurd to envisage Kallisthenes creating *ex nihilo* a story which in fact reflected some credit on the very polity, viz. Phokis, which Philip was busy dismantling on the ground of impiety towards the oracle and the sacred land.

These points are well taken, but do not begin to touch the main issues. There are two, one specific, another contextual. The specific issue is that we now have to ask of the material about this event the questions that always underlie a historical tradition: how did the sources know what they thought they knew? Specifically, whence did Aischines get his information? What materials did Aristotle and Kallisthenes have when they set about constructing their list of Pythian victors? If, as Plutarch tells us, Euanthes of Samos said that Solon was the commander of the Athenian forces, or if the Delphians' *hypomnemata* assert that Alkmaion was Commander-in-Chief, how did they know? It ought not to have needed Robertson's article to force such questions up the agenda, for they are implicit in any analysis of a body of source material about the past, especially (as here) when eyewitness accounts are wholly to seek and the entire source material is secondary. Now, at least, any response to Robertson has to seek and find a sensible answer to such questions if the historicity of the war is to be vindicated via that source material.

The contextual issue takes us beyond the historicity of one particular war towards a more general predicament, itself the product of two related trends in recent scholarship. The first of these is the pursuit of forged (or 'forged') documents, begun by Habicht's article of 1961. Triggered by the publication by M. H. Jameson in 1959 of the Themistokles decree, it argued that that decree, and eight other decrees which surfaced in the literary tradition soon after 350 and reflected creditably on Athenian actions during the Persian Wars, were fabrications, in the sense of being imaginative expansions of statements in Herodotus and elsewhere. Habicht's list has since been extended,[7] but subsequent published discussion on this theme has largely confined itself to the *bona fides* either of one document at a time (notably of course the Themistokles decree itself and the Peace of Kallias) or of a closely related

---

[7] Seager (1969): Perikles' Congress Decree; Davies (1971: 51): decrees granting assistance to Aristeides' children.

group, such as the 'constitutions' transmitted in *Ath. Pol.* 29–33.[8] Even
such discussion as there has been of larger groups has mostly confined
itself to the decrees or documents which Habicht singled out. However,
in a paper delivered in London in 1985 I argued that the state of play was
fundamentally unsatisfactory, for three reasons. First, to concentrate on
individual documents hindered the creation of a systematic methodology
for validating or invalidating such ostensibly documentary formulations.
Secondly, even those sceptics who had looked at groups of documents
had disagreed about the date and purpose of their 'forgery' or 'reconstruction' and had therefore not dug very deep into the history of ideas or
the implications for our understanding of Greek historiography. Thirdly,
and most basic, I claimed that the scale of the problem had been danger-
ously underestimated. At the time of that paper, in 1985, I had a file of
some 125 items transmitted on documents in the Greek literary, orator-
ical, historiographical, or antiquarian traditions, the reliability and
authenticity of which were at least open to question if not palpably
absent, and that file has grown since. Historiographically at least, we
have to do with a phenomenon of major proportions, which needs to be
dealt with as a unit. In my paper I tried to do that in a preliminary way, by
isolating certain common characteristics or 'signatures' which surfaced
repeatedly in the material: that it mostly concerned Athens and
Athenians, and set them in a favourable light; that some repeated
formulae, such as 'on a bronze stele' ($\dot{\epsilon}\nu\ \sigma\tau\dot{\eta}\lambda\eta\ \chi\alpha\lambda\kappa\hat{\eta}$), or repeated
subject-matter (honours to writers), were ways of asking questions about
several documents at a time; and that anachronisms of purportedly
official terminology, or shifts of focus or content, were also usable
criteria. It will not be irrelevant to the consideration of the First Sacred
War to report that the horizon at which this much wider group of
material began to surface in the literary material was not the 350s (as
Habicht had argued for *his* sub-set) but the 400s and the 390s, a good
generation earlier. The detailed reasons must await a separate exposition, but it is helpful and relevant to pick out the new historiographical
inputs which made this activity possible—the work of Nikomachos and
his colleagues in the archives from 410 till 399; the changing habits of
speechwriters (compare Andokides' lavish use of documents in his
speech 1 *De Mysteriis* with the predominance of non-evidential argument in Antiphon); the creation and use in the previous generation of
lists of eponyms for dating purposes; the need after 404 to restore

---

[8] Basic orientation in Rhodes (1981: 362 ff.) and Andrewes, *HCT* v (1981), 184–256.

battered Athenian collective self-esteem; and the new horizon created by
the use of documents by Herodotus and Thucydides.

The second relevant contextual trend in recent scholarship, summar-
ized in two recent books by Rosalind Thomas,[9] has been a growing
awareness that the gradual movement from oral towards written forms
of memory, record, and document affected historical material just as
much as it did literary material. The former category includes tales of the
acta of noble families, legends and myths (especially those linked to
festivals and cult spots), and genealogies or king-lists purporting to link
the present to a heroic or Homeric horizon. Such material was vulnerable
twice over, both by being subject to continuous change during oral transmission according to the political, cultural, or social needs of each
successive generation, and by being open to rationalization or codification when being committed to written form. Herodotus' use of his
materials reveals some of the processes involved, and it is not chance that
the debate over his methods and reliability has also taken on new life in
recent years.[10] Other processes are dimly visible through the surviving
works or fragments of those most responsible for committing such
material to written form, viz. the dithyrambic poets and the local historians of each *polis* or area from Pherekydes and Charon onwards.

The story of the First Sacred War has therefore now to be seen within the
new framework provided by such recent work. The consequences are far-
reaching but can be simply stated. *First*, we cannot any longer think like
Busolt, i.e. treat the various pieces of evidence as interlocking more or
less well to give us—admittedly with some empty spaces and ragged
edges—an approximately accurate picture of what took place.[11] Instead,
and secondly, we must accept the main thrust of the Habicht/Robertson/
Thomas transformation of the study of archaic Greek history, viz. that
our information is the production of a continually changing tradition
which is at least semi-oral, owes a great deal more to poetic utterance and
mythic modes of thought than to anything else, and is generated more by
the changing needs of a continually shifting present than by a scholarly
urge to reconstruct a static but receding past. In consequence, and
thirdly, we must conceptually separate the task of reconstructing a past
event, or sequence of events, from that of reconstructing a consciousness
(or a succession of consciousnesses) of that event or sequence, while none

---

[9] (1989 and 1992).
[10] Cf. among others Waters (1985); Gould (1989); and Fehling (1989).
[11] Though the most recent discussion (Tausend, 1992: 43–7, 161–6) does just that.

the less recognizing that the tasks are interdependent. Moreover, and fourthly, we must accept, probably for most events or sequences, and certainly for the First Sacred War, that our various pieces of evidence are pieces from different jigsaw puzzles, of different epochs and different levels of difficulty, which cannot be made to interlock.

That last point may be illustrated by identifying the periods when it is plausible to suppose that effort was being put in to creating the tradition. They do not coincide with the chronological list of extant or known source material set out above, for it is the period before the canonical story appears in full, in 343–330, that is of most interest, and just for that reason it is best to move backwards in time. One pre-343 horizon has to be 380, in view of the publication of the Amphiktyonic Law at Athens in that year (*CID* i. 10). A second is likely to have been the publication of Hellanikos' *Atthis* a generation before, partly because the late fifth-century preoccupation with Solon as a symbolic figure is a likely background for his imputed role in the war, partly because Athenian hostility towards a pro-Spartan Delphi, perceptible in the literature and historiography of the Peloponnesian War,[12] could well have served as a prototype for some form of the *logos*. The same consideration holds true with even more force for the (Second) Sacred War of the early 440s,[13] with liberation from the Phokians (on the part of the Spartans) and support for the Amphiktyony (on the part of the Athenians) equally at issue. That last motif in turn can be taken back not only to the 470s, when under Themistokles' influence Athens was actively manipulating the Amphiktyony (Plut. *Them.* 20. 3 etc.), but even more to the decade 520–510, when Athenian interest in, and influence at, Delphi is reliably attested for the first time, when Boiotia was turning itself (however embryonically) into a genuine political entity, when zones of influence in Sterea Hellas were exceptionally volatile as Thessalian predominance declined, and when a legitimating *logos* for a presence at Delphi would have been of particular value to several parties.

At least five such pre-343 horizons can therefore be identified, each providing different configurations of issues and alliances. Robertson's approach, therefore, though correct in principle, is misleadingly oversimple in concentrating on one horizon and in assuming that the canonical version of the war *logos* stems from that one horizon: things

---

[12] Notably in Euripides' *Ion* and *Elektra*, and *passim* in Aristophanes. See Hornblower (1992: 175 ff.) for Thucydides' underestimate of the 5th-cent. Amphiktyony.

[13] Thuc. 1. 112. 5 with Hornblower ad loc.; other sources in Robertson (1978: 38 n. 3).

are likely to have been much more complicated than that. All the more, then, do we have to review the basic question whether the *logos* contains any kernel of historic fact. In one important sense the answer has to be yes, for there are at least three pieces of evidence for conflict at Delphi, and many more for significant change there, in the period preceding the first in time of the horizons identified above. The question is whether they could safely engender the hypothesis of an archaic war on their own, even in the absence of the canonical *logos*.

First, the evidence for conflict. (*a*) Robertson himself identified as a component of the *logos* various local legends of Delphi, especially 'the original legend of a lair of bandits preying on Delphic pilgrims'.[14] More-over, there are ways of establishing an initial independence for com-ponents of this kind. For example, some aspects of the stories have only local relevance, such as the location of the second phase of the war on Mt. Kirphis or the use of the Mycenaean fortress of Ayios Georghios. Others are names, such as that of the Kra(u)gallidai, which are used but not exploited in the politicized version of the story, or involve commun-ities which are given no part in the war, such as Kos, Kalydon, or Aitolia in general.[15] Granted, the crystallization of such stories is impossible to date. Granted, also, they may be no less timeless and mythical than those of the Seven Against Thebes or those of Herakles' battles against various public nuisances. Granted, further; stories involving conflict and violence are never far to seek in the history or folk-tale of any archaic Greek com-munity. None the less, the very untidiness of the Delphian stories suggest that they pre-existed the partial systematization which was carried through as the canonical story developed.

(*b*) Representations in plastic art of a quarrel or fight between Herakles and Apollo over a tripod are found from the late eighth century BC onwards.[16] The earliest is on the leg of a bronze tripod from Olympia, but there is a long gap thence till the many representations in vase paint-ings (Brommer (1984) reports 230), which run from 560 to 460 BC. The proof that the conflict was located at Delphi fortunately does not depend on Apollodoros (2. 6. 2), but comes from an explicit reference at the end of the Hesiodic *Aspis*,[17] from a passing allusion in Pindar,[18] and of course fundamentally from the composition of the East Pediment of the

---

[14] Robertson (1978: 44, 68 ff.)　　　　　[15] References in Robertson (1978: 68–72).
[16] References in Parke and Boardman (1957: 278 ff.), Defradas (1972: 123 ff.), and Brommer (1984: 7–10).
[17] Lines 477–80. For the date see Guillon (1963) and West (1971: 305 f.).
[18] *Ol.* 9. 32 (4. 32 in error at Brommer, 1984: 8 n. 48).

Siphnian Treasury at Delphi *c*.525.[19] The sculptural group by Dipoinos and Skyllis at Sikyon representing Apollo, Artemis, Herakles, and Athene has been associated with the story of Herakles' attempt to purloin the tripod, and would be important evidence if Pliny's date for the sculptures (fiftieth Olympiad, i.e. 580–577 BC) could be trusted, but since the evidence does not predate Pliny (*NH* 36. 9–10) and 'much of the evidence about them proves to be anything but solid' (Griffin, 1982: 112), the temptation to read that group as an echo of the theme should probably be resisted.[20] Likewise, though Pausanias informs us that 'the poets took over the story and sing of the fight of Herakles with Apollo over the tripod' (10. 13. 8), his vagueness consigns his evidence to the end of a paragraph.

(*c*) The final lines of the *Homeric Hymn to Apollo* can hardly be read otherwise than as a warning, or more plausibly as a prophecy (presumably *post eventum*), that one regime at the shrine would be violently succeeded by another. Given that the whole poem as we have it is concerned to provide a divine charter for the oracle as Apollo's, and given that lines 388 ff. are concerned to provide the Kretan sailors who had been hijacked by Apollo's epiphany with a charter as priests of the sanctuary, these final lines are disconcerting. To propose the existence of a lacuna between lines 539 and 540 does not help much, for the final lines will still explicitly cancel the charter. Unless they belong to a wholly different poem, they have to be read, I think, as an addition, tacked on to accommodate a new reality as economically as possible. However, this is not the place to enter the debate over the unity of the hymn,[21] but rather to register, for what it may be worth, the information transmitted in the Pindaric scholia that the hymn was composed, and ascribed to Homer, by the rhapsode Kynaithos, active in the sixty-ninth Olympiad (504/3–501/0).[22] Whether, even if accurate, that information gives us a secure *terminus post quem non* for the hymn as we have it is undecidable.

Other phenomena are evidence of change, but not necessarily of conflict. For present purposes it is enough to take the Korinthian ascendancy at Delphi of the mid-seventh century as a datum line, and to list salient developments and realignments thereafter. Those known to us would include (again in rough chronological order):

---

[19] For reproduction see e.g. Boardman (1988: fig. 211).
[20] Griffin (1982: 114 ff.), against Parke and Boardman (1957: 279 ff.).
[21] Recent expositions e.g. in West (1975) and Clay (1989: 17 ff.).
[22] Schol. P. *Nem.* ii 1c, III p. 29 Drachmann, citing Hippostratos *FGrHist* 568 F 5.

1. The tapering-off of the Kretan artistic influence at Delphi,[23] *c.*620 or soon after.

2. The foundation of the Pythian Games. This can presumably be placed in the first quarter of the sixth century even if the elaborate tale of refoundations through the 590s and 580s carried by the Pindaric scholia suggests a struggle to reconcile incompatible traditions. Uncoupled from a hypothetical war, their foundation *c.*582/1 is still intelligible, but as one parallel to the foundation of the Isthmian Games in 581 and the Nemean Games in 573, i.e. as a shrewd response on the part of a Sanctuary and its management to the needs felt by the international aristocracy for a focus of display, prestige, and competition.

3. A shift in the ideology underlying oracular responses from one favouring Dorian powers to one more even-handed. The question is how much of this, Forrest's thesis of 1956, would survive if the hypothesis of a First Sacred War precipitating that shift were removed. Its main drift probably would, for there is still a sharp contrast between the attitudes of the seventh-century or early sixth-century responses in praise of Argos (PW ii no. 1 = Fontenrose (1978), Q 26), or in dispraise of Kleisthenes (PW ii no. 24 = Fontenrose Q 74), or in encouragement of the Spartan attack on Tegea (PW ii no. 31 = Fontenrose Q 88) on the one hand, and on the other the oracles of the 560s and 550s chartering the policies and initiatives in Spartan politics which are conventionally associated with Chilon. The distinction is rather that without the hypothesis of a Sacred War one would be more inclined to see a secular drift of attitude at the oracle, mirroring public opinion and changed power relationships rather than influencing them.

4. The incorporation of Delphi into the Pylaian Amphiktyony. This is not the place to explore the complexities of the problem, but merely to record two simple facts. (*a*) Alike from its name, from the title of some of its officials (*pylagorai*), and from the geographical distribution of the *ethne* which participated, the original cult-centre of the Amphiktyony must have been the sanctuary of Demeter at Anthela by Thermopylai (Hdt. 7. 200). (*b*) The first *secure* evidence that the Amphiktyons saw themselves as having power over—or responsibility for—the Apollo sanctuary at Delphi is their issuing the contract for rebuilding the temple after the fire of 548 (Hdt. 2. 180). Again, without the hypothesis of a First Sacred War the process of take-over will be undatable and untraceable.

---

[23] Guarducci (1943–6); Forrest (1956); Rolley (1977).

5. The absence of Krisa/Kirrha from the city-state map of classical Greece.

6. The status of the Krisaian Plain as 'holy' in the sense of being uncultivable.

Point 5, though in formal terms *ex silentio*, takes force from the detail of the narratives in Herodotus, Thucydides, Hellenika Oxyrhynchia, and Diodoros of events in Central Greece in the fifth and fourth centuries BC, especially when contrasted with the references to the town in epic poetry.[24] Point 6 in turn does seem to be a plain fact of the classical period, attested by the 380 version of the Amphiktyonic Law and confirmed for 404 by Isokrates' evidence in 373 (*Plat.* 14. 31). Given its fertility, in contrast to the surrounding landscape, the uncultivated status of the plain must reflect a deliberate decision, which clearly needed enforcement. Though not formally necessary, the hypothesis of a conflict over that decision and its enforcement is clearly plausible.

This is a paper about method in ancient and modern historiography, using the First Sacred War as an illuminating but intractable example, and not a paper strictly about the historicity and nature of that war. Nevertheless, the discourse cannot break off at this point. We have to ask if the components of conflict and of change listed above (i) help us to trace the crystallization of the canonical story, *and* (ii) require us to accept its substantial truth as history. The answer to (i) is certainly yes if poetic evidence of conflict between gods or between local heroes is taken as reflecting, however obscurely, a real conflict between human groups or communities. The creation of the canonical story will then have involved, alike for ancient and for modern historiography, the processes of (*a*) expanding that conflict, or of amalgamating others with it, (*b*) associating with that conflict changes which in reality unfolded over a far longer period or were even wholly unrelated, and (*c*) telescoping the pace of change so as to allow a single simple before-and-after structure. However, such an answer to (i) begs the major question of whether rehistoricizing mythic narratives in such a way is ever permissible.[25] If it is not, then we are left with only such evidence of change as requires formal public decisions to have been made. Items 2, 4, and 6 above are such, but do not have to belong with each other in time or political atmosphere. If

---

[24] *Iliad* 2. 520; *Hymn. Hom. Ap.* 282–5.
[25] For further discussion see Davies (1984).

they do, the idea of a Sacred War in the early sixth century remains a plausible hypothesis, but no more.

Such a conclusion is unsatisfactory, but it may be the best we can do. As Cassola wryly but justly observed in 1980 (1980: 422), 'Tutto sommato, credo che il fantasma della guerra crisea non possa essere esorcizzato; prevedo anzi che continuera a turbare i sonni degli storici.'

APPENDIX
Some Basic Texts

I. AISCHINES 3. 107–13

(107) Ἔστι γάρ, ὦ ἄνδρες Ἀθηναῖοι, τὸ Κιρραῖον ὠνομασμένον πεδίον καὶ λιμὴν ὁ νῦν ἐξάγιστος καὶ ἐπάρατος ὠνομασμένος. ταύτην ποτὲ τὴν χώραν κατῴκησαν Κιρραῖοι καὶ Κραγαλίδαι, γένη παρανομώτατα, οἳ εἰς τὸ ἱερὸν τὸ ἐν Δελφοῖς καὶ περὶ τὰ ἀναθήματα ἠσέβουν, ἐξημάρτανον δὲ καὶ εἰς τοὺς Ἀμφικτύονας. ἀγανακτήσαντες δ᾽ ἐπὶ τοῖς γιγνομένοις μάλιστα μέν, ὡς λέγονται, οἱ πρόγονοι οἱ ὑμέτεροι, ἔπειτα καὶ οἱ ἄλλοι Ἀμφικτύονες, μαντείαν ἐμαντεύσαντο παρὰ τῷ θεῷ, τίνι χρὴ τιμωρίᾳ τοὺς ἀνθρώπους τούτους μετελθεῖν. (108) καὶ αὐτοῖς ἀναιρεῖ ἡ Πυθία πολεμεῖν Κιρραίοις καὶ Κραγαλίδαις πάντ᾽ ἤματα καὶ πάσας νύκτας, καὶ τὴν χώραν αὐτῶν ἐκπορθήσαντας καὶ αὐτοὺς ἀνδραποδισαμένους ἀναθεῖναι τῷ Ἀπόλλωνι τῷ Πυθίῳ καὶ τῇ Ἀρτέμιδι καὶ τῇ Λητοῖ καὶ Ἀθηνᾷ Προναίᾳ ἐπὶ πάσῃ ἀεργίᾳ, καὶ ταύτην τὴν χώραν μήτ᾽ αὐτοὺς ἐργάζεσθαι μήτ᾽ ἄλλον ἐᾶν.

Λαβόντες δὲ τὸν χρησμὸν τοῦτον οἱ Ἀμφίκτυονες ἐψηφίσαντο Σόλωνος εἰπόντος Ἀθηναίου τὴν γνώμην, ἀνδρὸς καὶ νομοθετῆσαι δυνατοῦ καὶ περὶ ποίησιν καὶ φιλοσοφίαν διατετριφότος, ἐπιστρατεύειν ἐπὶ τοὺς ἐναγεῖς κατὰ τὴν μαντείαν τοῦ θεοῦ· (109) καὶ συναθροίσαντες δύναμιν πολλὴν τῶν Ἀμφικτυόνων, ἐξηνδραποδίσαντο τοὺς ἀνθρώπους καὶ τὸν λιμένα καὶ τὴν πόλιν αὐτῶν κατέσκαψαν καὶ τὴν χώραν καθιέρωσαν κατὰ τὴν μαντείαν. καὶ ἐπὶ τούτοις ὅρκον ὤμοσαν ἰσχυρόν, μήτ᾽ αὐτοὶ τὴν ἱερὰν γῆν ἐργάσεσθαι μήτ᾽ ἄλλῳ ἐπιτρέψειν, ἀλλὰ βοηθήσειν τῷ θεῷ καὶ τῇ γῇ τῇ ἱερᾷ καὶ χειρὶ καὶ ποδὶ καὶ φωνῇ καὶ πάσῃ δυνάμει. (110) καὶ οὐκ ἀπέχρησεν αὐτοῖς τοῦτον τὸν ὅρκον ὀμόσαι, ἀλλὰ καὶ προστροπὴν καὶ ἀρὰν ἰσχυρὰν ὑπὲρ τούτων ἐποιήσαντο. γέγραπται γὰρ οὕτως ἐν τῇ ἀρᾷ, "Εἴ τις τάδε," φησί, "παραβαίνοι ἢ πόλις ἢ ἰδιώτης ἢ ἔθνος, ἐναγής," φησίν, "ἔστω τοῦ Ἀπόλλωνος καὶ τῆς Ἀρτέμιδος καὶ τῆς Λητοῦς καὶ Ἀθηνᾶς Προναίας." (111) καὶ ἐπεύχεται αὐτοῖς μήτε γῆν καρποὺς φέρειν, μήτε γυναῖκας τέκνα τίκτειν γονεῦσιν ἐοικότα, ἀλλὰ τέρατα, μήτε βοσκήματα κατὰ φύσιν γονὰς ποιεῖσθαι, ἥτταν δὲ αὐτοῖς εἶναι πολέμου καὶ δικῶν καὶ ἀγορᾶς, καὶ ἐξώλεις εἶναι καὶ αὐτοὺς καὶ οἰκίας καὶ γένος ἐκείνων. "Καὶ μήποτε," φησίν, "ὁσίως θύσειαν τῷ Ἀπόλλωνι μηδὲ τῇ Ἀρτέμιδι μηδὲ τῇ Λητοῖ μηδ' Ἀθηνᾷ Προναίᾳ, μηδὲ δέξαιντο αὐτοῖς τὰ ἱερά." (112) ὅτι δ' ἀληθῆ λέγω, ἀνάγνωθι τὴν τοῦ θεοῦ μαντείαν. ἀκούσατε τῆς ἀρᾶς. ἀναμνήσθητε τῶν ὅρκων, οὓς ὑμῶν οἱ πρόγονοι μετὰ τῶν Ἀμφικτυόνων συνώμοσαν.

MANTEIA

[Οὐ πρὶν τῆσδε πόληος ἐρείψετε πύργον ἑλόντες,
πρίν γε θεοῦ τεμένει κυανώπιδος Ἀμφιτρίτης
κῦμα ποτικλύζῃ κελαδοῦν ἱεραῖσιν ἐπ' ἀκταῖς.]

ΟΡΚΟΙ. ΑΡΑ

(113) Ταύτης τῆς ἀρᾶς καὶ τῶν ὅρκων καὶ τῆς μαντείας ἀναγεγραμμένων ἔτι καὶ νῦν, οἱ Λοκροὶ οἱ Ἀμφισσεῖς, μᾶλλον δὲ οἱ προεστηκότες αὐτῶν, ἄνδρες παρανομώτατοι, ἐπηργάζοντο τὸ πεδίον, καὶ τὸν λιμένα τὸν ἐξάγιστον καὶ ἐπάρατον πάλιν ἐτείχισαν καὶ συνῴκισαν, καὶ τέλη τοὺς καταπλέοντας ἐξέλεγον, καὶ τῶν ἀφικνουμένων εἰς Δελφοὺς πυλαγόρων ἐνίους χρήμασι διέφθειρον, ὧν εἷς ἦν Δημοσθένης.

(107) There is, fellow citizens, a plain, called the plain of Cirrha, and a harbour, now known as 'dedicate and accursed'. This district was once inhabited by the Cirrhaeans and the Cragalidae, most lawless tribes, who repeatedly committed sacrilege against the shrine at Delphi and the votive offerings there, and who transgressed against the Amphictyons also. This conduct exasperated all the Amphictyons, and your ancestors most of all, it is said, and they sought at the shrine of the god an oracle to tell them with what penalty they should visit these men. (108) The Pythia replied that they must fight against the Cirrhaeans and the Cragalidae day and night, utterly ravage their country, enslave the inhabitants, and dedicate the land to the Pythian Apollo and Artemis and Leto and Athena Pronaea, that for the future it lie entirely uncultivated; that they must not till this land themselves nor permit another.

Now when they had received this oracle, the Amphictyons voted, on motion of Solon of Athens, a man able as a law-giver and versed in poetry and philosophy, to march against the accursed men according to the oracle of the god. (109) Collecting a great force of the Amphictyons, they enslaved the men, destroyed their harbour and city, and dedicated their land, as the oracle had commanded. Moreover they swore a mighty oath, that they would not themselves till the sacred land nor let another till it, but that they would go to the aid of the god and the sacred land with hand and foot and voice, and all their might. (110) They were not content with taking this oath, but they added an imprecation and a mighty curse concerning this; for it stands thus written in the curse: 'If any one should violate this,' it says, 'whether city or private man, or tribe, let them be under the curse,' it says, 'of Apollo and Artemis and Leto and Athena Pronaea.' (111) The curse goes on: That their land bear no fruit; that their wives bear children not like those who begat them, but monsters; that their flocks yield not their natural increase; that defeat await them in camp and court and market-place, and that they perish utterly, themselves, their houses, their whole race; 'And never,' it says, 'may they offer pure sacrifice unto Apollo, nor to Artemis, nor to Leto, nor to Athena Pronaea, and may the gods refuse to accept their offerings.' (112) As a proof of this, let the oracle of the god be read; hear the curse; call to mind the oaths that your fathers swore together with all the other Amphictyons.

THE ORACLE

[Ye may not hope to capture town nor tower,
Till dark-eyed Amphitrite's waves shall break
And roar against Apollo's sacred shore.]

THE OATHS. THE CURSE

(113) This curse, these oaths, and this oracle stand recorded to this day; yet the Locrians of Amphissa, or rather their leaders, most lawless of men, did till the plain, and they rebuilt the walls of the harbour that was dedicate and accursed, and settled there and collected port dues from those who sailed into the harbour; and of the deputies who came to Delphi they corrupted some with money, one of whom was Demosthenes.

2. PLUTARCH, *SOLON* 11

Ἤδη μὲν οὖν καὶ ἀπὸ τούτων ἔνδοξος ἦν ὁ Σόλων καὶ μέγας. ἐθαυμάσθη δὲ καὶ διεβοήθη μᾶλλον ἐν τοῖς Ἕλλησιν εἰπὼν ὑπὲρ τοῦ ἱεροῦ τοῦ ἐν Δελφοῖς, ὡς χρὴ βοηθεῖν καὶ μὴ περιορᾶν Κιρραίους ὑβρίζοντας εἰς τὸ μαντεῖον, ἀλλὰ προσαμύνειν ὑπὲρ τοῦ θεοῦ Δελφοῖς. πεισθέντες γὰρ ὑπ' ἐκείνου πρὸς τὸν πόλεμον ὥρμησαν οἱ Ἀμφικτύονες, ὡς ἄλλοι τε πολλοὶ μαρτυροῦσι καὶ Ἀριστοτέλης ἐν τῇ τῶν Πυθιονικῶν ἀναγραφῇ Σόλωνι τὴν γνώμην ἀνατιθείς. οὐ μέντοι στρατηγὸς ἐπὶ τοῦτον ἀπεδείχθη τὸν πόλεμον, ὡς λέγειν φησὶν Ἕρμιππος Εὐάνθη τὸν Σάμιον· οὔτε γὰρ Αἰσχίνης ὁ ῥήτωρ τοῦτ' εἴρηκεν, ἔν τε τοῖς τῶν Δελφῶν ὑπομνήμασιν Ἀλκμαίων, οὐ Σόλων, Ἀθηναίων στρατηγὸς ἀναγέγραπται.

Solon's reputation and authority had already been greatly increased by these events. But he became still more admired and celebrated throughout the Greek world when he spoke out on behalf of the temple at Delphi and declared that the Greeks must not allow the people of Cirrha to profane the oracle, but must come to its rescue and help the Delphians to ensure that Apollo was still honoured there. It was on his advice that the Amphictyonic Council went to war, as Aristotle, among others, confirms in his list of the victors at the Pythian games, where he gives Solon the credit for taking up this attitude. He was not, however, appointed general for this war, as Evanthes of Samos alleges (according to Hermippus). Certainly Aeschines the orator makes no such statement, and according to the records at Delphi it was Alcmaeon, not Solon, who commanded the Athenians.

3. PAUSANIAS 10. 37. 4–8

(4) ἐς δὲ Κίρραν τὸ ἐπίνειον Δελφῶν ὁδὸς μὲν σταδίων ἐξήκοντά ἐστιν ἐκ Δελφῶν· καταβάντι δὲ ἐς τὸ πεδίον ἱππόδρομός τέ ἐστι καὶ ἀγῶνα Πύθια ἄγουσιν ἐνταῦθα τὸν ἱππικόν. τὰ μὲν δὴ ἐς τὸν ἐν Ὀλυμπίᾳ Ταράξιππον ἐδήλωσέ μοι τὰ ἐς Ἠλείους τοῦ λόγου, ὁ δὲ ἱππόδρομος ἔοικε τοῦ Ἀπόλλωνος τάχα μέν που καὶ αὐτὸς τῶν ἱππευόντων τινὰ ἀνιᾶσαι, ἅτε ἀνθρώποις τοῦ δαίμονος ὁμοίως ἐπὶ ἔργῳ παντὶ καὶ ἀμείνω καὶ τὰ χείρω νέμοντος· οὐ μέντοι καὶ αὐτὸς ταραχὴν τοῖς ἵπποις ὁ ἱππόδρομος οὔτε κατὰ αἰτίαν ἥρωος οὔτε ἐπ᾽ ἄλλῃ πέφυκεν ἐργάζεσθαι προφάσει. (5) τὸ δὲ πεδίον τὸ ἀπὸ τῆς Κίρρας ψιλόν ἐστιν ἅπαν, καὶ φυτεύειν δένδρα οὐκ ἐθέλουσιν ἢ ἔκ τινος ἀρᾶς ἢ ἀχρεῖον τὴν γῆν ἐς δένδρων τροφὴν εἰδότες. λέγεται δὲ ἐς τὴν Κίρραν * * * καὶ ἀπὸ τῆς Κίρρας τὸ ὄνομα τὸ ἐφ᾽ ἡμῶν τεθῆναι τῷ χωρίῳ φασίν. Ὅμηρος μέντοι Κρῖσαν ἔν τε Ἰλιάδι ὁμοίως καὶ ὕμνῳ τῷ ἐς Ἀπόλλωνα ὀνόματι τῷ ἐξ ἀρχῆς καλεῖ τὴν πόλιν. χρόνῳ δὲ ὕστερον οἱ ἐν τῇ Κίρρᾳ ἄλλα τε ἠσέβησαν ἐς τὸν Ἀπόλλωνα καὶ ἀπετέμνοντο τοῦ θεοῦ τῆς χώρας. (6) πολεμεῖν οὖν πρὸς τοὺς Κιρραίους ἔδοξεν Ἀμφικτύοσι, καὶ Κλεισθένην τε Σικυωνίων τυραννοῦντα προεστήσαντο ἡγεμόνα εἶναι καὶ Σόλωνα ἐξ Ἀθηνῶν ἐπηγάγοντο συμβουλεύειν· χρωμένοις δέ σφισιν ὑπὲρ νίκης ἀνεῖπεν ἡ Πυθία·

> οὐ πρὶν τῆσδε πόληος ἐρείψετε πύργον ἑλόντες,
> πρίν κεν ἐμῷ τεμένει κυανώπιδος Ἀμφιτρίτης
> κῦμα ποτικλύζῃ κελαδοῦν ἐπὶ οἴνοπα πόντον.

ἔπεισεν οὖν ὁ Σόλων καθιερῶσαι τῷ θεῷ τὴν Κιρραίαν, ἵνα δὴ τῷ τεμένει τοῦ Ἀπόλλωνος γένηται γείτων ἡ θάλασσα. (7) εὑρέθη δὲ καὶ ἕτερον τῷ Σόλωνι σόφισμα ἐς τοὺς Κιρραίους· τοῦ γὰρ Πλείστου τὸ ὕδωρ ῥέον διὰ ὀχετοῦ σφισιν ἐς τὴν πόλιν ἀπέστρεψεν ἀλλαχόσε ὁ Σόλων. καὶ οἱ μὲν πρὸς τοὺς πολιορκοῦντας ἔτι ἀντεῖχον ἔκ τε φρεάτων καὶ ὕδωρ τὸ ἐκ τοῦ θεοῦ πίνοντες· ὁ δὲ τοῦ ἐλλεβόρου τὰς ῥίζας ἐμβαλὼν ἐς τὸν Πλεῖστον, ἐπειδὴ ἱκανῶς τοῦ φαρμάκου τὸ ὕδωρ ᾔσθετο ἔχον, ἀπέστρεψεν αὖθις ἐς τὸν ὀχετόν. καὶ — ἐνεφορήσαντο γὰρ ἀνέδην οἱ Κιρραῖοι τοῦ ὕδατος — [καὶ] οἱ μὲν ὑπὸ ἀπαύστου τῆς διαρροίας ἐξέλιπον οἱ ἐπὶ τοῦ τείχους τὴν φρουράν, (8) Ἀμφικτύονες δὲ ὡς εἶλον τὴν πόλιν, ἐπράξαντο ὑπὲρ τοῦ θεοῦ δίκας παρὰ Κιρραίων, καὶ ἐπίνειον Δελφῶν ἐστιν ἡ Κίρρα. παρέχεται δὲ καὶ ἐς θέαν Ἀπόλλωνος καὶ Ἀρτέμιδος καὶ Λητοῦς ναόν τε καὶ ἀγάλματα μεγέθει μεγάλα καὶ ἐργασίας Ἀττικῆς. ἡ δὲ Ἀδράστεια ἵδρυται μὲν ἐν τῷ αὐτῷ σφίσι, μεγέθει δὲ τῶν ἄλλων ἀποδέουσα ἀγαλμάτων ἐστίν.

(4) The length of the road from Delphi to Cirrha, the port of Delphi, is sixty stades. Descending to the plain you come to a race-course, where at the Pythian games the horses compete. I have told in my account of Elis the story of the Taraxippus at Olympia, and it is likely that the race-course of Apollo too may possibly harm here and there a driver, for heaven in every activity of man bestows either better fortune or worse. But the race-course itself is not of a nature to startle the horses, either by reason of a hero or on any other account. (5) The plain from Cirrha is altogether bare, and the inhabitants will not plant trees, either because the land is under a curse, or because they know that the ground is useless for growing trees. It is said that to Cirrha ... and they say that from Cirrha the place received its modern name. Homer, however, in the *Iliad*, and similarly in the hymn to Apollo, calls the city by its ancient name of Crisa. Afterwards the people of Cirrha behaved wickedly towards Apollo; especially in appropriating some of the god's land. (6) So the Amphictyons determined to make war on the Cirrhaeans, put Cleisthenes, tyrant of Sicyon, at the head of their army, and brought over Solon from Athens to give them advice. They asked the oracle about victory, and the Pythian priestess replied:

> You will not take and throw down the tower of this city,
> Until on my precinct shall dash the wave
> Of blue-eyed Amphitrite, roaring over the wine-dark sea.

So Solon induced them to consecrate to the god the territory of Cirrha, in order that the sea might become neighbour to the precinct of Apollo. (7) Solon invented another trick to outwit the Cirrhaeans. The water of the river Pleistus ran along a channel to the city, and Solon diverted it in another direction. When the Cirrhaeans still held out against the besiegers, drinking well-water and rain-water, Solon threw into the Pleistus roots of hellebore, and when he perceived that water held enough of the drug he diverted it back again into its channel. The Cirrhaeans drank without stint of the water, and those on the wall, seized with obstinate diarrhoea, deserted their posts, (8) and the Amphictyons captured the city. They exacted punishment from the Cirrhaeans on behalf of the god, and Cirrha is the port of Delphi. Its notable sights include a temple of Apollo, Artemis and Leto, with very large images of Attic workmanship. Adrasteia has been set up by the Cirrhaeans in the same place, but she is not so large as the other images.

BIBLIOGRAPHY

K. J. Beloch, *Griechische Geschichte*, i²(1) (De Gruyter; Berlin and Leipzig, 1924).

H. C. Bennett, 'On the Systematization of Scholia Dates for Pindar's Pythian Odes', *HSCP* 62 (1957), 61–78.

E. Bickermann and J. Sykutris, 'Speusipps' Brief an König Philipp', *SB Berlin* 80. 3 (1928).

J. Boardman, 'Herakles, Delphi and Kleisthenes of Sikyon', *Rev. Arch.* (1978), 227–34.

—— *Greek Sculpture: The Archaic Period* (Thames and Hudson; London, 1988).

K. Brodersen, 'Zur Datierung der ersten Pythien', *ZPE* 82 (1990), 25–31.

F. Brommer, *Herakles II: Die unkanonischen Taten des Helden* (Wiss. Buchgesellschaft; Darmstadt, 1984).

G. Busolt, *Griechische Geschichte*, i² (Gotha, 1893), 690–8.

F. Cassola, 'Nota sulla guerra crisea', in *Miscellanea E. Manni*, ii (Rome, 1980), 413–39.

J. K. Davies, *Athenian Propertied Families 600–300 BC* (Clarendon Press; Oxford, 1971).

—— 'The Reliability of the Oral Tradition', in L. Foxhall and J. K. Davies (edd.), *The Trojan War: Its Historicity and Context* (Bristol Classical Press, 1984), 87–110; repr. in C. Emlyn-Jones, L. Hardwick, and J. Purkis (edd.), *Homer: Readings and Images* (Duckworth; London, 1992), 211–25.

J. Defradas, *Les Thèmes de la propagande delphique* (Les Belles Lettres; Paris, 1972).

L. Dor, J. Jannoray, H. and M. van Effenterre, *Kirrha: Étude de préhistoire phocidienne* (Paris, 1960).

D. Fehling, *Herodotus and his 'Sources': Citation, Invention, and Narrative Art*, tr. J. G. Howie (ARCA Francis Cairns; Leeds, 1989).

J. Fontenrose, *The Delphic Oracle: Its Responses and Operations, with a Catalogue of Responses* (University of California Press; Berkeley, 1978).

G. Forrest, 'The First Sacred War', *BCH* 80 (1956), 33–52.

J. P. A. Gould, *Herodotus* (Weidenfeld and Nicolson; London, 1989).

A. Griffin, *Sikyon* (Clarendon Press; Oxford, 1982).

M. Guarducci, 'Creta e Delfi', *Studi e materiali di storia delle religioni*, 19–20 (1943–6), 85–114.

P. Guillon, *Études béotiennes: Le Boucher d'Héraclès et l'histoire de la Grèce centrale dans la période de la première guerre sacrée* (Aix en Provence, 1963).

C. Habicht, 'Falsche Urkunden zur Geschichte Athens im Zeitalter der Perserkriege', *Hermes* 89 (1961), 1–35.

S. Hornblower, 'The Religious Dimension to the Peloponnesian War', *HSCP* 94 (1992), 169–97.

J. Jannoray, 'Krisa, Kirrha, et la première guerre sacrée', *BCH* 61 (1937), 33–43.

G. A. Lehmann, 'Der "erste heilige Krieg"—eine Fiktion?', *Historia* 29 (1980), 242–6.

D. M. Lewis, 'An Aristotelian Publication-date', *CR*² 8 (1958), 18.

M. F. McGregor, 'Clisthenes of Sicyon and the Panhellenic Festivals', *TAPA* 72 (1941), 266–87.

S. G. Miller, 'The Date of the First Pythiad', *CSCA* 11 (1978), 127–58.

A. A. Mosshammer, 'The Date of the First Pythiad—again', *GRBS* 23 (1982), 15–30.

H. W. Parke and J. Boardman, 'The Struggle for the Tripod and the First Sacred War', *JHS* 77 (1957), 276–82.

—— and D. E. W. Wormell, *The Delphic Oracle*, i² (Oxford, 1956), 99–113.

M. Pohlenz, 'Philipps Schreiben an Athen', *Hermes* 64 (1929), 41–62.

H. Pomtow, 'Hippokrates und die Asklepiaden in Delphi', *Klio* 15 (1918), 303–38.

P. J. Rhodes, *A Commentary on the Aristotelian Athenaion Politeia* (Clarendon Press; Oxford, 1981).

N. Robertson, 'The Myth of the First Sacred War', *CQ*² 28 (1978), 38–73.

J. Roger and H. van Effenterre, 'Krisa-Kirrha', *Rev. Arch.*⁶ (1944), no. 21.

C. Rolley, *Les Trépieds à cuve clouée: Fouilles de Delphes*, v(3) (De Boccard; Paris, 1977).

R. Seager, 'The Congress Decree: Some Doubts and a Hypothesis', *Historia* 18 (1969), 129–41.

Wesley D. Smith (ed. and tr.), *Hippocrates: Pseudepigraphic Writings (Letters—Embassy—Speech from the Altar—Decree)*, Studies in Ancient Medicine, ed. J. Scarborough, vol. 2 (E. J. Brill; London–New York–Copenhagen–Cologne, 1990).

M. Sordi, 'La prima guerra sacra', *RFIC* 81 (1953), 320–46.

Jenny Strauss Clay, *The Politics of Olympus: Form and Meaning in the Major Homeric Hymns* (Princeton University Press; Princeton, NJ, 1989).

Klaus Tausend, *Amphiktyonie und Symmachie. Historia*, Einzelschrift 73 (Franz Steiner Verlag; Stuttgart, 1992).

Rosalind Thomas, *Oral Tradition and Written Record in Classical Athens* (University Press; Cambridge, 1989).

—— *Literacy and Orality in Ancient Greece* (University Press; Cambridge, 1992).

H. T. Wade-Gery, 'Kynaithos', in *Greek Poetry and Life* (Festschrift G. Murray) (Oxford, 1936); repr. in Wade-Gery, *Essays in Greek History* (Blackwell; Oxford, 1958), 17–36.

K. H. Waters, *Herodotus the Historian: His Problems, Methods, and Originality* (Croom Helm; London, 1985).

M. L. West, 'Stesichorus', *CQ*² 21 (1971), 302–14.

—— 'Cynaethus' Hymn to Apollo', *CQ*² 25 (1975), 161–70.

8

Diodorus and his Sources:
Conformity and Creativity

KENNETH S. SACKS

When in his great commentary on Thucydides Arnold Gomme comes to discuss the merits of Diodorus, he adds the expletive: alas![1] Gomme, of course, only echoes the judgement of most scholars. Diodorus' *Bibliotheke* extends from the mythological period before the Trojan War to Diodorus' lifetime in the first century BC, and its range has been its saving grace. We depend on it for much of what is known of the mid-fourth century, Alexander the Great, his Successors, and almost the entirety of Sicilian history. And throughout the *Bibliotheke*, even when the text is fragmentary, there are occasional bits of information which close important gaps in our knowledge.

But using the history requires great caution. Sensitive to its many factual errors and chronological blunders, scholars continually mine the *Bibliotheke* in the hopes of uncovering individual strata and attributing them to various sources. Determining whom Diodorus used is an obvious first step to establishing the worth of his narrative. Yet *Quellenforschung* brings its own dangers.

Over the past century and a half it has been virtually axiomatic that, aside from errors he inadvertently introduces, Diodorus is faithful to his sources. Thus he is called 'a mere copyist' and considered to be 'slavishly following his sources'. This should, ironically, be a compliment to an antiquarian historian, whose purpose it is to preserve the factual record. But the judgement is also applied to how Diodorus produced the sinews of the *Bibliotheke*—the philosophical, moral, and political judgements which establish for any work its intellectual unity. The assumption that Diodorus was incapable of imposing his own interpretations reflects a

---

[1] i. 51. Some of this paper is based on my *Diodorus Siculus and the First Century* (Princeton, NJ, 1990). Additional material was first presented at talks at Princeton, Oxford, Cambridge, and a meeting of the Association of Ancient Historians; I am grateful to all the audiences for helpful suggestions. I have kept documentation to a minimum where it can be found in my book.

lingering nineteenth-century approach to source criticism and is meth-
odologically weak: for the most part the corresponding narratives of the
original sources are no longer extant, so that there are few controls,
direct or indirect, over how much thematic material Diodorus has
borrowed from his sources. Indeed, once the belief in Diodorus'
incompetence is put aside, it is easy to establish his authorship of important concepts in the *Bibliotheke*. Here are two obvious examples.

An expert on Diodorus' Sicilian material, commenting on Diodorus
11. 26 and the phrase εὐεργέτης καὶ σωτήρ as applied to the tyrant
Gelon, states that it is a sentiment drawn directly from the historian
Timaeus, and refers to his analysis of the same expression at 16. 20.
When the reader turns to that, he finds as proof that it is Timaean a reference to the earlier passage, 11. 26. Far worse than the circularity is what
has been ignored: Diodorus uses εὐεργέτης and σωτήρ throughout the
mythological and historical sections, in passages where Timaeus could
not possibly be the author. They, in fact, form an important part of
Diodorus' emphasis on benefactors and culture-heroes.[2] Rather than
consider the possibility that Diodorus wrote the words himself and did so
as part of his larger view of history, scholars often focus narrowly on one
section of the *Bibliotheke*, boldly attributing phrases or *Tendenzen* to
historians largely unknown.

Another example concerns Diodorus' use of ethical judgements.
Throughout the *Bibliotheke*, it is stated that by the study of history the
reader can learn how to improve his own moral behaviour. To that end,
the *Bibliotheke* contains numerous moral assessments of historical
characters aimed at spurring the reader towards better living. Ephorus,
who is Diodorus' main source for the classical period, that is, for books
11–15/16, was strongly influenced by Isocrates, who in turn used historical *exempla* as a method for conveying moral principle. It is customarily
argued that if Isocrates used such an approach, so must have Ephorus.
Consequently, although little of Ephorus' work survives outside of what
is supposedly preserved in Diodorus' history, for over a century moral
judgements found in the *Bibliotheke* have been attributed to Ephorus.

Such easy explanations, however, run into problems. More recently, it
has been noted that similar ethical judgements are distributed throughout the entire *Bibliotheke*, certainly in parts for which Ephorus could not
be the source.[3] Typical of the ethical assessments is this statement:

---

[2] Sacks, *Diodorus*, 61–82.
[3] Initially argued by R. Drews, 'Diodorus and his Sources', *AJP* 83 (1962), 383–

Prusias, king of Bithynia, also came to congratulate the senate and the generals who had brought the conflict to a successful issue. This man's ignobility of spirit must not be allowed to go without comment. For when the virtue of good men is praised, many of later generations are guided to strive for a similar goal; and when the cowardice of bad men is held up to reproach, not a few who are taking the path of vice are turned aside. Accordingly the frank language of history should of set purpose be employed for the improvement of society.[4]

This passage is part of book 31, which covers the period two centuries after Ephorus' death. Further, it comes at a point where Diodorus' source is extant. Diodorus follows Polybius quite closely on the details of Prusias' embassy (30. 18), but introduces the discussion with his own statement on history's obligation to render ethical judgements. And, just as in the case with the expression εὐεργέτης and σωτήρ, the use of history as a moral force can be shown to play a significant role in Diodorus' historical outlook.[5]

Here, because Diodorus' source survives, his own creativity is demonstrable. Usually, however, as little is preserved of Diodorus' sources outside of what is found in the *Bibliotheke*, there is no certain way to establish the origins of these themes. Because of the long tradition that Diodorus is 'a mere copyist', scholars attempting to assert Diodorus' creativity bear the burden of proving the negative. Although there is very little evidence that he did plagiarize, neither is it easy to prove that he did not. In the instance cited above, however, the completely extant original provides certain evidence for Diodorus' intrusion.

Moreover, in the passage concerning Prusias, there are ways to assess Diodorus' originality independent of the preserved Polybian original. Diodorus' discussion of Prusias links thematically and verbally with several other passages and so provides important clues to his philosophy of history.[6] Because most of the original sources are lost, established unity of theme within the *Bibliotheke* is probably the best methodology for demonstrating Diodorus' creativity. This method is particularly appropriate where an historian is known to have drawn on a number of different traditions throughout his work. It is obviously far less effective when analysing an historian who is using only a few sources and where a

92. Similar judgements for the future moral improvement: e.g. 2. 2. 1–2; 10. 12; 11. 38. 5–6; 23. 15. 1; 37. 4; 37/38. 18. 1.

[4] 31. 15. 2; F. R. Walton (ed. and tr.), *Diodorus of Sicily*, xi (Loeb; Cambridge, Mass., 1957), modified.

[5] *Diodorus*, 23–35.

[6] On παρρησία see *Diodorus*, 34–5; on τοῦ κοινοῦ βίου see *Diodorus*, 11 n. 12.

thematic unity might be ascribed instead to the original source. In the case involving Prusias and ethical judgements, of course, one might argue that Diodorus was at least inspired by an Ephoran model found in the earlier books of his work. But, in fact, it can be shown that certainly Isocrates, and hence perhaps Ephorus, had in mind a notion of moral *didaxis* that is somewhat different than that usually found in the *Bibliotheke*.[7]

Despite such demonstrations, however, Diodorus gets very little credit for originality. It is a common belief that Diodorus drew the subjective elements of the work, just as the historical narrative, directly from his sources. Given the size and antiquarian nature of the work, undoubtedly at times he did so. But this assumption, dominant for over a century, has led to many unwarranted interpretations of the *Bibliotheke*. Here, I hope to demonstrate that the opposite may also be the case. I will suggest first that, rather than merely plagiarizing his sources, Diodorus intrudes into the text at important moments to emphasize his own philosophical and political beliefs. But at the end of the paper, I will point out the danger of using antiquated notions of Diodorus' methodologies in assuming that, when there are possible errors present, they are to be ascribed to Diodorus' methods. By concentrating on the single general theme of imperialism, I explore instances in which he imposes his own ideas on the subjective material and at least one highly controversial passage where he may well have preserved the narrative of his source accurately. My goal here is not to stand all assumptions of Diodorus on their head, but rather to emphasize that, because he has received so little attention in his own right, we still have much to learn about his methods.

The part of the *Bibliotheke* that covers classical Greece, books 11–16, is completely extant.[8] In those books there recurs a formulaic interpretation for the rise and fall of empires. In the fifth century, Athens suffered from the disaffection of her subject states because she changed her behaviour towards them. Initially, Athens had acted moderately (ἐπιεικῶς) and so her subjects complied with her rule. But by the 460s, after gaining so much power, Athens began treating her subjects harshly (βιαίως). As a consequence, many allies discussed rebellion and even began to act on their plans. Similarly, in the early fourth century the Spartans lost their empire. Their ancestors had gained power by acting

---

[7] See *Diodorus*, 26–9. As well, many of Diodorus' moral assessments are aimed at portraying women in a positive light, certainly a tendency more representative of the late Hellenistic period than of Ephorus' day.

[8] A fuller discussion of Diodorus on empires, with complete documentation, is found in *Diodorus*, 42–54.

moderately (ἐπιεικῶς), but the later generation behaved harshly (βιαίως). Hence when the Spartans were defeated at Leuctra, all their former subjects held them in contempt. Athens and Sparta built their empires on good will (ἐπιείκεια), but subsequently they turned to acting harshly towards their subjects; because of this new attitude, the subject states rebelled. Applied to Athenian and Spartan behaviour separately at least six times, the model is also used to explain the fall of Athens in conjunction with the rise of the short-lived Spartan empire: for example, in 421, the future of the two powers moved in different directions, as Sparta treated her allies moderately (ἐπιεικῶς), while Athens terrorized hers.

Ephorus is universally identified as Diodorus' main source for the classical period, and, were Diodorus merely a 'copy machine', we should be inclined to believe that this paradigm of empire derives from Ephorus. Yet no Ephoran material outside of that in Diodorus contains a similar model for empire, and, indeed, below I note that the Ephoran/Isocratean paradigm was probably somewhat different. More importantly, the formula is not limited to explaining the empires of Athens and Sparta. It is found in nearly two dozen other contexts in the *Bibliotheke*, where Ephorus is certainly not Diodorus' source; in passages ranging in content from mythology to Alexander's *Diadochoi*. Further, the key concept of moderate behaviour is a hallmark of Diodoran thought. He employs ἐπιείκεια and φιλανθρωπία nearly 300 times in assessing the behaviour of individuals or nations. And where a papyrus fragment of Ephorus provides a control, it is clear that Diodorus himself substitutes the notion. It so pervades the *Bibliotheke*, applied even to events involving Diodorus' home town of Agyrium (22. 2. 4), that it must reflect his own understanding of empires and not simply that of a specific source. The terms, moreover, connect directly with Diodorus' interest, mentioned earlier, in culture-heroes and benefactors. It was common practice in the Hellenistic age to attribute to culture-heroes and benefactors the characteristics of ἐπιείκεια and φιλανθρωπία.

This model of empire involving moderation turning to arrogance is also applied to Roman imperialism, but with an interesting modification. Book 32 begins:

> Those whose object it is to gain dominion over others use courage and intelligence to get it, moderation and consideration for others to extend it widely, and paralysing terror to secure it against attack. The proofs of these propositions are to be found in attentive consideration of the history of such empires as were created in ancient times as well as of the Roman domination that succeeded them. (32. 2; Walton trans.)

218 Kenneth S. Sacks

This analysis of the Roman empire, developed in detail a few chapters later, has sometimes in the past been attributed to Polybius, whom Diodorus followed for the narrative events in book 32. But now there is a growing belief that the actual analysis is Diodorus' own.[9] And indeed, the use of ἐπιείκεια should alert us to Diodoran intrusion. That Rome used terrorizing tactics without suffering a loss of empire is a variation Diodorus probably adapted because he lived when Roman power was still in place.

Though Roman behaviour is described here by the terminology of Diodorus' model, two books later a new element intrudes in material commonly attributed to Posidonius.[10] There, Cato the Censor and Scipio Nasica argue over the fate of Carthage. Cato urges that she be annihilated, while Nasica counters with two predictions: if Carthage is destroyed and there are no other strong forces in the Mediterranean, then, first, Roman magistrates will act rapaciously towards subject states and, second, Rome might suffer civil war at home. Cato wins the debate, and Nasica's warnings come true. The effects of the Third Punic War are amplified three books later when the Italian Social War occurs because 'the Romans exchanged their disciplined, simple, and controlled conduct through which they grew to such power for the ruinous pursuit of luxury (τρυφή) and licence' (37. 2. 1). Roman mores decline so greatly that the historian must supply vignettes of Romans who still maintain the older virtues.

This Posidonian analysis is far removed from the earlier simple model of empire based on clemency and terror. Central here is the consideration of the moral effects of imperialism on the ruling citizenry. Though such a sociological consideration is virtually absent in the *Bibliotheke*, similar material did exist in abundance in Diodorus' main sources for Rome, Polybius and Posidonius. They made the growing Roman decadence part of their themes. And many writers who are frequently thought to be Diodorus' sources for parts of his earlier narrative stressed τρυφή, including Timaeus, Theopompus, Duris, Agatharchides, and Herodotus. From what little of Ephorus survives, there are indications that he, too, used decadence as a dynamic force in history. Moreover, Isocrates, whose very words are at times found in Ephorus' history, argued that, after gaining their empires, Athens and Sparta fell into the pursuit of τρυφή and so declined as powers. But there is no mention of decadence

[9] To the bibliography in *Diodorus*, 45 n. 91, add Jean-Louis Ferrary, *Philhellénisme et impérialisme* (Rome, 1988), 334–9.
[10] 34/35. 33 = *FGrHist* 87 Posidonios F 112 = F 178 Theiler.

in the so-called Ephoran part of Diodorus, indicating that, if Ephorus stressed the relationship between τρυφή and empire, Diodorus chose to ignore it. Though the causal connection between empire and τρυφή is frequently emphasized by his sources, Diodorus himself generally minimized it.

This examination of imperialism leads to some interesting observations about Diodorus' methods. Although he offered his own model for the rise and fall of empires, upon close inspection it seems that Diodorus does not try to alter the narrative found in his sources. For example, Diodorus, probably following Polybius, condemns Perseus for starting the Third Macedonian War and effusively praises Rome for treating him leniently.[11] Later, however, when stressing the terrorizing tactics Rome employed in maintaining her empire, he himself makes the Macedonian king a victim (32. 4. 5). The contradiction in judgement occurs because in the latter instance Diodorus departs from his source in making his own point concerning Roman methods. The case of Athens is especially striking. Diodorus offers his explanation that Athens moved from clement behaviour to terror while describing events of the 460s (11. 70. 3–4). It is at that moment, however, that his source, Ephorus, is denouncing the enemies of Athens and extolling her achievement. Diodorus' pattern of decline creates a definite dissonance in the narrative, as Ephorus clearly meant to condemn Aegina, but Diodorus criticizes Athens.[12] In one of the rare instances in which survive apparently the original words of Ephorus, the reconstructed text describes the Athens of ten years earlier, in 470 BC, as δικαιοτάτην or a similar compliment.[13] Diodorus, however, substitutes ἐπιεικεστάτην (11. 59. 3). The change is minor, but, in declaring that Athens was 'most clement' instead of 'most just', Diodorus sets up his charge of a few chapters later that Athens ceased acting ἐπιεικῶς and resorted to ruling by terror. Diodorus follows the account and the general interpretation of his source, in this case Ephorus, but makes his own moral point.

More importantly, Diodorus' model for empire demonstrates his

---

[11] Diodorus 31. 8–9; Polyb. 36. 17. 13. Polybius' negative feelings on Perseus are also implied in the final sentence of 25. 3. 1–8.

[12] 11. 70. 2; cf. 78. 3 and 12. 28. 1 on the phrasing.

[13] *POxy.* xiii. 1610 FF 4–5 = *FGrHist* 70 F 191; J. Hornblower, *Hieronymus of Cardia* (Oxford, 1981), 28–9, 278. What can be read is [σο]φ[ωτάτην καὶ δικαι]οτά[την . . .], but its otherwise close relationship to Diodorus' account (see Grenfell–Hunt's comparison of the texts in *Oxyrhynchus Papyri*, xiii. 103) shows that Athens is being described, but not by the term ἐπιεικεστάτην. Isocrates, from whom Ephorus gets much of his material (see above), refers to Athens as δικαιοτάτην: e.g. *De Pace* 140.

independence in matters of interpretation. Most of his sources argued that ruling states grow weak as imperial power corrupts their social values. But Diodorus generally disregards this explanation, focusing instead on the dynamics between rulers and subjects. His message is that kind actions inspire loyalty, harsh ones disaffection. The pervasive use of the schema may have been intended as a subtle warning to Rome. I suggest below that Diodorus held mixed, but somewhat negative feelings for the imperial power. Here, he indicates that by 146 Rome reached the point of terrorizing her subjects. Yet he fails to draw on his model to make the obvious prophecy about what should follow. Diodorus perhaps felt more comfortable in applying his full paradigm to earlier empires and in allowing the reader to draw his own conclusion about Rome's present direction. Finally, it is clear from many points that Diodorus never saw Rome as the centre of the Mediterranean or of historical progress. With that as background, his preference for the ruler–subject paradigm of empire over the model involving internal decay is clarified. Diodorus was a provincial, with strong allegiance to Sicily; he was concerned less with the effects of imperialism on the Roman ruling class than with how the ruling power treated its subjects.

Diodorus' personal experience and its influence on shaping the *Bibliotheke* are rarely considered.[14] Yet, it was in Rome and during the late Republic, probably from 56 to around 30 BC, that Diodorus wrote his work. During all this time and for a longer period than any other area outside the Italian Peninsula, Diodorus' homeland was under Roman control. In Diodorus' lifetime, Verres shamelessly ravaged the island, Caesar and Antony extended the franchise, and Octavian and Sextus Pompey fought there, with Octavian exacting a harsh punishment, including the withdrawal of citizen rights, for Sicilian support of the younger Pompey. At the same time a series of civil wars transformed the Roman government from oligarchy to autocracy. No historian, not even a so-called copyist, could have ignored or remained unaffected by these events. Diodorus may have reproduced the narratives of his sources, but his feelings on the contemporary world must somehow also be reflected in the *Bibliotheke*.

Instructive is a comparison of his opening prologue with those of the two other Hellenistic historians whose works are still substantially preserved. Both Polybius and Dionysius of Halicarnassus, in the second chapters of their first *prooemia*, list the great empires of history and

---

[14] What follows, from here through the discussion of Erycinian Aphrodite, is found in greater detail in *Diodorus*, 117–59.

conclude that the most successful was the Roman. Such a σύγκρισις was a common convention in antiquity for justifying one's subject-matter, yet Diodorus, who devotes his early books to ancient oriental empires and thus had a natural setting for a σύγκρισις, fails to make the obvious comparison with Rome.

Throughout his opening prologue, in fact, Diodorus offers no laudation to the new world power. Dionysius begins his *Roman Antiquities* with fulsome praise of Rome, arguing that she achieved the greatest accomplishments and empire in history and calling her citizens 'god-like men'. The only time that Diodorus in his first proem mentions the Roman empire is to explain that, because of Rome's control of the entire οἰκουμένη, extensive research materials were available to him in the imperial city (1. 4. 3). Polybius, in his first prologue, describes the Roman empire as a manifestation of divine will (Τύχη). When Diodorus discusses Fate (πρόνοια) and history (1. 1. 3), he takes no note of Rome.

Other Greek historians who travelled to or resided in Rome also made sure to repay the city with direct praise, most notably Posidonius, Strabo, and Nicolaus. And the philosopher Philodemus, thankful for Roman patronage, wrote that Rome and Alexandria were equally the greatest cities of his day. Diodorus previously lived in Alexandria and was then a long-time resident of Rome. In reporting that 'some say' it is Alexandria which is the first city in the world, he may be handing the imperial capital an intentional insult (17. 52. 5).

Within the narrative, Rome's presence is quite uneven. In the books up to the First Punic War, discussions of Rome are insignificant, constituting less than three per cent of the extant material. Rome becomes substantially more important at the point of the First Punic War, for Diodorus' sources centre their accounts on the new organizing force in the Mediterranean. In particular, from book 32 until the final book, 40, Diodorus follows Polybius and Posidonius closely. Within those final, highly fragmentary books are found several judgements of Rome. Many of these can be attributed confidently or cautiously to Polybius and Posidonius. They are passing remarks reflecting the ambivalence of their original authors. The main charges are that Rome no longer fights only just wars and that her leaders are not of the same high calibre as before. But there are also statements about Rome which appear to be of Diodorus' own doing. Although a few are positive, most are negative. There are also a number of instances in which Diodorus appears to offer an extensive, rather than a passing personal opinion, and in each case he is critical of Rome. Here are two examples.

In the prologue to book 32, which many now consider to be of his own making, Diodorus notes that at one time the Romans were admired by their subjects for managing the empire with compassion and beneficence, but now 'they confirmed their power by terrorism and by the destruction of the most eminent cities' (32. 4. 5). His criticism of Roman imperialism is further articulated when narrating events of the Achaean War, found within the same book. In summing up the sack of Corinth, Diodorus emerges from behind Polybius:

[Corinth] was the city that, to the dismay of later ages, was now wiped out by her conquerors. Nor was it only at the time of her downfall that Corinth evoked great compassion from those that saw her; even in later times, when they saw the city levelled to the ground, all who looked upon her were moved to pity. . . . Nearly a hundred years later, Gaius Iulius Caesar (who for his great deeds was entitled *divus*), after viewing the site, restored the city.   (32. 27. 1; Walton trans.)

Although Diodorus followed Polybius closely on the Achaean War, the notice that Corinth was pitied even in later times and that the city was refounded by Caesar is certainly his own contribution. Indeed, in the same episode Diodorus must be altering the Polybian narrative by greatly exaggerating two points: Rome's brutality and the extent to which she deprived Greece of freedom. Diodorus writes:

The Greeks, after witnessing in person the butchery and beheading of their kinsmen and friends, the capture and looting of their cities, the abusive enslavement of whole populations, after, in a word, losing both their liberty and the right to speak freely [τὴν ἐλευθερίαν καὶ τὴν παρρησίαν], exchanged the height of prosperity for the most extreme misery.   (32. 26. 2; Walton trans.)

This is a strong indictment of Roman warfare and imperial rule, and nothing in the fragmentary Polybian text hints of such butchery or the suppression of Greek freedom. In the part that does remain, Polybius argues that L. Mummius acted leniently, given his extraordinary powers. If there was any brutal behaviour, the historian contends, it was due to Mummius' friends, but we are left to guess what that might have been. Nor do later writers confirm Diodorus' description. Other authors who may also have drawn on Polybius—Strabo, Pausanias, and Cassius Dio—emphasize the material, rather than human loss, and even that is limited to the city of Corinth; they also appear to suggest that Rome did not rob Greece of her freedom.

Diodorus declares that, as a result of the war, Greece went from great wealth to poverty. Polybius, on the other hand, notes specifically that Greece had already been poor (36. 17. 5), and Diodorus himself, probably in a passage inspired by Polybian material, speaks of the number of debtors before the war (32. 26. 3). Diodorus' statement about Greece's sudden fall from prosperity, then, reflects his attempt to widen the tragedy, as does his wrong-headed remark about the Greek loss of political liberty and freedom of speech. The strength of his emotional response is evident when he eulogizes Caesar for his re-establishment of Corinth, pronouncing that the Dictator: 'made amends for [the] unrelenting severity [of earlier Romans]' (Walton trans.).

In this case, Diodorus has exaggerated the effects of Roman behaviour. Here is another, more subtle example of how Diodorus can impose his own thoughts. This instance occurs in the early part of the *Bibliotheke* which covers the mythological prehistory of the Mediterranean. Diodorus declares that mythology has no factual basis but is important for supplying the reader with moral lessons. Thus here especially we might expect Diodorus to express some of his own philosophical or political sentiments.

In book 4, when describing the Sicilian shrine to Erycinian Aphrodite, Diodorus displays local pride, declaring that it is the only sanctuary in the world that has continued to flourish from the beginning of time to his own day. After noting that it was established by Aphrodite's son, Eryx, Diodorus lists the visitors who subsequently paid homage: first, Aeneas, also a son of the goddess, though by a different father, on his way to Italy anchored off Sicily and embellished the sanctuary; and then over the centuries the Sicanians, Carthaginians, and finally, in Diodorus' time, the Romans also come and contribute. The Romans are the most enthusiastic, tracing their ancestry and attributing their expansion to the goddess. All consuls, praetors, and Romans with any authority who come to the island visit the shrine. The Roman Senate even decreed the sanctuary special honours (4. 83).

Behind this casual description is an important story. The sanctuary to Erycinian Aphrodite was no mere local cult: it was in fact a significant part of the Roman foundation myth. In an attempt to win Greek support during the Second Punic War, the Romans stressed Aeneas' landing at Sicily. A temple, inspired by that at Eryx, was set within the *pomerium* at Rome, a rare honour. Aeneas' Sicilian experience was recalled during the late Republic in coins and literature (Cicero) and by the Augustan age it became central to the story, as told by Dionysius, Livy, and Virgil. The story illuminated the common cultural heritage of the Greeks and

Romans, and helped justify the Roman conquest of Sicily, Carthage, and the Greek East.

It is therefore striking that in the *Bibliotheke* the Aenean connection with Sicily is specifically minimized and the claim that he founded the Erycinian cult denied. According to Diodorus, Aeneas did not remain in Sicily, but anchored off shore in order to pay his respects at the sanctuary. The cult itself had been established much earlier by a different son of Aphrodite, Eryx, an indigenous Sicilian after whom the town and sanctuary were named. Eryx himself belonged to the prehistory of the island: he fought Heracles and lived in Sicily contemporaneously with the earliest Sicanians. Diodorus emphasizes the prehistoric and aboriginal status of Eryx, in contrast to his half-brother, Aeneas, who merely passed through later (ὕστερον).

Diodorus was not the first Greek to deny the Roman version of Aeneas. A century earlier Demetrius of Scepsis, a patriot of his native Troad, stressed the primacy of the story that Aeneas never went to Latium, but instead remained in Asia Minor. Having witnessed Roman incursions into his land, Demetrius rebutted the myth that connected Italy with the Greek East, implying that Romans ought to respect the true homeland of Aeneas. In standing against many contemporaries who embraced the Roman tradition, Demetrius used the Aeneas legend as part of resistance historiography.[15]

Diodorus, in making his own case against the Roman legend, even offers a coarse insult. While noting that all people who worshipped the goddess honoured her shrine with dedications, he adds that Roman praetors and consuls who visited the island 'laying aside the austerity of their authority . . . enter into sports and have intercourse (ὁμιλία) with women in a spirit of great gaiety, believing that only in this way will they make their presence there pleasing to the goddess' (4. 83. 6; Oldfather, modified). Certainly temple prostitutes were available to anyone who came to Eryx, but Diodorus connects them only with Romans, and provincial governors at that. The contrast with Strabo is illuminating: the Roman apologist mentions the temple prostitutes as being dedicated by 'both Sicilians and many foreigners' (6. 2. 6). Diodorus tells his story of Aeneas to defend his cultural heritage against Roman tradition and to offer a pointed barb at imperial rule.

Having proposed that Diodorus intrudes into the text with political, moral, and philosophical opinions, I now turn to an important episode in

[15] Emilio Gabba, 'Sulla valorizzazione politica della leggenda delle origini troiane di Roma fra III e II secolo a.C.', *Cont. Ist. st. ant.* 4 (1976), 84–101.

Roman imperialistic history where Diodorus is thought to have intentionally contaminated the narrative with unhistorical material. Here I suggest, conversely, that Diodorus may not have altered the account and that uncritical assumptions about his methodology have led scholars to minimize unduly the significance and accuracy of his account. Although Diodorus does contribute his own interpretations to the *Bibliotheke*, these interpretations are not intended to alter the factual narrative. In particular, when evaluating Diodorus' methods, it is important to distinguish between his treatment of narrative and set speeches.

In the spring of 198, Flamininus arrived in Greece to take up the war against Philip V of Macedon. Soon after, during a conference at the Aous River, Flamininus and Philip discussed possible conditions for peace. Diodorus (28. 11) and Livy (32. 10. 1–8) preserve the event, and scholars agree that they draw at least somewhat on Polybius, whose account for this is lost.

Diodorus and Livy both give the diplomatic exchange in a mixture of direct and indirect discourse. Polybius also frequently employs this style, which suggests that the later historians accurately reflect the general presentation of the original. Diodorus recounts: 'Flamininus held that Philip must completely evacuate Greece, which should thereafter be ungarrisoned and autonomous [Φλαμινῖνος μὲν ᾤετο δεῖν τὸν Φίλιππον ἐκχωρεῖν ἁπάσης Ἑλλάδος, ὅπως ἀφρούρητος ᾖ καὶ αὐτόνομος]. . . . But when Philip insisted that the evacuation of much of the territory might be submitted for arbitration but some of it was Antigonid by claim and tradition and hence not negotiable, Flamininus replied that he himself was under orders from the Senate to liberate not part, but all of Greece [καὶ διότι παρὰ τῆς βουλῆς ἐντολὰς ὅπως μὴ μέρος τῆς Ἑλλάδος ἀλλὰ πᾶσαν αὐτὴν ἐλευθεροῦν]. Philip retorted by asking: "What heavier condition would you have imposed if you had defeated me in war?", and with these words he departed in a rage' (Walton trans., modified).

According to Diodorus, Flamininus arrived armed with a senatorial order that Philip both evacuate and leave autonomous all of Greece—rather harsh demands given the lengthy Antigonid occupation of parts of the country. Appian's abbreviated narrative appears to be consistent with this tradition. There Flamininus told Philip 'to evacuate Greece' (ἐκστῆναι τῆς Ἑλλάδος: *Mac.* 5).

For much of this century, writers on Roman imperialism have credited Diodorus with accurately preserving the Polybian account. Recently, however, several scholars have doubted the historicity of Diodorus' version, arguing instead that details in Livy are more persuasive.[16] Livy's account, discussed below, has Flamininus request the Macedonian surrender of garrisons in Greek towns and the evacuation of Thessaly only. Livy does not explicitly mention a senatorial order, and there is no explicit demand by Flamininus for complete freedom of all mainland Greeks.

The difference between the two accounts is significant for understanding the development of Roman imperialism. Two years earlier when making its declaration of war, Rome merely insisted that Macedonia refrain from attacking others: Polybius gives no hint that the Senate was bent on liberating mainland Greece from Macedonian rule.[17] Yet, within months after the Aous Conference, Greeks considered, at least in some general fashion, that Rome's goal in the war was to free their country.[18] But did Flamininus bring such a commitment with him from the Senate or did the idea evolve through increased contact with the Greeks? Certainly the aims of the war could have changed and enlarged after Philip refused Roman demands and armies were dispatched to Greece. But how much did Roman aims expand and when?

Because of Diodorus' poor reputation, if important interpretations of Roman imperialism hang on a single episode of his work, there is bound to be controversy. Most relevant to Diodorus' veracity is Livy's failure to mention that Flamininus came to Greece armed with orders from the Senate and that those orders involved the explicit demand that Philip leave Greece ungarrisoned and autonomous. The condition of the Aous Conference expressed by Diodorus, that Philip leave Greece ungarrisoned and autonomous ($\dot{\alpha}\phi\rho o\acute{v}\rho\eta\tau o\varsigma$ $\kappa\alpha\grave{\imath}$ $\alpha\dot{v}\tau\acute{o}\nu o\mu o\varsigma$), resembles the settlement of the senatorial commission surrounding the Isthmian Declaration of 196. Critics of Diodorus suggest that he may here be following an annalist. Such a pro-Roman apologist, in trying to make Rome appear devoted to the Greek cause at the earliest possible moment, might have reworked Polybius by adding those details to the Aous Conference. Otherwise, critics suggest that perhaps Diodorus, prone to doublets anyway, himself applied the details, retrojecting them to the earlier event.

Here, rather than as copy machine, Diodorus is viewed as bumbler. That is, he has either followed an inferior, annalistic version of Polybius (when the original was clearly available to him), or contaminated the account himself. Certainly, Diodorus is frequently—and rightly—

---

[16] Summaries of the scholarship in A. M. Eckstein, 'Polybius, the Achaeans, and the "Freedom of the Greeks"', *GRBS* 31 (1990), 45–7.

[17] Polybius 16. 27. 2, 34. 3.   [18] Livy 32. 21. 35–7.

considered incompetent in preserving historical fact. But each passage must be considered in its context and with full consideration for Diodorus' methods and the surrounding historiographical tradition.

The surrounding tradition consists of a possible annalistic source and Livy, and they are taken first. Because Appian frequently drew on Roman annalists for his material and here appears to agree with Diodorus, some argue that Diodorus followed a tradition contaminated by annalistic intrusion. But it will not do to attribute the episode in Diodorus to the creative skills of such an intermediary. Put simply, for the many instances involving Rome and the East where parallel accounts of Diodorus and Polybius exist, there is no indication that Diodorus drew on Polybius through an annalistic filter. Where we have both accounts, their close correlation makes the suggestion of an intermediary completely unnecessary.[19] To impugn the material, therefore, one must argue that Diodorus himself altered the Polybian account of the Aous Conference by introducing material from the later Isthmian Declaration. Livy must be preferred to Diodorus.

In such circumstances, however, Livy does not inspire confidence. He has been amply shown to rewrite for rhetorical effect[20] and abbreviate the Polybian original. He does so with especial frequency when recording diplomatic exchanges.[21] There are interesting examples in surrounding events. At the conference at Lysimachea, held in 196 between Antiochus and Roman envoys, Livy shortens some of the debate. Further, the single point of substance mentioned in both Polybian speeches which Livy omits is the Roman demand that Antiochus should not interfere with the autonomous cities of the north-east Aegean.[22] But perhaps most telling, it is clear that Livy seriously misunderstands the diplomatic conditions for the Roman settlement established just prior to the Isthmian Declaration.[23] These are sufficiently similar to the Aous Conference that we should not use Livy's silence on diplomatic details as proof that they are not historical.

[19] Still useful is H. Nissen, *Kritische Untersuchungen über die Quellen der vierten und fünften Dekaden des Livius* (Berlin, 1863), 112–13, on Diodorus' use of Polybius through a cross-check of common passages in Diodorus and Livy. Indeed, the attempt by P. Meloni (*Il valore storico e le fonti del libro Macedonico di Appiano* (Rome, 1955), 55–6) to show that Diodorus, Appian, and even Livy drew on annalistic sources for the Aous Conference is unsuccessful: E. Gabba, review of Meloni, *Rivista storica italiana* 68 (1956), 103.

[20] Nissen, *Kritische Untersuchungen*, 21–6.

[21] T. J. Luce, *Livy: The Composition of his History* (Princeton, NJ, 1977), 205–21.

[22] τῶν αὐτονόμων ἀπέχεσθαι πόλεων; P. 18. 49. 7=L. 33. 39. 7; P. 18. 51. 9=L. 33. 39. 2. Cf. Luce, *Livy*, 211–12.

[23] Polybius 18. 44. 2–3=Livy 33. 30. 2; cf. J. Briscoe, *A Commentary on Livy: Books XXXI–XXXIII* (Oxford, 1973), 304–5.

Indeed, Livy's version of Flamininus' Aous speech may be more compatible with Diodorus' than is frequently supposed. In outline, they are quite similar; in choice of words (allowing for different languages) they are at times nearly identical. Most importantly, regarding Roman demands on Philip there is in Livy sufficient ambiguity to permit interpretation. At the beginning of the Livian account, just as in the *Bibliotheke*, the consul requires that Philip withdraw his garrisons from the cities. But at the end of the episode, Livy states: 'Then, when they came to discuss what states were to be set free [*quae civitates liberandae essent*], the consul named the Thessalians before all the rest [*Thessalos primos omnium*]', at which point Philip angrily broke off negotiations. 'Thessalos primos omnium' appears to mean that the liberation of Thessaly was simply at the top of the list. That was particularly galling to Philip, since it had been an Antigonid possession for 150 years and was strategically critical for Macedonian defence. But it need not have been all that Livy found in Polybius: Livy may have not been so interested in all the legal implications as in portraying the key demand that caused negotiations to fail. It would appear, then, that using Livy as a corrective has its own problems. First, Livy is not especially reliable in preserving the diplomatic details found in his source; second, his account may after all contain an echo of what Diodorus articulated fully.

The more sources are examined, the more complex the problem appears. To widen our scope, Polybius' version of the Isthmian settlement, it is claimed, provided the model for Diodorus' language at the Aous Conference. Yet its key condition is itself a precise echo of Greek demands made a generation earlier and also found in Polybius.[24] No one doubts that Polybius got right the substance of the Isthmian Declaration. Why then should scholars discard what is in Diodorus? Moreover, the notion of leaving Greece ἀφρούρητος καὶ αὐτόνομος is found four other times in the *Bibliotheke* in highly specialized diplomatic contexts. In each case, there is no suggestion that Diodorus misrepresented the issue or created a doublet by repeating what he found in his sources in an inappropriate context.[25] And for a somewhat similar diplomatic inter-

[24] 18. 46. 4 = 4. 25. 6, for the Symmachy before the Social War; Walbank, *HCP* ii. 612. Most recently discussed by Eckstein (n. 16 above), 66.

[25] At 15. 38. 2 Diodorus' language of the Common Peace of 375 is confirmed precisely by Isocrates *De Pace* 16. On that and 15. 5. 1, on the King's Peace, see G. L. Cawkwell, 'The King's Peace', *CQ* 31 (1981), 69–83. At 19. 61. 3 is Antigonus Monophthalmos' declaration of 315 at Tyre, which is accepted as accurate: most recently, R. A. Billows, *Antigonos the One-eyed and the Creation of the Hellenistic State* (Berkeley, 1990), 114. The passage at 20. 99. 3, on Rhodian terms of surrender to Demetrius, cannot be confirmed or denied by Plutarch, *Demetrius* 22. 4, but is generally accepted: E. S. Gruen, *The Hellenistic*

change, for which there exists an inscription as a control, Diodorus is given high marks for accurately preserving the meaning and the language.[26] Even those who dismiss Diodorus' version of the Aous Conference acknowledge that the sentiment of ἀφρούρητος καὶ αὐτόνομος was by then current. Livy notes that soon after the Aous Conference, Greeks believe that the intent of the Roman army was to liberate Greece (32. 21. 36). Polybius records that, a few months later, at a meeting between Flamininus and Philip at Nicaea, Flamininus demands Philip's complete withdrawal from Greece.[27]

A closer look at Diodorus' methods produces greater confidence in his account. Before arguing that Diodorus has carried Polybian phrasing from the Isthmian Declaration back to the Aous Conference, one must consider the following. Certainly Diodorus commits numerous doublets and other factual errors throughout the *Bibliotheke*. But here is the critical consideration: they occur in the narrative, not in the speeches. At times, Diodorus intervenes in speeches to express his own philosophical and moral points, but those additions are epideictic, rather than factual.[28] And Diodorus is quite accurate in recounting Polybian speeches. In the more than a dozen instances where the Diodoran speech and the Polybian original survive,[29] in no case does he transpose Polybius' words from one oration to another or significantly alter the text. In one speech, Diodorus does make many factual blunders, but those come from a misreading of the Polybian original and not from the intrusion of outside material.[30] In several other instances, Diodorus leaves out details.[31] One of the episodes involves the conference at Lysimachea (28. 12), in which, it was noted above, Livy also omits detail. Omission of details is expected in the *Bibliotheke*, for Diodorus had to condense the hundreds of books of his sources into forty of his own. Additionally, most of the Polybian material in Diodorus is preserved by

World and the Coming of Rome (Berkeley, 1984), 136. Because Diodorus is such an important source for the diplomatic language and actions of the *Diadochoi*, impugning his veracity at 28. 11 would imply that he cannot be trusted in other such circumstances. That would call into question our fundamental understanding of early Hellenistic diplomacy.

[26] Gruen, *Hellenistic World*, 135 n. 20.

[27] 18. 1. 12; cf. 2. 6, 5. 5, 9. 1; cf. Livy 32. 32. 2; Diodorus is not preserved here.

[28] *Diodorus* 93–108.

[29] DS 28. 6=P. 16. 34. 2–7; 28. 12=P. 18. 50–1; 29. 8=P. 21. 15; 29. 17=P. 22. 7–9; 29. 21=P. 23. 14; 29. 22=P. 24. 5; 29. 11=P. 21. 18–24; 29. 12=P. 18. 49–51; 30. 5=P.30. 4; 30. 23=P. 29. 20; 31. 1=P. 27. 6; 31. 5=P. 30. 4; 31. 15. 1–2=P. 30. 18; 32. 22=P. 38. 8. 4–12; 32. 23=P. 38. 20. 1–3.

[30] DS 29. 17=P. 22. 7–9.

[31] DS 28. 12=P. 18. 50–1; 29. 11=P. 21. 18–24; 29. 12=P. 18. 49–51; 30. 5=P. 30. 4; 31. 1=P. 27. 6; 31. 5=P. 30. 4; 32. 22=P. 38. 8. 4–12; 32. 23=P. 38. 20. 1–3.

the Constantinian excerptors, and there is the possibility that they further abbreviated what they found in the _Bibliotheke_. But the question at present is not whether Diodorus omitted details found in Polybius, for that issue pertains to Livy's account. The question here is whether Diodorus _added_ material not in the Polybian original. Using the more than a dozen examples where the Diodoran and Polybian versions of speeches are extant, it is clear that Diodorus does not add factual material to Polybius' speeches. Given these considerations of Diodorus' methods and the fact that there may even be an echo of the autonomy clause in the Livian account, it seems difficult to accept that Diodorus introduced the phrase from a later Polybian oration.

Indeed, what is especially striking in the Diodoran episode—and particularly hard to dismiss—is Flamininus' claim that he brought orders from the Senate. It is argued that Livy, a Roman apologist, would not omit this fact, because it is obviously favourable to Rome. But on the other hand, why would Diodorus go to the trouble to invent or retroject the claim from the Isthmian Declaration unless he is engaging in pro-Roman propaganda? Yet, if Diodorus does have a political bias, it is mildly anti-Roman.[32]

Not long after the Aous Conference, the warring states hold a conference at Nicaea, for which the Polybian account is completely preserved. One of the most frequently voiced arguments against the historicity of the senatorial order at the Aous Conference is the belief that Flamininus does not again appeal to that authority at the Nicaean Conference. Yet at Nicaea, Polybius paraphrases Flamininus: 'The Roman general said that what it was his duty to say was simple and obvious [ὁ δὲ τῶν Ῥωμαίων στρατηγὸς αὐτῷ μὲν ἁπλοῦν τινα λόγον ἔφη καθήκειν καὶ φαινόμενον]. He demanded that Philip should withdraw from the whole of Greece. ...'[33] Polybius' construction, καθήκειν and the pronoun in the dative, indicates, as Paton here translates, that Flamininus was obliged to demand complete autonomy for Greece.[34] But was that a moral, diplomatic, or legal obligation? If the last, it would confirm Diodorus' earlier notice at the Aous River that the consul was issued senatorial orders. Certainly, the statement allows sufficient room for interpretation.

[32] See above and, for fuller detail, _Diodorus_, 117–203. The question is extremely complex and probably will never be completely settled. But certainly for the middle Republic, and especially in the Greek military theatre, there is no evidence that Diodorus acted as a Roman apologist.

[33] 18. 1. 12: tr. W. S. Paton, _Polybius_, 6 vols. (Loeb; Cambridge, Mass., 1922–7).

[34] Cf. Walbank, _HCP_ ii. 550–1.

How would acceptance of Diodorus' version fit into the immediately subsequent events? Deliberations from the Aous Conference through Nicaea and its resulting meetings in the Senate are complex and beyond the scope of this discussion (Polybius 18. 1–12). But all details easily accommodate the interpretation of a Roman commander, armed with a senatorial ultimatum, who, through inexperience, political acumen, or simple flexibility, eventually allows Philip to negotiate the conditions by referring him to the Senate. In any case, to reject Diodorus' version because it may not fit some modern assumptions about ancient diplomacy of course fails to privilege the unique experiences we study.

Certainly I do not here propose to settle the question of whether, already in early 198, the Senate demanded the complete autonomy of Greece—or whether it even sent Flamininus to Greece with a specific diplomatic agenda. Indeed, even if Diodorus' version is accepted, the demands made by the Senate may well be quite different from the later notion of 'Freedom of the Greeks', which could have evolved as Romans gained closer contact with Greeks and circumstances changed.[35] What I want to emphasize here is how the sources have been handled. Polybius, of course, stands supreme for this period, and it takes a very clever argument to impeach his authority. Where his account is lost, we rely on Livy. Dozens of studies of Livy have revealed various *Tendenzen*, often contradictory, which can be used at will by scholars. Implicit in the decision of many to favour Livy over Diodorus here is the belief that Livy is a Roman apologist who would not have intentionally omitted an earlier senatorial decision to free Greece. The so-called 'best' authors have drawn nearly all our attention, because they shape the major historical questions that we investigate. The result, in this instance, is that a considerable amount of scholarship on Livy and a few uncritical assumptions about Diodorus have been marshalled to provide an answer. When Diodorus is also studied in his own right, however, different possibilities emerge.

In conclusion, there is no sure way to decide what in the *Bibliotheke* derives from Diodorus' sources and what is his own intrusion. But the growing belief that the work does indeed have some intellectual unity allows us to discern features that are far more likely of Diodorus' authorship than of any of his sources. Generally speaking, Diodorus appears to add his own philosophical and political principles, and to do so fairly consistently. But, generally speaking, he does not intentionally alter the

---

[35] See esp. the excellent discussion by Eckstein (n. 16 above), 45–71.

narrative to conform to those philosophical and political interpretations. These observations are, of course, preliminary. With further investigation of Diodorus and other minor authors, we will get a much fuller notion—for a notion is all that it can be—of their methods and *Weltanschauung*.

9

Symbol of Unity? The Persian-Wars Tradition in the Roman Empire

ANTONY SPAWFORTH

INTRODUCTION

The Roman empire was a multi-cultural conquest-state, the long-term survival of which depended on the creation of political consensus between rulers and subjects (and subject-élites in particular).[1] Historians recognize that shared symbols played their part in this process, as with the imperial cult or (for parts of the empire) the figure of Alexander the Great.[2] This paper focuses on another touchstone in the historical memory of subject-Greeks, the tradition about the Persian wars, and seeks to answer two questions. (1) What historical significance should we attach to the blending of this tradition into the official ideology of Rome from Augustus onwards? (2) What relationship, if any, was there between this development and the Persian-wars 'mania' which is such a striking feature of Greek culture in the first three centuries AD? What follows is in three parts. In Section I I begin by re-examining the Parthenon-inscription in honour of the emperor Nero. Once its full context emerges, this epigraphic monument not only underscores how much the Persian wars constituted a living tradition in first-century Greece, but also suggests how this tradition could draw some of its vitality from official attitudes at Rome. In Section II I briefly discuss the Roman state's linkage of these wars with Parthian policy from Augustus onwards and try to answer the question 'why?'. In Section III I consider

[1] My warm thanks to Simon Hornblower for inviting me to address the Greek Historiography seminar in November 1991. Versions of this paper have also been read at seminars in Leeds, Newcastle-upon-Tyne, and the Institute of Classical Studies, London; I am grateful to the audiences on all these occasions for helpful comments, and to Donald Hill for subsequent discussion; faults, of course, are mine.

[2] K. Hopkins, *Conquerors and Slaves* (Cambridge, 1978), ch. 5 (imperial cult). The Alexander-bibliography is predictably vast, but note now *Neronia 4. Alejandro Magno, modelo de los emperadores romanos: Actes du IVᵉ colloque international de la SIEN* (Coll. Latomus 209; Brussels, 1990); C. Bourazelis, Θεία δωρεά: Μελέτες πάνω στὴν πολιτικὴ τῆς δυναστείας τῶν Σεβήρων καὶ τὴν Constitutio Antoniniana (Athens, 1989), 41–51.

the effectiveness of the Persian wars as a symbol of unity for the Greek
half of the empire.

I. THE PARTHENON-INSCRIPTION AGAIN

On the east architrave of the Parthenon the eye can easily see a row of
groups of holes in the marble for the attachment of bronze (and no doubt
gilded) letters, each about fourteen centimetres high. These settings
permit an unproblematic reconstruction of an honorific inscription for
the emperor Nero:[3]

Translation: The Council of the Areopagus, the Council of the Six Hundred, and
the Athenian people (honour) the Greatest Emperor Nero Caesar Claudius
Augustus Germanicus, son of a God, when Ti(berius) Claudius Novius, son of
Philinus, was Hoplite-General for the eighth time, as well as Epimelete and
Nomothete, and the priestess (of Athena) was Paulina, daughter of Capito.

Since Kevin Carroll's careful restudy of this text, we know that it does
not signify a rededication of the Parthenon to Nero, but should be seen as
an honorific gesture in its own right (which of course is not to deny that a
cult for Nero *may* have been installed in the Parthenon). Carroll clarified
the context of this honour, which is firmly dated by the office of Tib.
Claudius Novius to the Athenian archon-year 61/2, when Nero's
generals were waging an aggressive war with Parthia over control of
Armenia: he argued that, by setting this inscription where they did, the
Athenians presented the viewer with an honorific analogy between
Nero's Armenian war and the great tradition of Greek and Hellenistic
triumphs over eastern barbarism commemorated on the Acropolis. Lest
this seem a woolly proposition, it must be remembered that 'symbolic
juxtaposition', whereby a monument or work of art disclosed its full
meaning only when viewed (literally or imaginatively) in conjunction
with others in the vicinity, was a time-honoured practice in the setting up
of Greek dedications and honours.[4] The Parthenon-inscription was set
into a veritable grid of such visual cross-references. The later Athenians
thought of the Parthenon itself as paid for by Persian spoils (Dem. 22.
13); the inscription was punctuated by a row of shields forming part of

---

[3] *IG* ii². 3277 = K. Carroll, *The Parthenon Inscription* (Greek, Roman and Byzantine
Monographs 9; Durham, NC, 1982); R. Sherk (ed.), *The Roman Empire: Augustus to
Hadrian* (Cambridge, 1988), no. 78.

[4] At Roman Athens note Cass. Dio 47. 20. 4; also L. Robert, *Hellenica*, iv (Paris, 1948),
42 (statues of Roman governors).

Alexander's dedication of Persian trophies to Athena after the Granicus;[5] and it overlooked a Pergamene dedication by (probably) Attalus I of 200 BC, in which four sculptural tableaux juxtaposed Pergamene troops fighting Galatians with the time-honoured trinity of Marathon, Theseus against the Amazons, and the Gods versus the Giants.[6] This particular corner of the Acropolis, in fact, offered a visualization of a universal mythology of struggle against eastern barbarism which by the imperial age had come to accommodate not only the Persian wars, but also the feats of Alexander, the Attalids—and now Nero.

That the inscription for Nero consciously sought to exploit these associations is confirmed by the identity of its sponsor, whose personal familiarity with the tradition about the Persian wars can be documented. With little doubt this sponsor can be recognized in the Athenian notable, Tib. Claudius Novius, whose eighth term as hoplite-general ostentatiously replaces the eponymous archon in the dating formula. Novius, whose career was studied recently by D. Geagan, was a leading light in the Athens of his day; in a city with a disastrous recent history of disloyalty to Rome, his success in local politics must have owed a lot to his carefully cultivated *Romanitas*, manifested *inter alia* in his terms as high-priest of the Athenian imperial cult and agonothete of the first Athenian celebration of Sebastan games.[7] His familiarity with the Persian wars is vouchsafed by an inscription (*IG* ii². 1990) linking him closely with the commemorative cults and ceremonies at Plataea. As we learn from the heading of this Athenian list of ephebes likewise dated to 61/2, Novius combined his eighth term as hoplite-general in Athens with the Plataean post of 'high-priest of Nero Claudius Caesar Germanicus and Zeus Eleutherius by appointment from the Hellenes'.[8] These 'Hellenes' were the body known officially as the 'common council of the Hellenes who congregate at Plataea' (*Syll*³ 393), and their chief function was to celebrate the pentaeteric 'Freedom' games (Eleutheria) in memory of 479 BC; the appointment of Novius is part of a body of evidence showing that for long periods in the Hellenistic and imperial ages these Plataean cults were effectively controlled by Athens (unsurprisingly, in the light of both

---

[5] L. Schneider and C. Höcker, *Die Akropolis von Athen: antikes Heiligtum und modernes Reiseziel* (Cologne, 1990), 223.

[6] J. J. Pollitt, *Art in the Hellenistic Age* (Cambridge, 1986), 90–5.

[7] D. Geagan, 'Tib. Claudius Novius, the Hoplite-Generalship and the Epimeleteia of the Free City of Athens', *AJP* 100 (1979), 279–87.

[8] Confirmation of this dating (*pace* V. Rosivach, *P del P* 42 (1987), 277 n. 44); S. Follet, *BÉp.* 1989, no. 388.

history and geography).[9] The priestly title of Novius shows that by 61/2 a cult of the reigning emperor had been installed at Plataea alongside that of Zeus Eleutherius—probably not on the initiative of Novius himself, since he is not hailed as 'first-ever' incumbent; even so the cult was probably recent, given Achaia's general slowness to establish permanent institutions of emperor-worship above city level; the Achaean League only got round to doing so in the years around 54,[10] and the Plataean cult may have followed this Peloponnesian lead.

The Plataean ties of Novius were much older than his high-priesthood. The same Athenian text gives him (lines 5–6) the title 'best of the Greeks', by which winners in the men's race in hoplite armour at the Eleutherian games, run over a course which began at the trophies of 479 BC, were proclaimed;[11] Novius must have been a champion runner in his youth. He also married into an upper-class family from Sparta, the other major city involved in organizing the Plataean ceremonies at this date, since it has so far gone unnoticed that his wife 'Damosthenia of Marathon, daughter of Lysinicus', already known from a Delian inscription (*IDélos* 1629), turns up in the following Spartan (funerary?) inscription copied at Mistra (*IG* v(1), 509):

$$\Pi(\acute{o}\pi\lambda\iota os)\ M\acute{\epsilon}(\mu\mu\iota os)\ \Lambda\upsilon\sigma\acute{\iota}\nu\epsilon\iota\kappa[os]$$
$$\phi\iota\lambda\acute{a}\delta\epsilon\lambda\phi os\ \kappa a[\grave{\iota}]$$
$$\Delta a\mu o\sigma\theta\acute{\epsilon}\nu\epsilon\iota a$$
$$\Lambda\upsilon\sigma\iota\nu\epsilon\acute{\iota}\kappa ou\ \Pi\acute{o}(\pi\lambda\iota o\nu)\ M\acute{\epsilon}(\mu\mu\iota o\nu)\ \Gamma o[\rho\text{-}]$$
$$5\quad \gamma\iota\pi\pi\acute{\iota}\delta a\nu\ \phi\iota\lambda\acute{a}\delta\epsilon\lambda\phi o\nu$$
$$\Lambda\upsilon\sigma\iota\nu\epsilon\acute{\iota}\kappa ou\ \upsilon\acute{\iota}\grave{o}\nu\ \tau\grave{o}[\nu]$$
$$[\acute{a}\delta\epsilon\lambda\phi\acute{o}\nu\text{-----}]$$

Translation: Publius Memmius Lysinicus, brother-lover, and Damosthenia, daughter of Lysinicus, (honour) Publius Memmius Gorgippidas, brother-lover, the son of Lysinicus, their [brother-----].

Like Novius, this Spartan family of P. Memmii enjoyed the patronage of P. Memmius Regulus, governor of Greece between 35 and 44, who brokered a grant of Roman citizenship for the two brothers. The text confirms the suspicion of Geagan[12] that Damosthenia was not an

---

[9] N. Robertson, 'A Point of Precedence at Plataia: The Dispute between Athens and Sparta over Leading the Procession', *Hesperia* 55 (1986), 88–102.

[10] See my paper, 'Corinth, Argos and the Imperial Cult: A Reconsideration of Pseudo-Julian, *Letters* 198 Bidez', *Hesperia* (forthcoming).

[11] J. Strubbe, 'Gründer kleinasiatischer Städte: Fiktion und Realität', *Anc. Soc.* 15–17 (1984–6), 283–4.

[12] Geagan (n. 7 above), 288. Kolbe (to *IG* v(1), 509) read $\Lambda\upsilon\sigma\iota\nu\epsilon\acute{\iota}\kappa[\eta s]$ in line 1; 'Lysinices' is an otherwise unattested name at Roman Sparta; for 'Lysinicus' see *IG* v(1), 141. 26.

Athenian by birth (since the Delian inscription gives an Athenian demotic to her but not her father); conceivably it was at Plataea that her husband-to-be first came into contact with this Spartan family (for instance, her brothers might also have competed there). In Novius, a picture can now be pieced together of a type of civic notable by no means peculiar to Roman Athens: wealthy, but not from the established political and social élite of his city (his *cognomen*, in fact a Roman *nomen*, may have derived from family links with eastern *negotiatores*), and knowledgeable about and attached to local tradition in the way of the enthusiastic social outsider.[13] In 61/2, his background made him extremely well placed to propound to his fellow Athenians the special appropriateness of honouring Nero in the 'Persian-wars corner' of the Acropolis.

Carroll saw the Parthenon-inscription as a piece of Athenian 'flattery'. This is indeed a legitimate way, favoured by the Romans themselves, of evaluating the honorific practices of the Greek city.[14] But its contemporary significance is by no means thereby exhausted. To start with, we know now that Athens was not alone in the Roman east in giving monumental expression to provincial hopes for the success of Nero's Armenian war: at much the same date the Aphrodisians incorporated into their 'Sebasteum' a dramatic relief of the personified Armenia slumped against the striding figure of the emperor.[15] The Athenian gesture reveals a characteristically Greek way of comprehending the Parthians, by equating them with the Persian bogymen of the Classical past and, in this case, literally inscribing them into local historical tradition. It can also be argued, however, that this provincial Greek perception was deliberately fostered by the central administration and that the Parthenon-inscription should be seen as a sign of subject-Greeks responding favourably to a theme in imperial ideology. To examine the evolution of this theme it is necessary to go now to Rome itself.

II. PERSIANS AND PARTHIANS IN IMPERIAL IDEOLOGY

This section briefly gathers together, in chronological order, the evidence known to the writer for the official representation of Rome's Parthian adversaries as the reincarnation of the Persians.

[13] See my comments on C. Vibius Salutaris of Ephesus in *CR* 42 (1992), 383–4, reviewing G. Rogers, *The Sacred Identity of Ephesus* etc. (London, 1991).

[14] See A. Wallace-Hadrill, 'Roman Arches and Greek Honours: The Language of Power at Rome', *PCPhS* 216 (1990), 143–81 at 153–4.

[15] R. R. R. Smith, 'The Imperial Reliefs from the Sebasteion at Aphrodisias', *JRS* 77 (1987), 88–138 at 117–20.

1. From its remnants in museums in Naples and Copenhagen a lavish imperial monument has recently been reconstructed by R. Schneider, who argues that Augustus commissioned it to commemorate his 'victory' over Parthia in 21 BC. A sculptural group over four metres high is envisaged, comprising three kneeling barbarians (identified as 'eastern' by the use of Phrygian marble), who supported a massive bronze tripod. Schneider sees the inspiration for this monument as Greek, not only in its use of barbarian supports, but also in the tradition of tripod-trophies to which it belonged, one descending from the famous tripod-column dedicated by the Greeks at Delphi after the battle of Plataea. Likewise, on Schneider's view, this Augustan monument was an offering to Apollo, set up in his Palatine sanctuary, close to the residence of Augustus himself.[16]

2. In 2 BC Augustus staged a mock sea-battle in a specially excavated arena by the Tiber. Lavish enough to merit inclusion in the *Res Gestae* (23), the occasion saw 'thirty beaked ships, triremes or biremes, and many smaller vessels' joining in a re-enactment of the (mainly) Athenian victory at Salamis, as we learn from Cassius Dio, who adds the telling details (55. 10. 7) that the opposing sides were called 'Athenians' and 'Persians' and that 'on this occasion too the Athenians won'; Ovidian panegyric (*Ars amat.* 1. 171–2) linked this entertainment with the imminent departure of Gaius Caesar on an aggressive campaign of revenge (as Ovid presents it) against the 'usurper' Phraates V, king of Parthia.

3. In 39 Gaius staged a great pageant in the Bay of Naples, which he spanned by means of a bridge of boats; this acted as the focus for two days of military manœuvres and a quasi-triumphal procession, in which the emperor paraded in the alleged breastplate of Alexander the Great and displayed beside him the Parthian 'hostage' Darius, son of King Artabanus II. At the time this bridge of boats was said to be in imitation of its famous Persian predecessor, built by Xerxes across the Hellespont for the invasion of Europe.[17]

4. In 57 or 58 Nero staged a naumachia in Rome which, like that of Augustus, pitted 'Athenians' against 'Persians'. Again recalling Augustan precedent, this spectacle coincided with an imperial initiative in the east, in this case the launching of Nero's Armenian war in the winter of 57/8.[18]

[16] R. Schneider, *Bunte Barbaren* (Worms, 1986).

[17] Imitation of Xerxes: Suet. *Gaius* 19; *Brev. Vit.* 18. 5; other accounts: Cass. Dio 59. 17; Jos. *AJ* 19. 5–6. For the ambiguous meaning of *obses*, 'hostage', see D. Braund, *Rome and the Friendly King* (Beckenham, 1984), 12–13.

[18] Cass. Dio 61. 9. 6; Suet. *Nero* 12.

5. Albeit with less clarity, Persian-wars reminiscences can be heard in the milieu of Hadrian's new organization of Greek cities, the Athens-based Panhellenium. Pausanias (1. 18. 8) saw in the Olympieum a tripod-monument in the form of Persians supporting a bronze tripod. The date of this monument, and the identity of its unknown donor, are controversial, although a date no earlier than Hadrian is favoured by the fact that this Athenian sanctuary stood unfinished until 131/2, when Hadrian dedicated it in person.[19] In form the monument was a twin of the Augustan tripod-dedication reconstructed by Schneider, who may well be right in assuming an imperial donor for the Athenian offering too, although for the reason just given Hadrian seems a likelier candidate than Augustus (as Schneider proposed[20]). The Olympieum, moreover, has been shown to be closely linked with the Panhellenium, which was founded in the same year, and two officials of the organization, both Asian Greeks, are found combining the priesthood of the imperial founder ('Hadrian Panhellenius') with those of Zeus Eleutherius and Greek Concord at Plataea.[21] A case can be made out, therefore, for seeing an interest on the part of the Panhellenium's imperial founder in associating the symbolism of the Persian wars with his initiative.

6–7. In 161 the emperor Lucius Verus was accompanied on his Persian expedition by a contingent of Spartan auxiliaries, presumably recruited when he passed through Achaia on his way to the front. I have argued elsewhere that the symbolic character of this gesture emerges when set beside the similar action of Caracalla, who in 214 likewise levied Spartan auxiliaries for his Parthian war.[22] This time our sources make the Persian-wars echoes explicit: the Spartan levy was styled the 'Laconian and Pitanate *lochos*', and Caracalla paired it with a levy of provincial Macedonians armed in the style of Alexander the Great's phalanx.[23]

8. In 235 the emperor Gordian III staged a Greek agonistic festival at Rome in honour of the protectress of the Athenian army at Marathon, Athena Promachus. As with the naumachiae of Augustus and Nero, the spectacle in the capital preceded an imperial campaign on the eastern

[19] D. Willers, *Hadrians panhellenisches Programm* (Basle, 1990), 35–6.

[20] Schneider, *Bunte Barbaren*, 82–90 (rejected in his review of Schneider by R. Cohon), *JRA* 3 (1990), 264–70).

[21] A. Spawforth and S. Walker, 'The World of the Panhellenion', *JRS* 75 (1985), 92; with a different emphasis: A. R. R. Sheppard, '*Homonoia* in the Greek Cities of the Roman Empire', *Anc. Soc.* 15–17 (1984–6), 238–9.

[22] P. Cartledge and A. Spawforth, *Hellenistic and Roman Sparta: A Tale of Two Cities* (London, 1989), 115.

[23] Cartledge and Spawforth, op. cit. 118–19.

frontier, in this case against the Sassanian Persians, the successor-state to Parthia.[24]

There is enough here to recognize a persistent theme in the Roman representation of eastern policy under the empire. The key figure here looks like Augustus himself, whose innovations provided models for his successors. The historical significance of this theme crystallizes around the following possibilities.

By the time of Augustus the Hellenized Roman élite was sufficiently familiar with Greek history (its knowledge gleaned from the rhetorical schools, from travels in Greece, and from a budding Greek historiography in Latin[25]) for the equation of Parthians and Persians to be almost a cultural reflex—as early as 57 BC Cicero could refer to the Parthians as 'Persae' (*de domo sua* 60). Among Roman admirers of Greek culture this equation was highly flattering to Rome, since it absorbed her stand against Parthia into a universal myth-historical tradition of struggle against barbarism stretching back to the war between Gods and Giants (a theme already exploited in the Augustan presentation of Actium[26]). It was a comparison which would have seemed especially apt in the late first century BC, following the Roman defeat at Carrhae in 53 BC and the Parthian breakthrough to the Mediterranean in 40 BC: Augustan Rome lived with the uncomfortable realization that the Parthians were 'a formidable foe, able and willing to challenge Roman domination of the east',[27] and the Persian wars offered a resounding symbol of western success in repelling eastern invaders. From this point of view the official Persian/Parthian metaphor was a form of 'sociological propaganda',[28] slipping almost spontaneously into the official thinking of the regime because in fact it drew on the dominant cultural values of the Roman upper class.

The metaphor also had more overtly political resonances. Augustus had broken with the aggressive stance against Parthia of Caesar and

[24] L. Robert, 'Deux concours grecs à Rome . . .', *CRAI* 1970, 11–17.

[25] A. E. Wardman, *Rome's Debt to Greece* (London, 1976), chs. 4–5; A. C. Dionisotti, 'Nepos and the Generals', *JRS* 78 (1988), 35–49, esp. 37 (Rome's 'fresh and vigorous interest' in Greek history by the mid-50s BC).

[26] P. Zanker, *The Power of Images in the Age of Augustus*, tr. A. Shapiro (Ann Arbor, 1988), esp. 84; G. Sauron in X. Lafon and G. Sauron (eds.), *L'Art décoratif à Rome à la fin de la république et au début du principat* etc. (Collection de l'École française de Rome 55; Rome, 1981), 286–94.

[27] R. Frye, *The History of Ancient Iran* (Handbuch der Altertumswissenschaft 3. 17; Munich, 1984), 233.

[28] For this category I am indebted to D. Kennedy, *Liverpool Classical Monthly* 9. 10 (December 1984), 158 (citing J. Ellul, *Propaganda: The Formation of Men's Attitudes* (New York, 1973)).

Antony in favour of more or less peaceful coexistence.[29] Official commemoration of the diplomatic success of 21 BC, when Augustus negotiated the return of the captured Roman standards, assumed distinctly martial overtones,[30] and the Persian-wars parallelism will have reinforced this interpretation. Likewise in 2 BC the Salaminian naumachia emitted the message that the expedition of Gaius was to be understood as a military campaign. In general the parallelism had the effect of magnifying the exploits of Augustus against Parthia (just as it did in the case of Attalus I, whose Galatian victory-monuments at Delphi and Athens implicitly glorified the king's military prowess—the ultimate basis of royal legitimacy in the Hellenistic age[31]), and reinforced the emperor's self-representation as world-conqueror.[32]

The official representation of eastern conflict in this way must also be seen in the light of the claims by the Parthian kings to be the successors of both the Achaemenids and Alexander, and their professed ambitions to recover the western satrapies of the old Persian empire. Although we first hear of Parthian ambassadors addressing these claims to a Roman emperor only in 35 (Tac. *Ann.* 6. 31), they probably were a part of Parthian royal ideology long before the late first century BC, when Rome would have had some justification for taking seriously the Parthian capacity to implement them.[33] Almost two generations ago, Balsdon argued that Gaius laid on the spectacle of the bridge of boats primarily to impress high-ranking Parthian 'hostages', including Darius, and that the whole pageant may have been staged to mark the arrival in Italy of the 'hostages' given to Rome by Artabanus II in 37. This attempt at a 'rational' explanation of Gaius' grandiose gesture deserves more serious attention than it has received.[34] In the light of Parthian pretensions, a Roman counterpart to Xerxes' famous bridge of boats would have made a nicely pointed display of imperial power and resources. Recent work emphasizes the importance of foreign courts and ambassadors as targets

[29] H. Sonnabend, *Fremdenbild und Politik: Vorstellungen der Römer von Ägypten und dem Partherreich in der späten Republik und frühen Kaiserzeit* (Frankfurt am Main–Berne–New York, 1986), 197–8.

[30] Zanker, *Power of Images*, 183–92.

[31] Delphi: H.-J. Schalles, *Untersuchungen zur Kultur-politik der pergamenischen Herrscher im dritten Jahrhundert vor Christus* (Istanbuler Forschungen 36; Tübingen, 1985), 104–23.

[32] C. Nicolet, *Space, Geography and Politics in the Early Roman Empire*, tr. H. Leclerc (Ann Arbor, 1991), ch. 2.

[33] J. Wolski, 'Les Achéménides et les Arsacides', *Syria* 43 (1966), 65–89, esp. 72–6.

[34] J. P. V. D. Balsdon, *The Emperor Gaius* (Oxford, 1934), 50–4 (not discussed by A. Barrett, *Caligula: The Corruption of Power* (London, 1989), 211–12).

for the 'image-making' of early modern monarchs, notably Louis XIV.[35] To judge from a passage in Strabo, Parthian ambassadors were a frequent presence at Augustan Rome; under Augustus and his immediate successors Rome was the residence of various princely Arsacid 'hostages' and their families; and our sources confirm that Parthian royals and Parthian ambassadors attended public shows in the capital.[36] From returned 'hostages' and ambassadors reports of the latest imperial spectacles and monuments could eventually have filtered back, and given food for thought, to the Parthian court.

Apropos the spectacle of 2 BC, the late R. Syme was the first to promulgate what has since become almost an orthodoxy: 'this piece of pageantry advertised Rome as the champion of Hellas against the Orient.'[37] The Augustan regime, moreover, had its opponents in the eastern provinces, one strand in whose thinking was relish at Rome's discomfiture before Parthian arms.[38] There was a much earlier precedent too for the use of the Persian-wars motif in the presentation of Roman policy to Greeks, if the modern view is sound that during the second Macedonian war Flamininus sought to compare the Macedonian threat to Greek freedom with the Persian.[39] On the other hand, when Alexander and the Attalids sought to align themselves with the Greek tradition of struggle against Persia, they did so by making dedications in central Greek sanctuaries. But the Roman and Italian setting of the gestures by Augustus, Gaius, and Nero does not suggest that they were *chiefly* designed to reach out to the Greek-speaking provinces so much as to glorify the Roman emperor and state; the provincial children present by invitation at the Augustan naumachia came from all over the empire, not just the east (Ovid, *Ars Amat.* 1. 173–4).

A change can be discerned, however, in the high empire of the second century. By this date, with Parthia on the defensive, earlier Roman fears about her power could be seen as groundless and the Persian-wars parallel less apt; no mention of Classical Greek history is heard in connection with the offensive Parthian war of Trajan, for whom Alexander

[35] P. Burke, *The Fabrication of Louis XIV* (New Haven, Conn. and London, 1992), 162 (note too his interesting comparison of Augustus and Louis XIV, 193–7).

[36] Strab. 16. 1, C 648–9; Suet. *Aug.* 43; *Claud.* 25.

[37] R. Syme, *SBAW* 1974, 15 = *Roman Papers*, iii (Oxford, 1984), 922; G. W. Bowersock, 'Augustus and the East: The Problem of the Succession', in F. Millar and E. Segal (eds.), *Caesar Augustus: Seven Aspects* (Oxford, 1984; corr. edn. 1990), 175–6.

[38] G. W. Bowersock, *Augustus and the Greek World* (Oxford, 1965), 108–9, discussing Livy's famous passage (9. 17–19).

[39] F. W. Walbank, 'Alcaeus of Messene, Philip V and Rome, part I', *CQ* 36 (1942), 145 n. 1; J.-L. Ferrary, *Philhellénisme et impérialisme* (Rome, 1988), 86.

the conqueror offered a more suitable symbol of triumph over the east. A chauvinistic reaction to the recent magnification of Classical Greek history can now be discerned in some Roman writers, the historian Florus in particular, who in a striking piece of polemic against Athens claimed that Rome too had met her Xerxes—in Antiochus III—and won her Salamis at Ephesus (an obscure naval engagement which Appian presents as a Roman defeat).[40] Florus wrote under Hadrian, however, and it was precisely then that the central administration came to favour the more systematic promotion of Greek interests within the empire; the perception that Greeks were now privileged subjects can be found in Greek writers of the Antonine age, notably the sophist Aelius Aristides, but perhaps too in the historian Arrian's vision (*Anab.* 7. 11. 9) of subject-Persians as 'partners in rule' with their Macedonian overlords. The levies of Spartans by Lucius Verus and Caracalla belong to this more advanced political climate; for educated Greeks who remembered that Alexander the Great failed to win over the Spartans (Arr. *Anab.* 1. 1. 2), they would have advertised the greater success of Roman emperors in achieving united Greek support for their eastern wars.

III. SUBJECT-GREEKS AND THE PERSIAN WARS

In the light of what has gone before, this last section considers the pervasive celebration of the Persian wars in Greek culture under the empire, and asks how far imperial ideology contributed to the dynamic of this phenomenon. For cities in mainland Greece like Athens, Sparta, or Delphi, the 'official' standing of the Persian-wars tradition helps to explain the assiduity with which they cultivated their commemorative landscape of monuments and mementoes from the Persian wars, of which a striking example comes from Megara, which boasted in the fourth century of its continuous maintenance 'until our own day' of the annual bull-sacrifice at the cenotaph of the Megarians who fell at Plataea.[41] Another example comes from Athens itself—the city's reacquisition of the island of Salamis through the generosity of a foreign benefactor, whom the Athenians promptly dubbed a 'new

---

[40] *Epit.* 1. 24. 3; App. *Syr.* 24. Tacitean irritation with the historiographical self-absorption of subject-Greeks: *Ann.* 2. 88; F. Goodyear (ed.), *The Annals of Tacitus*, ii (Cambridge, 1981), 448–9.

[41] Tod i², no. 20, line 15; C. Fornara (ed.), *Archaic Times to the End of the Peloponnesian War*² (Cambridge, 1983), no. 60. See too Cartledge and Spawforth (n. 22 above), 190–3; Robertson, art. cit. (n. 9 above).

Themistocles'.[42] The civic perception that a place in this tradition could provide political currency in dealings with the ruling power is suggested by the Parthenon-inscription no less than by the embarrassment of the Argives at their ancestral medism in an approach to a Roman official in the late first century AD, when they felt obliged to argue that leadership of the Trojan War conferred no less glory than the part played by Athens and Sparta in the Persian wars.[43]

But subject-Greek attachment to the memory of the Persian wars was more deep-seated than this. The terror which the irruption of barbarians into Greek lands could strike, and the value of a knowledge of the Persian-wars tradition in fortifying contemporary Greek resolve in the face of such threats, is underlined by Polybius in the second century BC, writing about the Gauls (2. 35). That the Parthians aroused ancient fears is suggested by the gratitude for imperial protection implicit in the provincial monuments at Athens and Aphrodisias discussed in Section I and by the great Antonine altar at Ephesus, now more or less firmly identified as an Asian thank-offering for the Parthian victory of Lucius Verus in 166.[44] Ingrained cultural habit made subject-Greeks view the Parthians almost inevitably through the Persian 'frame'. This emerges no less in the setting of the Parthenon-inscription than in the flurry of Herodotean and Thucydidean reminiscences among the Greek historians who followed Lucius Verus on his Parthian campaigns and whose literary pretensions the satirist Lucian mocked so effectively.[45]

These pretensions are a reminder of course that for Greek men of letters under the empire the Persian wars were attractive as a theme for more narrowly literary reasons, and it was the desire to imitate the old masters, no doubt, which contributed to the popularity of rehashes of 'the deeds of the Assyrians and Medes' among the nameless Greek historians of the Flavian age rebuked by Josephus (*BJ* 1. pr. 13). If we turn to the sophists of the imperial age, we find that the Persian wars provided them with a mere twelve per cent or so of their known themes from Greek history.[46] On the other hand, Philostratus leaves us in no doubt that these wars inspired some of the most popular and frequently

---

[42] For the disputed date of this episode (Augustus or AD 61/2) see C. P. Jones, 'Three Foreigners in Attica', *Phoenix* 32 (1978), 222–8; Geagan (n. 7 above), 281 n. 13; T. L. Shear, 'Athens: From City-State to Provincial Town', *Hesperia* 50 (1981), 366–7 (evidently unaware of Jones).

[43] Pseudo-Julian, *Letters* 198, 407b–c Bidez (n. 10 above for the date).

[44] On the date, and on eastern attitudes to the victory of Verus: C. P. Jones, *Culture and Society in Lucian* (Cambridge, Mass., 1986), 161–6.

[45] Jones, op. cit., ch. 6.

[46] Figure based on D. A. Russell, *Greek Declamation* (Cambridge, 1983), 107.

performed material. Part of their appeal lay with their histrionic possibilities—Scopelian of Clazomenae would 'shake his body as if in a
Bacchic frenzy' in his representations of Darius and Xerxes (*VS* 5 1 9–20).
But these tastes in rhetoric and historiography also had much deeper
well-springs. As well as belonging to a wider tendency to relive the past in
Greek culture under the empire, they drew strength from the continued
intensity of Greek ethnic self-consciousness, which led Livy to the
remarkable observation that 'all Greeks are and always will be at war
with foreigners and barbarians' (31. 29. 15).[47] Intentionally or not,
imperial ideology played on the old ethnocentrism of the Greeks, and it
may well be here that the real strength of the Persian wars as a unifying
symbol should be sought, especially since the Romans themselves, as
they came into contact with the 'uncivilized' peoples on their frontiers,
Parthians included, in due course constructed their own brand of
'barbarology' on Greek lines.[48]

A passage in Plutarch's *Politika Parangelmata* (8 1 4e) warning Greek
city-politicians not to use the Persian wars to excite 'the many' (τοὺς
πόλλους) deserves comment at this point, since it might seem to show
that in the years around 100 Marathon and Plataea could be used by
cynical politicians to conjure up fond memories of successful Greek
resistance to foreign empires; one might then wonder if the official
Roman appropriation of these victories was not an attempt to neutralize
this subversive thread in the civic discourse of subject-Greeks. This
aspect of the symbolism of the Persian wars had, after all, a long history,
its potency stemming from the long struggle of the Greek *poleis* in the
fourth and third centuries BC to defend their freedoms against Macedonian encroachment. Appearing as it does in a work addressed to a
Sardian notable, Plutarch's passage certainly suggests that by his day a
general knowledge of the Persian wars had worked its way well into the
Greek historical memory on both sides of the Aegean (these wars did,
after all, culminate in the liberation of the Asian Greek cities from Persian
control). But Plutarch is unlikely to have been thinking here of serious
attempts by local politicians to stir up resistance to Roman rule as such.
First, it needs to be emphasized that under the empire the politically
active *demos* in Greek cities to whom local politicians addressed their
rhetoric was less likely to be the full citizen-body than a much smaller

---

[47] E. Hall (*Inventing the Barbarian* (Oxford, 1989), 195) drew my attention to this
passage.
[48] See Y. Dauge, *Le Barbare: Recherches sur la conception romaine de la barbarie et de la
civilisation* (Coll. Latomus 176; Brussels, 1981).

group of 'privileged ecclesiasts', who alone enjoyed the right to conduct public business and who, as men of property, had a stake in maintaining the political status quo (as at Roman Delphi, a city which Plutarch knew extremely well[49]). Secondly, provincial Greek cities were constantly faced with unpalatable Roman decisions in petty matters of local administration and did their best to resist them, usually by 'diplomatic' channels (appeals to the governor, for instance); I suggest that more hotheaded manifestations of non-compliance, and their serious consequences for those concerned, lay behind Plutarch's warning. That the Persian wars retained a subversive resonance for some Greeks in Plutarch's day is quite possible. But the case of Tib. Claudius Novius suggests that civic notables—the only provincial social stratum which mattered much politically—could find pride in the Persian wars entirely compatible with loyalty to Rome.

If cherished cultural values meant that educated Greek provincials were predisposed to inject the memory of the Persian wars with contemporary meaning, the incorporation of these wars into official ideology is likely to have flattered the obsessive pride of subject-Greeks in their past and to have been valued in particular by the cities of the ethnic Greek heartlands, and by those further afield—in Asia Minor above all— which claimed kinship with Athens and Sparta, the leaders of the resistance to Persia; a note of Athenian self-congratulation can probably be heard in the Parthenon-inscription. Moreover, it was a Roman institution, the Panhellenium, which fostered the marked Athenocentricity characteristic of Greek culture in the Antonine age—the background against which the Asian sophist Aelius Aristides in his Panathenaic speech is found paying tribute to the role of Classical Athens as protectress of the Greeks, and Asiatic Greeks served as priests at Plataea (above).[50] But the limitations of old Greek victories as a universal symbol for the Hellenized east must also be recognized: the nearer the eastern frontier, the weaker the ethnic basis of local 'Greekness' and the greater the competition from other historical traditions[51]—the contempt of the Jewish historians Josephus for the Persian-wars 'mania' of his day has already been encountered. Another example, relevant here, of the

---

[49] C. Vatin, 'Damiurges et épidamiurges à Delphes', *BCH* 85 (1961), 248–50 (Delphi); G. E. M. de Ste Croix, *The Class Struggle in the Ancient Greek World* (London, 1981), 518–37 (generally).

[50] For remarkable new evidence from Phrygian Aezani for 'flag-waving' for Athens by an enthusiastic ex-Panhellene see M. Wörrle, 'Neue Inschriftenkunde aus Aizanoi I', *Chiron* 22 (1992), 337–70.

[51] For these other traditions in Roman Asia Minor see e.g. Strubbe (n. 11 above), 267–8.

cultural diversity masked by the veneer of Hellenism in the Roman east comes in a new inscription from Pamphylian Selge which reveals a local lady of Roman consular stock called Aurelia Volussia Quirina Atossa; suffice it to say that no notable of Roman Athens would have dreamt of naming his daughter after the mother of Xerxes.[52]

---

[52] J. Nollé and F. Schindler, *Die Inschriften von Selge* (IGSK 37; Bonn, 1991), no. 20 (but her husband was a Pericles!).

Bibliography

ACKROYD, P., *Dickens* (London, 1990).

ALGRA, K., review of Kidd, *Posidonius*, CR 41 (1991), 316–19.

ALONSO-NUÑEZ, J. M. (ed.), *Geschichtsbild und Geschichtsdenken im Altertum* (Darmstadt, 1991).

ALTY, J., 'Dorians and Ionians', *JHS* 102 (1982), 1–14.

ALY, W., *Volksmärchen, Sage und Novelle bei Herodot und seine Zeitgenossen*[2] (Göttingen, 1969).

ANDERSON, G., *The Second Sophistic: A Cultural Phenomenon in the Roman Empire* (London, 1993).

ANDREWES, A., 'Sparta and Arcadia in the Early Fifth Century', *Phoenix* 6 (1952), 1–5.

—— 'Thucydides and the Persians', *Historia* 10 (1961), 1–18.

—— 'The Arginusai Trial', *Phoenix* 28 (1974), 112–22.

—— 'Spartan Imperialism?', in P. Garnsey and C. Whittaker (eds.), *Imperialism in the Ancient World* (Cambridge, 1978), 90–102, 302–6.

—— 'The Hoplite *Katalogos*', in G. Shrimpton and D. McCargar (eds.), *Classical Contributions: Studies in Honor of M. F. McGregor* (Locust Valley, NY, 1981), 1–3.

—— 'Notion and Kyzikos: The Sources Compared', *JHS* 102 (1982), 15–25.

—— see Abbreviations (above p. xi) under *HCT*.

ARAFAT, K. W., 'Pausanias' Attitude to Antiquities', *BSA* 87 (1992), 387–409.

ARMAYOR, O. K., 'Did Herodotus Ever Go to the Black Sea?', *HSCP* 82 (1978), 45–62.

—— *Herodotus' Autopsy of the Fayoum: Lake Moeris and the Labyrinth of Egypt* (Amsterdam, 1985).

BADIAN, E., 'Greeks and Macedonians', in *Studies in the History of Art* 10 (1982), 33–51.

—— *From Plataea to Potidaea: Studies in the History and Historiography of the Pentecontaetia* (Baltimore, 1993).

—— 'The Peace of Callias', in *From Plataea to Potidaea*, 1–72.

—— 'Toward a Chronology of the Pentecontaetia down to the Renewal of the Peace of Callias', in *From Plataea to Potidaea*, 73–107.

—— 'Thucydides and the Outbreak of the Peloponnesian War: A Historian's Brief', in *From Plataea to Potidaea*, 125–62.

—— 'Thucydides and the *Arche* of Philip', in *From Plataea to Potidaea*, 171–85.

BAL, M., *Narratology: Introduction to the Theory of Narrative* (Toronto–Buffalo–London, 1985).

BALCER, J. M., 'Persian Occupied Thrace (Skudra)', *Historia* 37 (1988), 1–21.

BALSDON, J. P. V. D., *The Emperor Gaius* (Oxford, 1934).

BARBER, G. L., *The Historian Ephorus* (Cambridge, 1935).

BARLOW, S. (ed.), *Euripides: Trojan Women* (Warminster, 1986).

BARRETT, A., *Caligula: The Corruption of Power* (London, 1989).

BARTHES, R., 'The Discourse of History', in M. Lang (ed.), *Structuralism: A Reader* (London, 1970), 145–55; also in R. Barthes, *The Rustle of Language* (Oxford, 1986), 127–41.

BARTOLETTI, V. (ed.), *Hellenica Oxyrhynchia* (Leipzig, 1959).

BAUSLAUGH, R., 'The Text of Thucydides iv.8.6 and the South Channel at Pylos', *JHS* 99 (1979), 1–6.

BEAGON, M., *Roman Nature: The Thought of Pliny the Elder* (Oxford, 1992).

BEERBOHM, M., *A Peep into the Past*, ed. R. Hart-Davis (London, 1972).

BELLEMORE, J. (ed.), *Nicolaus of Damascus: Life of Augustus* (Bristol, 1984).

BELOCH, K. J., *Griechische Geschichte*², 4 vols. in 8 (Strasburg–Leipzig–Berlin, 1912–27).

BENNETT, H. C., 'On the Systematization of Scholia Dates for Pindar's Pythian Odes', *HSCP* 62 (1957), 61–78.

BERGER, H., *Die Geographische Fragmente des Eratosthenes* (Leipzig, 1880).

BERGER, S., 'Seven Cities at Thuc. 6.20.2–3', *Hermes* 120 (1992), 421–4.

BERNSTEIN, C., see Woodward, B.

BERTRAC, P., see Chamoux, F.

BICKERMAN(N), E. J., 'Origines gentium', *CP* 47 (1952), 65–81.

—— and SYKUTRIS, J., 'Speusipps Brief an König Philipp', *SB Berlin* 80. 3 (1928).

BILLOWS, R. A., *Antigonos the One-eyed and the Creation of the Hellenistic State* (Berkeley, 1990).

BLOCH, H., 'Studies in the Historical Literature of the Fourth Century B.C.', in *Athenian Studies presented to W. S. Ferguson* (*HSCP* suppl. 1; 1940), 303–76.

BOARDMAN, J., 'Herakles, Delphi, and Kleisthenes of Sikyon', *Rev. Arch.* 1978, 227–34.

—— *Greek Sculpture: The Archaic Period* (London, 1988).

—— and see Parke, H. W.

BOEDEKER, D. (ed.), *Herodotus and the Invention of History*, *Arethusa* 20 (1978) (special number).

BOOTH, W., *The Rhetoric of Fiction*² (Harmondsworth, 1983).

BORZA, E. N., *In the Shadow of Olympus: The Emergence of Macedon* (Princeton, NJ, 1990).

BOSWORTH, A. B., 'Philip II and Upper Macedonia', *CQ* 21 (1971), 93–105.

—— 'Arrian and the Alexander-vulgate', in E. Badian (ed.), *Alexandre le grand: Image et réalité* (Entretiens Hardt 22; Geneva, 1976), 1–46.

—— *A Historical Commentary on Arrian's History of Alexander*, i (Oxford, 1980).

—— *From Arrian to Alexander* (Oxford, 1988).

BOURAZELIS, C., Θεῖα δωρεά: Μελέτες πάνω στὴν πολιτικὴ τῆς δυναστείας τῶν Σεβήρων καὶ τὴν Constitutio Antoniniana (Athens, 1989).

Bowersock, G. W., *Augustus and the Greek World* (Oxford, 1965).

—— *Greek Sophists in the Roman Empire* (Oxford, 1969).

—— 'Introduction to the Greek Renaissance', in Bowersock (ed.), *Approaches to the Second Sophistic* (Pennsylvania, 1974).

—— review of Pritchett, *Dionysius on Thucydides*, CP 71 (1976), 361–4.

—— *Roman Arabia* (Cambridge, Mass. and London, 1983).

—— 'Augustus and the East: The Problem of the Succession', in F. Millar and E. Segal (eds.), *Caesar Augustus: Seven Aspects* (Oxford, 1984), 169–88.

—— 'Gibbon's Historical Imagination', *The American Scholar* (Winter 1988), 33–47.

Bowie, E. L., 'Greeks and their Past in the Second Sophistic', in M. I. Finley (ed.), *Studies in Ancient Society* (London, 1974), 166–209.

—— 'Early Greek Elegy, Symposium and Public Festival', *JHS* 106 (1986), 13–38.

Braccesi, L., *Grecità adriatica*² (Bologna, 1979).

Braudel, F., *The Mediterranean World in the Age of Philip II*, 2 vols. (London, 1972).

Braund, D., *Rome and the Friendly King* (Beckenham, 1984).

Bretzl, H., *Botanische Forschungen des Alexander-Zugen* (Leipzig, 1903).

Briscoe, J., *A Commentary on Livy: Books XXXI–XXXIII* (Oxford, 1973).

Brodersen, K., *Appians Abriss der Seleukidengeschichte: Syriake 45.232–70.369. Text und Kommentar* (Munich, 1989)=*Münchener Arbeiten zur alten Geschichte*, i.

—— 'Zur Datierung der ersten Pythien', *ZPE* 82 (1990), 25–31.

—— *Appians Antiochike: Syriake 1.1–44.232. Text und Kommentar* (Munich, 1991)=*Münchener Arbeiten zur alten Geschichte*, ii.

—— 'Appian und sein Werk' in *ANRW* 2. 34. 2 (Berlin and New York, 1973), 339–63.

—— 'Appian' in *OCD*³ (forthcoming).

Brommer, F., *Herakles II: Die unkanonischen Taten des Helden* (Darmstadt, 1984).

Brown, T. S., *Timaeus of Tauromenium* (Berkeley and Los Angeles, 1958).

—— *The Greek Historians* (Lexington–Toronto–London, 1973).

—— 'Echoes from Herodotus in Xenophon's Hellenica', *Ancient World* 21 (1990), 97–101.

Brunt, P. A., 'Marcus Aurelius in his *Meditations*', *JRS* 64 (1974), 1–20.

—— *Arrian*, 2 vols. (Loeb; London, 1976–83).

—— 'On Historical Fragments and Epitomes', *CQ* 30 (1980), 477–94; repr. in J. M. Alonso-Nuñez, *Geschichtsbild und Geschichtsdenken im Altertum* (Darmstadt, 1991), 334–62.

—— *Studies in Greek History and Thought* (Oxford, 1993).

—— 'The Bubble of the Second Sophistic' (*BICS* 39 (1994)).

Buraselis, K., see Bourazelis, C.

BURKE, P., *The Fabrication of Louis XIV* (New Haven, Conn. and London, 1992).

BURKERT, W., *Homo Necans* (Berkeley, 1983).

—— 'Herodot als Historiker fremden Religionen', in G. Nenci and O. Reverdin (eds.), *Hérodote et les peuples non-grecs* (Entretiens Hardt 35; Vandœuvres–Geneva, 1990), 1–39.

BURY, J. B., *Ancient Greek Historians* (London, 1909).

BUSOLT, G., *Griechische Geschichte*, 4 vols. in 3 (=i², ii², iii (1) and (2); Gotha, 1893–1904).

CAIRNS, F., 'Cleon and Pericles: A Suggestion', *JHS* 102 (1982), 203–4.

CALEY, E. R., and RICHARDS, J. F. C., *Theophrastus on Stones: Introduction, Greek Text, English Translation and Commentary* (Columbus, 1956).

CAMERON, AVERIL, ed., *History as Text: The Writing of History* (London, 1989).

CAPELLE, W., 'Der Garten des Theophrast', in *Festschrift für F. Zucker* (Berlin, 1954), 47–82.

—— 'Theophrast in Kyrene?', *Rh. Mus.* 97 (1954), 169–89.

—— 'Theophrast in Ägypten', *Wiener Studien* 69 (1956), 173–86.

CARLYLE, T., *History of the French Revolution* (London, 1837).

CARROLL, K., *The Parthenon Inscription* (*GRBM* 9; Durham, NC, 1982).

CARTLEDGE, P., *Sparta and Lakonia: A Regional History 1300–362 BC* (London, 1979).

—— and SPAWFORTH, A., *Hellenistic and Roman Sparta: A Tale of Two Cities* (London, 1989).

CASSOLA, F., 'Nota sulla guerra crisea', *Miscellanea E. Manni*, ii (Rome, 1980), 413–39.

CAWKWELL, G. L. (ed.), *Xenophon: A History of my Times* (Harmondsworth, 1979).

—— 'The King's Peace', *CQ* 31 (1981), 69–83.

CHAMBERS, M. H., GALLUCCI, R., and SPANOS, P., 'Athens' Alliance with Egesta in the Year of Antiphon', *ZPE* 83 (1990), 38–63.

CHAMOUX, F., 'Pausanias historien', in *Mélanges A. Tuilier* (Paris, 1988), 37–45.

—— 'De Pyrrhus à Marc Antoine: Considérations sur la biographie historique', in *Mélanges P. Lévêque* (Besançon, 1989), 83–8.

—— and BERTRAC, P., 'Introduction générale' in *Diodore de Sicile, Bibliothèque historique, Livre I* (Paris, 1993).

CHRIST, M. R., 'Theopompus and Herodotus: A Reassessment', *CQ* 43 (1993), 47–52.

CLANCHY, M., *From Memory to Written Record* (London, 1979).

CLAY, JENNY STRAUSS, *The Politics of Olympus: Form and Meaning in the Major Homeric Poems* (Princeton, NJ, 1989).

COBET, J., 'Herodotus and Thucydides on War', in I. S. Moxon, J. D. Smart, and A. J. Woodman (eds.), *Past Perspectives: Studies in Greek and Roman Historical Writing* (Cambridge, 1986), 1–18.

COHEN, R., review of R. Schneider, *JRA* 3 (1990), 264–70.

CONNOR, W. R., 'Narrative Discourse in Thucydides', in M. H. Jameson (ed.), *The Greek Historians: Literature and History. Papers Presented to A. E. Raubitschek* (Stanford, Calif., 1985), 1–17.

COOPER, G. L., 'Intrusive Oblique Infinitives in Herodotus', *TAPA* 104 (1974), 23–76.

CORNFORD, F. M., *Thucydides Mythistoricus* (London, 1907).

CRAWLEY, R., *Thucydides: History of the Peloponnesian War* (London, 1876, and reprints).

DAUGE, Y., *Le Barbare: Recherches sur la conception romaine de la barbarie et de la civilisation* (Coll. Latomus 176; Brussels, 1981).

DAVIDSON, J., 'The Gaze in Polybius' Histories', *JRS* 81 (1991), 10–24.

DAVIES, J. K., *Athenian Propertied Families 600–300 BC* (Oxford, 1971).

—— 'The reliability of the oral tradition', in L. Foxhall and J. K. Davies (eds.), *The Trojan War: Its Historicity and Context* (Bristol, 1984), 87–110; repr. in C. Emlyn-Jones, L. Hardwick, and J. Purkis (eds.), *Homer: Readings and Images* (London, 1992), 211–25.

DEFRADAS, J., *Les Thèmes de la propaganda delphique* (Paris, 1972).

DE JONG, see under Jong, I. J. F. de.

DE ROMILLY, see under Romilly, J. de.

DEROW, P. S., 'Polybius, Rome and the East', *JRS* 69 (1979), 1–15.

—— 'Polybius', in J. Luce (ed.), *Ancient Writers* (New York, 1982), I 525–39.

DE STE CROIX, see under Ste Croix, G. E. M. de.

DIELS, H., 'Herodot und Hekataios', *Hermes* 22 (1887), 411–44.

DIONISOTTI, A. C., 'Nepos and the Generals', *JRS* 78 (1988), 35–49.

DOR, L., JANNORAY, J., and VAN EFFENTERRE, H. and M., *Kirrha: Étude de pré-histoire phocidienne* (Paris, 1960).

DORAN, R., '2 Maccabees and Tragic History', *Hebrew Union College Annual* 50 (1979), 107–14.

DOVER, K. J., 'La colonizzazione della Sicilia', *Maia* 6 (1953), 1–20; Ger. tr. 'Über die Kolonisierung Siziliens bei Thukydides', in H. Herter (ed.), *Thukydides* (Darmstadt, 1968), 344–68.

—— 'Thucydides' Historical Judgment: Athens and Sicily', in Dover, *The Greeks and their Legacy: Collected Papers*, ii: *Prose Literature, History, Society, Transmission, Influence* (Oxford, 1988), 74–82.

—— see Abbreviations (above, p. xi) under *HCT*.

DRACHMANN, A. B., *Atheism in Pagan Antiquity* (Copenhagen and London, 1922).

DREWS, R., 'Diodorus and his Sources', *AJP* 83 (1962), 383–92.

—— *The Greek Accounts of Eastern History* (Washington, DC, 1973).

EAGLETON, T., *Literary Theory: An Introduction* (Oxford, 1983).

ECKSTEIN, A. M., 'Polybius, the Achaeans, and the "Freedom of the Greeks"', *GRBS* 31 (1990), 45–71.

EDELSTEIN, L., see Kidd, I.

254 *Bibliography*

EDMONDSON, J., *Dio, the Julio-Claudians: Selections from Books 58–63 of the Roman History of Cassius Dio* (*Lactor* 15; London, 1992).

EDWARDS, M., *The Iliad: A Commentary*, v: *Books 17–20* (Cambridge, 1991).

EFFENTERRE, H. and M., see Dor, L.

—— and see Roger, J.

EHRMAN, J., *The Younger Pitt*, 2 vols. (London, 1969–83).

EICHHOLZ, D. E., *Theophrastus: De Lapidibus* (Oxford, 1965).

EINARSON, B., 'The Manuscripts of Theophrastus' *Historia Plantarum*', CP 71 (1976), 67–76.

—— and LINK, G. K. K., *Theophrastus: De Causis Plantarum*, 3 vols. (Loeb; Cambridge, Mass., 1976–90).

ELLUL, J., *Propaganda: The Formation of Men's Attitudes* (New York, 1973).

ELSNER, J., 'Pausanias: A Greek Pilgrim in the Roman World', *Past and Present* 135 (1992), 3–29.

ERBSE, H., *Studien zum Verständnis Herodots* (Berlin, 1992).

ERRINGTON, R. M., 'Bias in Ptolemy's History of Alexander', CQ 19 (1969), 233–42.

—— 'Alexander the Philhellene and Persia', in H. J. Dell (ed.), *Ancient Macedonian Studies in Honor of Charles F. Edson* (Thessaloniki, 1981), 139–43.

ERSKINE, A., *The Hellenistic Stoa: Political Thought and Action* (London, 1990).

EVERETT, B., *Poets in their Time: Essays on English Poetry from Donne to Larkin* (Oxford, 1991).

FABIA, P., *Les Sources de Tacite dans les histoires et les annales* (Paris, 1893).

FEENEY, D. C., *The Gods in Epic* (Oxford, 1991).

FEHLING, D., *Herodotus and his Sources: Citation, Invention and Narrative Art*, tr. J. G. Howie (Liverpool, 1989).

FERGUSON, W. S., *Hellenistic Athens* (London, 1911).

FERRARY, J.-L., *Philhellénisme et impérialisme* (Rome, 1988).

FINLEY, J. H., *Three Essays on Thucydides* (Cambridge, Mass., 1967).

FINLEY, M. I., 'Myth, Memory and History', *History and Theory* 4 (1965), 281–302=Finley, *The Use and Abuse of History*[2] (London, 1975), 11–33=Alonso-Nuñez (ed.), *Geschichtsbild und Geschichtsdenken im Altertum*, 9–38.

FLOWER, H., 'Delphic Traditions about Croesus', in M. Flower and M. Toher (eds.), *Georgica: Greek Studies in Honour of George Cawkwell* (*BICS* suppl. 58; London, 1991), 57–77.

—— 'Thucydides and the Pylos Debate', *Historia* 41 (1992), 40–57.

FOLLET, S., 'Chronologie et calendriers', in *Bulletin épigraphique* 1989, *RÉG* 102 (1989), 361–509 at 386–9.

FONTENROSE, J., *The Delphic Oracle: Its Responses and Operations, with a Catalogue of Responses* (Berkeley, 1978).

FORNARA, C. W., *Herodotus: An Interpretative Essay* (Oxford, 1971).

—— (ed.), *Archaic Times to the End of the Peloponnesian War*[2] (Cambridge, 1983).

FORREST, G., 'The First Sacred War', *BCH* 80 (1956), 33–52.

FORTENBAUGH, W. W. (*et al.*), *Theophrastus of Eresus: Sources for his Life, Writings, Thought and Influence*, 2 vols. (Leiden, 1992).

FOWLER, D., 'Deviant Focalisation in Virgil's *Aeneid*', *PCPhS* 36 (1990), 43–63.

—— 'Narrate and Describe: The Problem of Ekphrasis', *JRS* 81 (1991), 25–35.

FOX, ROBIN LANE, see Lane Fox, R.

FRAENKEL, E. (ed.), *Aeschylus: Agamemnon*, 3 vols. (Oxford, 1950).

FRASER, P. M., review of Treves, *Euforione*, *Gnomon* 28 (1956), 578–86.

—— *Ptolemaic Alexandria*, 3 vols. (Oxford, 1972).

—— 'The Son of Aristonax at Kandahar', *Afghan Studies* 2 (1979), 9–21.

FREUDENTHAL, A., *Alexander Polyhistor und die von ihm erhaltenem Reste judäischer und samaritanischer Geschichtswerk* (*Hellenistischer Studien* 1–2; Breslau, 1875).

FRYE, R., *The History of Ancient Iran* (Munich, 1984).

GABBA, E., review of Meloni, *Riv. stor. it.* 68 (1956), 100–6.

—— (ed.), *Polybe* (Entretiens Hardt 20; Geneva, 1974).

—— 'Sulla valorizzazione politica della leggenda delle origini troiane di Rome fra III et II secolo a.C.', in M. Sordi (ed.), *I canali della propaganda nel mondo antico* (*Contributi dell'istituto di storia antica* 4; Milan, 1976), 84–101.

—— 'True History and False History in Classical Antiquity', *JRS* 71 (1981), 50–62.

—— 'The Historians and Augustus', in F. Millar and E. Segal (eds.), *Caesar Augustus: Seven Aspects* (Oxford, 1984), 61–88.

GARDNER, H., *The Art of T. S. Eliot*[3] (London, 1964).

GARLAN, Y., *Slavery in Ancient Greece*, tr. J. Lloyd (Ithaca, NY, 1988).

GEAGAN, D., 'Tib. Claudius Novius, the Hoplite-Generalship and the Epimeleteia of the Free City of Athens', *AJP* (1979), 279–87.

GEFFCKEN, J., *Timaios' Geographie des Westens* (*Philologische Untersuchungen* 13; Berlin, 1892).

GENETTE, G., *Figures III* (Paris, 1972).

—— *Narrative Discourse* (Oxford, 1980) = Eng. tr. by J. E. Lewin of *Figures III*, 67–286.

—— *Fiction et diction* (Paris, 1991).

GERA, DEBORAH L., *Xenophon's Cyropaedia: Style, Genre, and Literary Technique* (Oxford, 1993).

GOMME, A. W., *The Greek Attitude to Poetry and History* (Berkeley and Los Angeles, 1954).

—— see Abbreviations (above, p. xi) under *HCT*.

GOODMAN, M., see Schürer, E.

GOODYEAR, F. (ed.), *The Annals of Tacitus Books 1–6*, i: *Annals 1. 1–54*; ii: *Annals 1. 55–81 and Annals 2* (Cambridge, 1972–81).

GOULD, J., 'On Making Sense of Greek Religion', in P. E. Easterling and J. V. Muir (eds.), *Greek Religion and Society* (Cambridge, 1985), 1–33, 219–21.

GOULD, J., *Herodotus* (London, 1989).

GRAFTON, A., *Forgers and Critics* (Princeton, NJ, 1990).

GRAY, V. J., 'Two Different Approaches to the Battle of Sardis', *CSCA* 12 (1979), 183–200.

—— *The Character of Xenophon's Hellenica* (London, 1989).

GREEN, P., review of Momigliano's *Classical Foundations of Modern Historiography*, *TLS*, 19 July 1991, 3–4.

GRENE, D. (tr.), *Herodotus: The History* (Chicago, 1987).

GRIFFIN, A., *Sikyon* (Oxford, 1982).

GRIFFIN, J., *Homer on Life and Death* (Oxford, 1980).

—— 'Words and Speakers in Homer', *JHS* 106 (1986), 36–57.

—— *The Odyssey* (Cambridge, 1987).

GRIFFITH, G. T., 'Some Habits of Thucydides when Introducing Persons', *PCPhS* 187 (1961), 21–33.

—— and see Hammond, N. G. L.

GRUEN, E. S., *The Hellenistic World and the Coming of Rome*, 2 vols. (Berkeley, 1984).

GUARDUCCI, M., 'Creta e Delfi', *Studi e materiali di storia delle religioni* 19–20 (1943–6), 85–114.

GUILLON, P., *Études béotiennes: Le Boucher d'Héraclès et l'histoire de la Grèce centrale dans la période de la première guerre sacrée* (Aix en Provence, 1963).

HABICHT, C., 'Falsche Urkunden zur Geschichte Athens im Zeitalter der Perserkriege', *Hermes* 89 (1961), 1–35.

—— 'Royal Documents in Maccabees II', *HSCP* 80 (1976), 1–18.

—— *Pausanias' Guide to Ancient Greece* (Berkeley–Los Angeles–London, 1985).

HAINSWORTH, B., *The Iliad: A Commentary*, iii: *Books 9–12* (Cambridge, 1993).

HALL, C. M. (ed.), *Nicolaus of Damascus: Life of Augustus* (Smith Classical Studies 4; Northampton, Mass., 1923).

HALL, E., *Inventing the Barbarian: Greek Self-Definition through Tragedy* (Oxford, 1989).

HAMMOND, N. G. L., *Three Historians of Alexander the Great* (Cambridge, 1983).

—— *Sources for Alexander the Great: An Analysis of Plutarch's Life and Arrian's Anabasis Alexandrou* (Cambridge, 1993).

—— and GRIFFITH, G. T., *A History of Macedonia*, ii (Oxford, 1979).

HANSEN, M. H., *Demography and Democracy: The Number of Athenian Citizens in the Fourth Century BC* (Herning, 1986).

HANSON, V. D., 'Thucydides and the Desertion of Attic Slaves during the Decelean War', *Classical Antiquity* 11 (1992), 210–28.

HARRISON, S. J., *Commentary on Aeneid 10* (Oxford, 1991).

HARTOG, F., *The Mirror of Herodotus: The Representation of the Other in the Writing of History*, tr. J. Lloyd (Berkeley–Los Angeles–London, 1988).

HARVEY, F. D., 'The Political Sympathies of Herodotus', *Historia* 15 (1966), 254–5.

HEMER, C. J., *The Book of Acts in the Setting of Hellenistic History* (Winona Lake, Ind., 1990).

HERSHBELL, J. P., 'Plutarch and Herodotus: The Beetle in the Rose', *Rh. Mus.* 136 (1993), 143–63.

HERTER, H. (ed.), *Thukydides* (Darmstadt, 1968).

HOBSBAWM, E., and RANGER, T. (eds.), *The Invention of Tradition* (Cambridge, 1983).

HÖCKER, C., see Schneider, L.

HOFFMANN, W., *Rom und die griechische Welt* (*Philologus* Suppbd. 27(1); Leipzig, 1934).

HOLFORD-STREVENS, L., *Aulus Gellius* (London, 1988).

HOPKINS, K., *Conquerors and Slaves* (Cambridge, 1978).

HORNBLOWER, J., *Hieronymus of Cardia* (Oxford, 1981).

HORNBLOWER, S., review of Stadter, *Arrian*, *CR* 31 (1981), 12–14.

—— review of Hammond, *Three Historians*, *CR* 34 (1984), 261–4.

—— *Thucydides* (London, 1987).

—— review of Gould, *Herodotus*, *TLS*, 10–16 November 1989, 1237.

—— review of Weiskopf, *CR* 40 (1990), 363–5.

—— 'When was Megalopolis founded?' *BSA* 85 (1990), 71–7.

—— *A Commentary on Thucydides*, i: *Books I–III* (Oxford, 1991).

—— review of Starr, *Birth of Athenian Democracy*, *CR* 41 (1991), 388–90.

—— 'The Religious Dimension to the Peloponnesian War, Or, What Thucydides Does not Tell Us', *HSCP* 94 (1992), 169–97.

—— 'Thucydides' Use of Herodotus', in J. Sanders (ed.), *ΦΙΛΟΛΑΚΩΝ: Lakonian Studies in Honour of Hector Catling* (Athens, 1992), 141–54.

—— 'Sources and their Uses', in D. M. Lewis, J. Boardman, S. Hornblower, and M. Ostwald (eds.), *Cambridge Ancient History*, vi²: *The Fourth Century BC* (1994), ch. 1.

—— 'Epilogue', ibid., ch. 18.

—— 'Thucydides and Boiotia' in *Proceedings of the Second International Congress of Boiotian Studies* (forthcoming).

—— 'The Reception of Thucydides' (*JHS* 115 (1995), forthcoming).

HORT, A. F., *Theophrastus: Enquiry into Plants*, 2 vols. (Loeb; Cambridge, Mass., 1916).

HÜBER, L., 'Herodots Homerverständnis', in H. Flashar and K. Gaiser (eds.), *Synusia: Festgabe für W. Schadewaldt* (Pfullingen, 1965), 29–52.

HUDE, C., *Scholia in Thucydidem* (Leipzig, 1927).

HULSCH, F., *Griechische und römische metrologie* (Berlin, 1882).

HUNTER, V., *Thucydides the Artful Reporter* (Toronto, 1973).

JACOBY, F., *Atthis: The Local Chronicles of Ancient Athens* (Oxford, 1949).

—— *Griechische Historiker* (Stuttgart, 1956).

JANKO, R., *The Iliad: A Commentary*, iv: *Books 13–16* (Cambridge, 1992).

JANNORAY, J., 'Krisa, Kirrha et la première guerre sacrée', *BCH* 61 (1937), 33–43.
—— and see Dor, L.

JEFFERY, L. H., rev. A. W. Johnston, *The Local Scripts of Archaic Greece* (Oxford, 1990).

JONES, A. H. M., 'Slavery in the Ancient World', in M. I. Finley (ed.), *Slavery in Classical Antiquity* (Cambridge, 1960), 1–16.

JONES, C. P., *Plutarch and Rome* (Oxford, 1971).
—— 'Three Foreigners in Attica', *Phoenix* 32 (1978), 222–8.
—— *Culture and Society in Lucian* (Cambridge, Mass., 1986).

JONG, I. J. F. DE, *Narrators and Focalizers: The Presentation of the Story in the Iliad* (Amsterdam, 1987).
—— 'Homeric Words and Speakers: An Addendum', *JHS* 108 (1988), 188–9.
—— *Narrative in Drama: The Art of the Euripidean Messenger-Speech* (*Mnemosyne* suppl. 116; Leiden, 1991).
—— 'Studies in Homeric Denomination', *Mnemosyne* 46 (1993), 289–306.

KEANEY, J. J., 'The Early Tradition of Theophrastus' Historia Plantarum', *Hermes* 96 (1968), 293–8.
—— *The Composition of Aristotle's Athenaion Politeia: Observation and Explanation* (New York and Oxford, 1992).
—— 'Theophrastus on Ostracism and the Character of his NOMOI', in M. Piérart (ed.), *Aristote et Athènes* (Paris, 1993), 261–78.

KENNEDY, D., review of Woodman and West, *Poetry and Politics in the Age of Augustus*, *LCM* 9 (1984), 157–60.

KENNEY, E., 'Books and Readers in the Classical World', in Kenney (ed.), *Cambridge History of Classical Literature* 2: *Latin Literature* (Cambridge, 1982), 3–32.

KENT, R. G., *Old Persian: Grammar, Texts, Lexicon*² (New Haven, Conn., 1950).

KIDD, I. (and EDELSTEIN, L.), *Posidonius*, i²: *The Fragments* (Cambridge, 1989).
—— *Posidonius*, ii (1) and (2): *The Commentary* (Cambridge, 1988).
—— 'Posidonius as Philosopher-Historian', in M. Griffin and J. Barnes (eds.), *Philosophia Togata: Essays on Philosophy and Roman Society* (Oxford, 1989), 38–50.
—— 'Posidonius', in *OCD*³ (forthcoming).

KIRK, G. S., *The Iliad: A Commentary*, i: *Books 1–4*; ii: *Books 5–8* (Cambridge, 1985 and 1990).
—— and RAVEN, J. S., *The Presocratic Philosophers*, 2nd edn. by M. Schofield (Cambridge, 1984).

KITTO, H. D. F., *Poiesis: Structure and Thought* (Berkeley–Los Angeles–London, 1966).

KROLL, W. (ed.), *Cicero: Orator* (1913; repr. Berlin, 1958).

KUHRT, A., see Sherwin-White, S. M.

LABARBE, J., *L'Homère de Platon* (Liège, 1949).

LAMB, W. R. M., *Clio Enthroned: A Study of Prose- Form in Thucydides* (Cambridge, 1914).

LANE FOX, R., 'Theopompus of Chios and the Greek World 411–322 BC', in J. Boardman and C. Vaphopoulou-Richardson (eds.), *Chios: A Conference Held at the Homereion in Chios 1984* (Oxford, 1986), 106–20.

—— *The Unauthorized Version: Truth and Fiction in the Bible* (London, 1991).

LANG, M., *Herodotean Narrative and Discourse* (Cambridge, Mass., 1984).

—— 'Unreal Conditions in Homeric Narrative', *GRBS* 30 (1989), 5–26.

LATEINER, D., *The Historical Method of Herodotus* (*Phoenix* suppl. 23; Toronto–Buffalo–London, 1989).

LEHMANN, G., 'Der "erste heilige Krieg"—eine Fiktion?', *Historia* 29 (1980), 242–6.

—— 'Der "lamische Krieg" und die "Freiheit der Hellenen": Überlegungen zur Hieronymianischen Tradition', *ZPE* 73 (1988), 121–49.

—— 'The "Ancient" Greek History in Polybios' *Historiae*: Tendencies and Political Objectives', *SCI* 10 (1989–90), 66–77.

LESKY, A., 'Homeros', *RE* Suppbd. xi (1968), cols. 687–846.

LEVENE, D., 'Sallust's *Jugurtha*: An "historical fragment"', *JRS* 82 (1992), 53–70.

LEWIS, D. M., 'An Aristotelian Publication-Date', *CR* 8 (1958), 108.

—— *Sparta and Persia* (Leiden, 1977).

—— 'The Archidamian War', in D. M. Lewis, J. Boardman, J. K. Davies, and M. Ostwald (eds.), *Cambridge Ancient History*, v²: *The Fifth Century BC* (Cambridge, 1992), 370–432.

—— 'Antony Andrewes', *PBA* 80 (1993), 221–31.

LINFORTH, I., 'Herodotus' Avowal of Silence in his Account of Egypt', *UCPCPh* 7 (1919–24) (1924), 269–92.

—— 'Greek Gods and Foreign Gods in Herodotus', *UCPCPh* 9 (1926–9) (1926), 1–25.

—— 'Named and Unnamed Gods in Herodotus', *UCPCPh* 9 (1926–9) (1928), 210–43.

LINK, G. K. K., see Einarson, B.

LITTMANN, R., 'The Strategy of the Battle of Cyzicus', *TAPA* 99 (1968), 265–72.

LUCE, T. J., *Livy: The Composition of his History* (Princeton, NJ, 1977).

McGREGOR, M. F., 'Clisthenes of Sicyon and the Panhellenic Festivals', *TAPA* 72 (1941), 266–87.

MACLEOD, C. W. (ed.), *Iliad XXIV: Commentary* (Cambridge, 1982).

—— *Collected Essays* (Oxford, 1983).

McNEAL, R. A., 'On Editing Herodotus', *L'Ant. class.* 52 (1983), 110–29.

MACVE, R., 'Some Glosses on de Ste Croix's "Greek and Roman Accounting"', in P. Cartledge and D. Harvey (eds.), *CRUX: Essays Presented to G. E. M. de Ste Croix on his 75th Birthday* (Exeter, 1985), 233–64.

MALITZ, J., *Die Historien des Poseidonius* (*Zetemata* 79; Munich, 1983).

MARTIN, R. P., *The Language of Heroes: Speech and Performance in the Iliad* (Ithaca, NY, and London, 1989).

MASSON, O., 'Le Culte ionien d'Apollon Oulios, d'après des données onomastiques nouvelles', *Journal des savants* (1988), 173–83.

MEIER, C., 'Historical Answers to Historical Questions: The Origins of History', in *Arethusa* 20 (1987) (see Boedeker), 41–57.

MEISTER, K., 'The Role of Timaeus in Greek Historiography', *SCI* 10 (1989–90), 55–65.

—— *Die griechische Geschichtsschreibung* (Cologne, 1990).

MELONI, P., *Il valore storico e le fonti del libro Macedonico di Appiano* (Rome, 1955).

MILLAR, F., *A Study of Cassius Dio* (Oxford, 1964).

—— 'The Problem of Hellenistic Syria', in A. Kuhrt and S. Sherwin-White (eds.), *Hellenism in the East: The Interaction of Greek and Non-Greek Civilizations from Syria to Central Asia after Alexander* (London, 1987), 110–33.

—— 'Polybius between Greece and Rome', in J. T. A. Koumoulides (ed.), *Greek Connections: Essays on Culture and Diplomacy* (Notre Dame, Ind., 1987), 1–18.

—— and see Schürer, E.

MILLER, S. G., 'The Date of the First Pythiad', *CSCA* 11 (1978), 127–58.

MOMIGLIANO, A., *Filippo il Macedone* (Florence, 1934).

—— *Studies in Historiography* (London, 1966).

—— *The Development of Greek Biography* (Harvard, 1971).

—— *Alien Wisdom: The Limits of Hellenization* (Cambridge, 1975).

—— *Essays in Ancient and Modern Historiography* (Oxford, 1977).

—— *Sesto contributo alla storia degli studi classici e del mondo antico* (Rome, 1980).

—— *Settimo contributo alla storia degli studi classici e del mondo antico* (Rome, 1984).

—— 'The origins of Rome', in F. W. Walbank, A. Astin, M. Frederiksen, and R. M. Ogilvie (eds.), *Cambridge Ancient History*, vii²(2): *The Rise of Rome to 220 BC* (Cambridge, 1989), 52–112.

—— *The Classical Foundations of Modern Historiography* (Berkeley and Los Angeles, 1990).

MORGAN, C., *Athletes and Oracles: The Transformation of Olympia and Delphi in the Eighth Century BC* (Cambridge, 1990).

MOSSHAMMER, A. A., 'The Date of the First Pythiad—again', *GRBS* 23 (1982), 15–30.

—— (ed.), *Georgius Syncellus, Ecloga Chronographica* (Leipzig, 1984).

MOSSMAN, J. M., 'Tragedy and Epic in Plutarch's *Alexander*', *JHS* 108 (1988), 83–93.

MOXON, I. S., SMART, J. D., and WOODMAN, A. J. (eds.), *Past Perspectives: Studies in Greek and Roman Historical Writing* (Cambridge, 1986).

MÜNSCHER, K., *Xenophon in der griechisch-römischen Literatur* (*Philologus* suppl. 13(2); Leipzig, 1920).

MURRAY, O., 'Hecataeus and Pharaonic Kingship', *JEA* 56 (1970), 141–71.

—— 'Herodotus and Hellenistic Culture', *CQ* 22 (1972), 200–13.

—— 'The Greek Symposium in History', in E. Gabba (ed.), *Tria corda: Scritti in onore di A. Momigliano* (Como, 1983), 257–72.

—— 'The Symposium as Social Organisation', in R. Hägg (ed.), *The Greek Renaissance of the 8th Century BC: Tradition and Innovation* (Stockholm, 1983), 195–9.

—— 'Herodotus and Oral History', in H. Sancisi-Weerdenburg and A. Kuhrt (eds.), *Achaemenid History*, ii: *The Greek Sources* (Leiden, 1987), 93–115.

—— 'The Ionian Revolt', in J. Boardman, N. G. L. Hammond, D. M. Lewis, and M. Ostwald (eds.), *Cambridge Ancient History*, iv²: *Persia, Greece, and the Western Mediterranean c.525–479 BC* (Cambridge, 1988), ch. 8.

—— (ed.), *Sympotica* (Oxford, 1990).

—— 'Arnaldo Momigliano in England', in M. P. Steinberg (ed.), *The Presence of the Historian: Essays in Memory of Arnaldo Momigliano* (*History and Theory*, Beiheft 30; Chicago, 1991), 49–64.

—— 'Thucydides and Local History' (forthcoming).

NAGY, G., 'Herodotus the Logios', in *Arethusa* 20 (1987) (see Boedeker), 175–201.

—— *Pindar's Homer* (Baltimore, 1991).

Neronia 4. Alejandro Magno, modelo de los emperadores romanos: Actes du IVᵉ colloque international de la SIEN (Coll. Latomus 209; Brussels, 1990).

NESSELRATH, H.-G., *Ungeschehenes Geschehen: 'Beinahe-Episoden' im griechischen und römischen Epos von Homer bis zum Spätantike* (Stuttgart, 1992).

NICOLET, C., *Space, Geography and Politics in the Early Roman Empire*, tr. H. Leclerc (Ann Arbor, 1990).

NISSEN, H., *Kritische Untersuchungen über die Quellen der vierten und fünften Dekaden des Livius* (Berlin, 1863).

NOCK, A. D., 'Posidonius', *JRS* 49 (1959), 1–15 = Z. Stewart (ed.), *Arthur Darby Nock: Essays on Religion and the Ancient World* (Oxford, 1972), 853–76.

NOLLE, J., and SCHINDLER, F., *Die Inschriften von Selge* (Inschriften griechischer Städte aus Kleinasien 37; Bonn, 1991).

NORDEN, E., *Die antike Kunstprosa vom VI Jhdt. v. Chr. bis in die Zeit der Renaissance*, 2 vols. (Leipzig, 1898).

NUTTALL, A. D., *Openings: Narrative Beginnings from the Epic to the Novel* (Oxford, 1992).

OLDFATHER, C. H. (ed. and tr.), *Diodorus of Sicily*, vi (Loeb; Cambridge, Mass., 1939).

OSTWALD, M., 'The Reform of the Athenian State by Cleisthenes', in J. Boardman, N. G. L. Hammond, D. M. Lewis, and M. Ostwald (eds.), *Cambridge Ancient History*, iv² : *Persia, Greece, and the Western Mediterranean c.525–479 BC* (Cambridge, 1988), 303–46.

PACKMAN, Z. M., 'The Incredible and the Incredulous: The Vocabulary of Disbelief in Herodotus, Thucydides and Xenophon', *Hermes* 119 (1991), 399–414.

PAGE, D. L., *History and the Homeric Iliad* (Berkeley and Los Angeles, 1963).

PARKE, H. W., and BOARDMAN, J., 'The Struggle for the Tripod and the First Sacred War', *JHS* 77 (1957), 276–82.

—— and WORMELL, D. E. W., *The Delphic Oracle*, i² (Oxford, 1956).

PARKER, R., 'The Festivals of the Attic Demes', *Boreas* 15 (1987; symposium called *Gifts for the Gods*), 137–47.

PARRY, A., *The Language of Achilles and Other Papers*, ed. H. Lloyd-Jones (Oxford, 1989).

PATER, W., *The Renaissance* (London, 1910; repr. Chicago, 1977).

PATON, W. S., *Polybius*, 6 vols. (Loeb; Cambridge, Mass., 1922–7).

PAUL, G. M., *A Historical Commentary on Sallust's Bellum Jugurthinum* (Liverpool, 1984).

PEARSON, L., *Early Ionian Historians* (Oxford, 1939).

—— *Local Historians of Attica* (Philadelphia, 1942).

—— *Lost Histories of Alexander the Great* (New York, 1960).

—— 'The Pseudo-History of Messenia and its Authors', *Historia* 11 (1962), 397–426.

—— *The Greek Historians of the West* (Atlanta, 1987).

PÉDECH, P., *Historiens compagnons d'Alexandre* (Paris, 1984).

PELLING, C. B. R., 'Plutarch's Method of Work in the Roman Lives', *JHS* 99 (1979), 74–96.

—— 'Plutarch's Adaptation of his Source-Material', *JHS* 100 (1980), 127–40.

—— (ed.), *Plutarch: Life of Antony* (Cambridge, 1988).

—— 'Plutarch and Thucydides', in P. A. Stadter (ed.), *Plutarch and the Historical Tradition* (London, 1991), 10–40.

—— 'Thucydides' Archidamus and Herodotus' Artabanus', in M. Flower and M. Toher (eds.), *Georgica: Greek Studies in Honour of George Cawkwell* (*BICS* suppl. 58; London, 1991), 120–42.

PFEIFFER, R., *A History of Classical Scholarship from 1300 to 1850* (Oxford, 1976).

PFISTER, F., 'Das Alexander-Archiv und die hellenistisch-römische Wissenschaft', *Historia* 10 (1961), 30–67.

POHLENZ, M., 'Philipps Schreiben an Athen', *Hermes* 64 (1929), 41–62.

POLLITT, J. J., *Art in the Hellenistic Age* (Cambridge, 1986).

Pomtow, H., 'Hippokrates und die Asklepiaden in Delphi', *Klio* 15 (1918), 303–38.

Powell, E., *Lexicon to Herodotus* (Cambridge, 1939).

Pringsheim, F., *Greek Law of Sale* (Weimar, 1950).

Pritchett, W. K. (ed.), *Dionysius of Halicarnassus: On Thucydides* (Berkeley–Los Angeles–London, 1975).

Rajak, T., *Josephus* (London, 1983).

Ranger, T., see Hobsbawm, E.

Raven, J. S., see Kirk, G. S.

Rawlings, H. H., *The Structure of Thucydides' History* (Princeton, NJ, 1981).

Redfield, J., *Nature and Culture in the Iliad: The Tragedy of Hector* (Chicago, 1975).

—— 'Herodotus the Tourist', *CP* 80 (1985), 97–118.

Regenbogen, O., 'Theophrastos', *RE* Suppbd. vii (1940), cols. 1354–1562.

Rhodes, P. J., *A Commentary on the Aristotelian Athenaion Politeia* (Oxford, 1981).

—— *Défense et illustration des historiens grecs* (pamphlet; Liège, 1992).

Rich, J. W., *Cassius Dio: The Augustan Settlement (Roman History 53–59.9)* (Warminster, 1990).

Richards, J. F. C., see Caley, E. R.

Richardson, N. J., *The Iliad: A Commentary*, vi: *Books 21–24* (Cambridge, 1993).

Richardson, Scott, *The Homeric Narrator* (Nashville, 1990).

Riemann, K. A., *Das herodoteische Geschichtswerk in der Antike* (Diss.; Munich, 1967).

Ritchie, C. E., jun., 'The Lyceum, the Garden of Theophrastus and the Garden of the Muses: A Topographical Re-evaluation', in Φίλια Ἔπη (*Studies in Honour of G. Mylonas*), iii (Athens, 1989), 250–60.

Robert, L., *Hellenica*, iv (Paris, 1948).

—— 'Une bilingue gréco-araméenne d'Asoka, II, observations sur l'inscription grecque', *JA* (1958), 7–18=*OMS* iii (1969), 551–62.

—— *Documents d'Asie mineure* (Paris, 1987).

—— 'De Delphes à l'Oxus: Inscriptions grecques nouvelles de la Bactriane', *CRAI* (1968), 416–57=*OMS* v (1989), 510–51.

—— 'Deux concours grecs à Rome (Antoninia Pythia sous Elagabal et concours d'Athéna Promachos depuis Gordien III)', *CRAI* (1970), 6–27=*OMS* vi (1989), 647–68.

—— *Opera minora selecta* (=*OMS*) (Amsterdam, 1969–90).

Robertson, N., 'The Myth of the First Sacred War', *CQ* 28 (1978), 38–73.

—— 'A Point of Precedence at Plataia: The Dispute between Athens and Sparta over Leading the Procession', *Hesperia* 55 (1986), 88–102.

Roger, J., and Effenterre, H. van, 'Krisa-Kirrha', *Rev. Arch.* 1944, no. 21.

ROISMAN, J., 'Ptolemy and his Rivals in his History of Alexander', *CQ* 34 (1984), 373–85.

ROLLEY, C., *Les Trépieds à cuve clouée: Fouilles de Delphes*, v(3) (Paris, 1977).

ROMILLY, J. DE, *Histoire et raison chez Thucydide* (Paris, 1956).

ROSIVACH, V. J., 'The Cult of Zeus Eleutherios at Athens', *P del P* 42 (1987), 262–85.

ROUECHÉ, C., and SHERWIN-WHITE, S. M., 'Some Aspects of the Seleucid Empire: The Greek Inscriptions from Failaka in the Arabian Gulf', *Chiron* 15 (1985), 1–39.

RUBHARDT, J., and REVERDIN, O. (eds.), *Le Sacrifice dans l'antiquité* (Entretiens Hardt 27; Geneva, 1981).

RUBINCAM, C. REID, 'Qualification of Numerals in Thucydides', *AJAH* 4 (1979), 77–95.

—— 'Thucydides 1.74.1 and the Use of 'ΕΣ with Numbers', *CP* 74 (1979), 327–37.

—— 'The Theban Attack on Plataia: Herodotus 7.233.2 and Thucydides 2.2.1 and 5.8–9', *LCM* 6 (1981), 47–9.

RUSCHENBUSCH, E., 'Der Endpunkt der Historien des Poseidonius', *Hermes* 121 (1993), 70–6.

RUSSELL, D. A., *Greek Declamation* (Cambridge, 1983).

RUSSO, J., FERNANDEZ-GALIANO, M., and HEUBECK, A., *A Commentary on Homer's Odyssey*, iii: *Books XVII–XXIV* (Oxford, 1992).

RUTHERFORD, R. B., 'What's New in Homeric Studies?', *JACT Review* 11 (1992), 15–17.

—— *Odyssey XIX–XX: Commentary* (Cambridge, 1992).

RYLE, G., *Plato's Progress* (Cambridge, 1966).

SACKS, K. S., *Polybius on the Writing of History* (Berkeley, 1981).

—— 'Rhetoric and Speeches in Hellenistic Historiography', *Ath* 64 (1986), 383–95.

—— *Diodorus Siculus and the First Century BC* (Princeton, NJ, 1990).

STE CROIX, G. E. M. DE, 'Aristotle on History and Poetry (*Poetics* 9, 1451[a]36–[b]11)', in B. M. Levick (ed.), *The Ancient Historian and his Materials: Essays in Honour of C. E. Stevens on his Seventieth Birthday* (Farnborough, 1975), 45–58; repr. in A. Rorty (ed.), *Essays on Aristotle's Poetics* (Princeton, NJ, 1992), 23–32.

—— *The Class Struggle in the Ancient Greek World* (London, 1981).

SAUGÉ, A., *De l'épopée à l'histoire: Fondement de la notion de l'historié* (Frankfurt am Main–Berne–New York–Paris, 1992).

SAURON, G., 'Aspects du néo-atticisme à la fin du Ier S. av. J.-C.: formes et symboles', in X. Lafou and G. Sauron, *L'Art décoratif à Rome à la fin de la république et au début du principat* (Collection de l'École française de Rome 55; Rome, 1981), 286–94.

SCAIFE, R., 'Alexander I in the Histories of Herodotus', *Hermes* 117 (1989), 129–37.

SCARDIGLIO, B., *Nicolao di Damasco: Vita di Augusto* (Florence, 1983).

SCHACHERMEYR, F., 'Athen als Stadt des Großkönigs', *Grazer Beiträge* 1 (1973), 211–20.

SCHALLES, H.-J., *Untersuchungen zur Kultur-politik der pergamenischen Herrscher im dritten Jahrhundert vor Christus* (Istanbuler Forschungen 36; Tübingen, 1985).

SCHAMA, S., *Dead Certainties (Unwarranted Speculations)* (London and New York, 1991).

SCHINDLER, F., see Nolle, J.

SCHMID, W., and STÄHLIN, O., *Geschichte der griechischen Literatur*, i(2) (Munich, 1934).

SCHNEIDER, C., *Information und Absicht bei Thukydides* (Göttingen, 1974).

SCHNEIDER, L., and HÖCKER, C., *Die Akropolis von Athen: Antikes Heiligtum und modernes Reiseziel* (Cologne, 1990).

SCHNEIDER, R., *Bunte Barbaren* (Worms, 1986).

SCHÜRER, E., rev. G. Vermes and F. Millar, *History of the Jewish People in the Age of Jesus Christ (175 BC–AD 135)*, i (Edinburgh, 1973).

—— rev. G. Vermes, F. Millar, and M. Goodman, *History of the Jewish People in the Age of Jesus Christ (175 BC–AD 135)* iii(1) (Edinburgh, 1986).

SCHWARZ, E., *Griechische Geschichtsschreiber* (Leipzig, 1959).

SCHWARZ, F., 'Daimachos von Plataia: zum geistesgeschichtlichen Hintergrund seiner Schriften', in R. Stiehl and R. Stier (eds.), *Beiträge zur alten Geschichte (F. Altheim Festschrift)*, i (Berlin, 1969), 293–304.

SCULLY, S., *Homer and the Sacred City* (Ithaca, NY, 1990).

SEAGER, R., 'The Congress Decree: Some Doubts and a Hypothesis', *Historia* 18 (1969), 129–41.

SEGAL, C., 'Bard and Audience in Homer', in R. Lamberton and J. Keaney (eds.), *Homer's Ancient Readers: The Hermeneutics of Greek Epic's Earliest Exegetes* (Princeton, NJ, 1992), 3–29.

SENN, G., *Die Pflanzenkunde des Theophrast von Eresos* (Basle, 1956).

SHAHBAZI, A. SH., 'Darius in Scythia and Scythians at Persepolis', *AMI* 15 (1982), 189–235.

SHEAR, T. L., 'Athens: From City-State to Provincial Town', *Hesperia* 50 (1981), 356–77.

SHEPPARD, A. R. R., 'Homonoia in the Greek Cities of the Roman Empire', *Anc. Soc.* 15–17 (1984–6), 238–9.

SHERK, R., *The Roman Empire: Augustus to Hadrian* (Cambridge, 1988).

SHERWIN-WHITE, S. M., and KUHRT, A., *From Samarkhand to Sardis: New Perspectives on the Seleucid Empire* (London, 1983).

SHOTTER, D. C. A. (ed.), *Tacitus: Annals IV* (Bristol, 1989).

SHRIMPTON, G. S., *Theopompus the Historian* (Montreal–London–Buffalo, 1991).

SHUCKBURGH, E. S. (tr.), *The Histories of Polybius*, 2 vols. (London and New York, 1889).

SIEVEKING, F., 'Die Funktion geographischer Mitteilungen im Geschichtswerk des Thukydides', *Klio* 42 (1964), 73–179.

SMART, J. D., 'Thucydides and Hellanicus', in I. S. Moxon, J. D. Smart, and A. J. Woodman (eds.), *Past Perspectives: Studies in Greek and Roman Historical Writing* (Cambridge, 1986), 19–35.

SMITH, A., ed. J. Lothian, *Lectures on Rhetoric and Belles Lettres* (London, 1963).

SMITH, R. R. R., 'The Imperial Reliefs from the Sebasteion at Aphrodisias', *JRS* 77 (1987), 88–138.

SMITH, WESLEY, D. (ed. and tr.), *Hippocrates: Pseudepigraphic Writings (Letters—Embassy—Speech from the Altar—Decree)* (London and New York, 1990).

SOLLENBERGER, M. G., *Theophrastus of Eresus, his Life and Work* (New Brunswick and Oxford, 1985).

SONNABEND, H., *Fremdenbild und Politik: Vorstellungen der Römer von Ägypten und dem Partherreich in der späten Republik und frühen Kaiserzeit* (Frankfurt am Main–Berne–New York, 1986).

SORDI, M., 'La prima guerra sacra', *RIFC* 81 (1953), 320–46.

SPAWFORTH, A. J. S., review of O. Andrei, *JRS* 76 (1986), 328.

—— review of G. Rogers, *CR* 42 (1992), 383–4.

—— 'Corinth, Argos and the Imperial Cult: A Reconsideration of Pseudo-Julian, Letters 198 Bidez', *Hesperia* (forthcoming).

—— and WALKER, S., 'The World of the Panhellenion', *JRS* 75 (1985), 78–104, and 76 (1986), 88–105.

—— and see under Cartledge, P.

SPENCE, I., 'Perikles and the Defence of Attika during the Peloponnesian War', *JHS* 110 (1990), 91–109.

SQUIRES, J. T., *The Plan of God in Luke–Acts* (Cambridge, 1993).

STADTER, P., *Arrian of Nicomedia* (Chapel Hill, NC, 1980).

STAHL, H.-P., *Thukydides: die Stellung des Menschen im geschichtlichen Prozess* (*Zetemata* 40; Munich, 1966).

STEINMETZ, P., *Die Physik des Theophrast* (Bad Homburg–Zurich–Berlin, 1964).

STERN, M. (ed.), *Greek Authors on Jews and Judaism*, i (Jerusalem, 1974).

STEVENSON, R. B., 'Lies and Inventions in Deinon's Persica', in H. Sancisi-Weerdenburg and A. Kuhrt (eds.), *Achaemenid History*, ii: *The Greek Sources* (Leiden, 1987), 27–35.

—— *Persica* (forthcoming).

STINTON, T. C. W., '*Si credere dignum est*: Some Expressions of Disbelief in Euripides and Others', *PCPhS* 22 (1976), 60–89 (= *Collected Papers on Greek Tragedy* (Oxford, 1990), 236–64).

STRASBURGER, H., review of Tarn's *Alexander*, *Bibliotheca Orientalis* 9 (1952), 202–11.

—— 'Herodot und das perikleische Athen', *Historia* 4 (1955), 1–25 = *Studien*, 592–626.

—— 'Posidonius on Problems of the Roman Empire', *JRS* 55 (1965), 40–53; Ger. tr. in *Studien*, 920–45.

—— *Studien zur alten Geschichte*, ed. W. Schmitthenner and R. Zoepffel (Hildesheim and New York, 1982).

STRATTON, G. M., *Theophrastus and Greek Physiological Psychology before Aristotle* (London and New York, 1917; repr. Iowa, n.d.).

STREBEL, H. G., *Wertung und Wirkung des Thukydideischen Geschichstwerk in der griechisch-römischen Literatur* (Diss.; Munich, 1935).

STROUD, R., 'Thucydides and the Battle of Solygeia', *CSCA* 4 (1971).

—— 'State Documents in Archaic Athens', in W. Childs (ed.), *Athens Comes of Age: From Solon to Salamis* (Princeton, NJ, 1978), 20–42.

STRUBBE, J., 'Gründer kleinasiatischer Städte: Fiktion und Realität', *Anc. Soc.* 15–17 (1984–6), 253–304.

SYKUTRIS, J., see Bickermann, E. J.

SYME, R., *Tacitus*, 2 vols. (Oxford, 1958).

—— *Sallust* (Cambridge, 1964).

—— *Roman Papers*, iii (Oxford, 1984).

TAPLIN, O., *Homeric Soundings: The Shaping of the Iliad* (Oxford, 1992).

TARN, W. W., *Alexander the Great*, 2 vols. (Cambridge, 1948).

TAUSEND, K., *Amphktyonie und Symmachie* (*Historia* Einzelschr. 73; Stuttgart, 1992).

TAYLOR, A. J. P., *The Struggle for Mastery in Europe, 1848–1918* (Oxford, 1954).

TAYLOR, G., *Reinventing Shakespeare: A Cultural History from the Restoration to the Present* (London, 1990).

THEILER, W., *Posidonius: Die Fragmente*, 2 vols. (Berlin, 1982).

THOMAS, R., *Oral Tradition and Written Record in Classical Athens* (Cambridge, 1989; paperback, 1992).

—— *Literacy and Orality in Ancient Greece* (Cambridge, 1992).

TODD, S., *The Shape of Athenian Law* (Oxford, 1993).

TUPLIN, C. J., 'Pausanias and Plutarch's *Epaminondas*', *CQ* 34 (1984), 346–58.

—— *The Failings of Empire: A Reading of Xenophon's Hellenica 2.3.11–7.5.27* (*Historia* Einzelschr. 76; Stuttgart, 1993).

USHER, S., *The Historians of Greece and Rome* (London, 1969).

VANDIVER, E., *Herodotus and Heroes: The Interaction of Myth and History* (Frankfurt am Main, 1991).

VANSINA, J., *Oral Tradition as History* (London, 1985).

VATIN, C., 'Damiurges et épidamiurges à Delphes', *BCH* 85 (1961), 236–55.

VERMES, G., see Schürer, E.

VEYNE, P., *Writing History* (Manchester, 1984).

WADE-GERY, H. T., 'Kynaithos', in *Greek Poetry and Life: Essays Presented to G. Murray on his Seventieth Birthday* (Oxford, 1936), 56–78; repr. in Wade-Gery, *Essays in Greek History*, 17–36.

WADE-GERY, H. T., *The Poet of the Iliad* (Cambridge, 1952).
—— *Essays in Greek History* (Oxford, 1958).
—— 'The "Rhianos-Hypothesis"', in E. Badian (ed.), *Ancient Society and Institutions: Studies Presented to V. Ehrenberg on his 75th Birthday* (Oxford, 1966), 289–302.
WALBANK, F. W., 'Alcaeus of Messene, Philip V and Rome, part I', *CQ* 36 (1942), 134–45.
—— 'Tragic History: A Reconsideration', *BICS* 2 (1955), 4–14.
—— *Historical Commentary on Polybius*, 3 vols. (Oxford, 1957–79).
—— *Polybius* (Berkeley–Los Angeles–London, 1972).
—— *Selected Papers* (Cambridge, 1985).
—— 'Timaeus' Views on the Past', *SCI* 10 (1989–90), 41–54; Ger. tr., *Timaios und die westgriechische Sicht der Vergangenheit* (Xenia 29; Konstanz, 1992).
WALLACE-HADRILL, A., 'Roman Arches and Greek Honours: The Language of Power at Rome', *PCPhS* 216 (1990), 143–81.
WALSER, G., *Die Völkerschaften auf den Reliefs von Persepolis: Historische Studien über den sogenannten Tributzug an der Apadanentreppe* (Teheraner Forschungen 2; Berlin, 1966).
WALTON, F. R. (ed. and tr.), *Diodorus of Sicily*, xi (Loeb; Cambridge, Mass., 1957).
WARDMAN, A. E., *Rome's Debt to Greece* (London, 1976).
WATERS, K. H., *Herodotus the Historian: His Problems, Methods and Originality* (London, 1985).
WAUGH, E., *The Letters of Evelyn Waugh*, ed. M. Amory (London, 1980).
WEISKOPF, M., *The So-called 'Great Satraps' Revolt', 366–60 BC: Concerning Local Instability in the Achaemenid Far West* (Historia Einzelschr. 63; Stuttgart, 1989).
WENSKUS, O., review of S. Hornblower, *Thucydides*, *Gnomon* 37 (1990), 577–9.
WEST, M. L., 'Stesichorus', *CQ* 21 (1971), 302–14.
—— 'Cynaethus' Hymn to Apollo', *CQ* 25 (1975), 161–70.
—— 'Simonides Redivivus', *ZPE* 98 (1993), 1–14.
WEST, S., 'Herodotus' Portrait of Hecataeus', *JHS* 111 (1991), 144–60.
—— review of Lateiner, *CR* 41 (1991), 23–5.
WESTERMANN, W., 'Athenaeus and the Slaves of Athens', in M. I. Finley (ed.), *Slavery in Classical Antiquity* (Cambridge, 1960), 73–92.
WESTLAKE, H. D., 'The Sources of Plutarch's *Pelopidas*', *CQ* 33 (1939), 11–22.
—— *Essays on the Greek Historians and Greek History* (Manchester, 1969).
—— *Studies in Thucydides and Greek History* (Bristol, 1989).
WHEELDON, M., '"True Stories": The Reception of Historiography in Antiquity', ch. 2 in Averil Cameron (ed.), *History as Text: The Writing of Ancient History* (London, 1989).
WHITE, H., *Metahistory* (Baltimore, 1973).
—— *Tropics of Discourse: Essays in Cultural Criticism* (Baltimore, 1978).

WHITEHEAD, D. (ed.), *Aineias the Tactician: How to Survive under Siege* (Oxford, 1990).

WHYBRAY, R. N., *The Succession Narrative* (London, 1968).

WIKÉN, E., *Die Kunde der Hellenen von dem Lande und den Völkern der Apenninenhalbinsel bis 300 v. Chr.* (Lund, 1937).

WILAMOWITZ-MOELLENDORFF, U. VON, *Aristoteles und Athen*, 2 vols. (Berlin, 1893).

—— *Der Glaube der Hellenen*, 2 vols. (Berlin, 1931–2).

—— *A History of Classical Scholarship*, ed. H. Lloyd-Jones, tr. A. Harris (London, 1982).

WILL, W., 'Die griechische Geschichtschreibung des 4. Jahrhunderts: Eine Zusammenfassung', in J. M. Alonso-Nuñez (ed.), *Geschichtsbild und Geschichtsdenken im Altertum* (Darmstadt, 1991), 113–35.

WILLERS, D., *Hadrians panhellenische Programm* (Basle, 1990).

WIMMER, F., *Theophrasti Eresii Opera*, 3 vols. (Leipzig, 1854–62).

WOLSKI, J., 'Les Achéménides et les Arsacides', *Syria* 43 (1966), 65–89.

WOODMAN, A. J., *Rhetoric in Classical Historiography* (Beckenham, 1988).

—— and see Moxon, I. S.

WOODWARD, B., and BERNSTEIN, C., *Final Days* (London, 1976).

WORMELL, D. E. W., see under Parke, H. W.

WÖRRLE, M., 'Neue Inschriftenfunde aus Aizanoi I', *Chiron* 22 (1992), 337–70.

ZANKER, P., *The Power of Images in the Age of Augustus*, tr. A. Shapiro (Ann Arbor, 1988).

ZIEGLER, K., 'Polybios', *RE* xxi (1952), cols. 1440–1578.

Index